D0406894

River Ecology and Man

ENVIRONMENTAL SCIENCES

An Interdisciplinary Monograph Series

EDITORS

DOUGLAS H. K. LEE
National Institute of
Environmental Health Sciences
Research Triangle Park
North Carolina

E. WENDELL HEWSON
Department of
Atmospheric Science
Oregon State University
Corvallis, Oregon

DANIEL OKUN
University of North Carolina
Department of Environmental
Sciences and Engineering
Chapel Hill, North Carolina

ARTHUR C. STERN, editor, AIR POLLUTION, Second Edition. In three volumes, 1968

L. FISHBEIN, W. G. FLAMM, and H. L. FALK, CHEMICAL MUTAGENS: Environmental Effects on Biological Systems, 1970

DOUGLAS H. K. LEE and DAVID MINARD, editors, PHYSIOLOGY, ENVIRONMENT, AND MAN, 1970

KARL D. KRYTER, THE EFFECTS OF NOISE ON MAN, 1970

R. E. MUNN, BIOMETEOROLOGICAL METHODS, 1970

M. M. KEY, L. E. KERR, and M. BUNDY, PULMONARY REACTIONS TO COAL DUST: "A Review of U. S. Experience," 1971

DOUGLAS H. K. LEE, editor, METALLIC CONTAMINANTS AND HUMAN HEALTH, 1972

DOUGLAS H. K. LEE, editor, ENVIRONMENTAL FACTORS IN RESPIRATORY DISEASE, 1972

H. ELDON SUTTON and MAUREEN I. HARRIS, editors, MUTAGENIC EFFECTS OF ENVIRONMENTAL CONTAMINANTS, 1972

RAY T. OGLESBY, CLARENCE A. CARLSON, and JAMES A. McCANN, editors, RIVER ECOLOGY AND MAN, 1972

In preparation

MOHAMED K. YOUSEF, STEVEN M. HORVATH, and ROBERT W. BULLARD, PHYSIOLOGICAL ADAPTATIONS: Desert and Mountain

DOUGLAS H. K. LEE and PAUL KOTIN, editors, MULTIPLE FACTORS IN THE CAUSATION OF ENVIRONMENTALLY INDUCED DISEASE

River Ecology and Man

EDITED BY

Ray T. Oglesby
Department of Natural Resources
Cornell University
Ithaca, New York

Clarence A. Carlson
New York Cooperative Fisheries Unit
Cornell University
Ithaca, New York

James A. McCann
Massachusetts Cooperative Fisheries Unit
University of Massachusetts
Amherst, Massachusetts

Proceedings of an International
Symposium on River Ecology and
the Impact of Man, Held at the
University of Massachusetts,
Amherst, Massachusetts,
June 20-23, 1971

Academic Press 1972 New York and London

ACADEMIC PRESS, INC.
111 Fifth Avenue, New York, New York 10003

United Kingdom Edition published by
ACADEMIC PRESS, INC. (LONDON) LTD.
24/28 Oval Road, London NW1

LIBRARY OF CONGRESS CATALOG CARD NUMBER: 75-182634

PRINTED IN THE UNITED STATES OF AMERICA

CONTENTS

ADVISORS TO SYMPOSIUM COMMITTEE ix
SYMPOSIUM COMMITTEE xi
CONTRIBUTORS xiii
FOREWORD xv
ACKNOWLEDGMENTS xvii

Introduction 1
    John Bardach

I. WHAT IS A RIVER?
Ruth Patrick, Session Chairman

Rivers — A Geomorphic and Chemical Overview 9
    Robert R. Curry

What Is a River? — A Zoological Description 33
    Kenneth W. Cummins

Plant Ecology in Flowing Water 53
    John L. Blum

Commentary 67
    Ruth Patrick

II. USES OF A RIVER — PAST AND PRESENT
E. D. LeCren, Session Chairman

Man's Impact on the Columbia River 77
    Parker Trefethen

The Delaware River — A Study in Water Quality Management 99
    Robert V. Thomann

Man and the Illinois River 131
    William C. Starrett

CONTENTS

The Nile River — A Case History 171
    Desmond Hammerton

Case History — The River Thames 215
    Kenneth H. Mann

Uses of the Danube River 233
    Reinhard Liepolt

Commentary 251
    E. D. LeCren

III. EFFECTS OF RIVER USES
Theodore P. Vande Sande, Session Chairman

Regulated Discharge and the Stream Environment 263
    Jack C. Fraser

Morphometric Changes 287
    T. Blench

Sedimentation (Suspended Solids) 309
    H. A. Einstein

A Comparative Assessment of Thermal Effects
in Some British and North American Rivers 319
    T. E. Langford

Effects of Pesticides and Industrial Waste on Surface Water Use . . 353
    William A. Brungs

Radionuclides in River Systems 367
    Daniel J. Nelson, S. V. Kaye, and R. S. Booth

Nutrients 389
    Walter M. Sanders, III

Commentary 417
    Theodore P. Vande Sande

CONTENTS

IV. RATIONALIZATION OF MULTIPLE USE
David J. Allee, Session Chairman

Rationalization of Multiple Use of Rivers   421
   John Cairns, Jr.

Multiple Use of River Systems — An Economic Framework   431
   James Crutchfield

Political Aspects of Multiple Use   441
   John D. Dingell

Summary of Symposium   455
   Justin W. Leonard

SUBJECT INDEX   461

ADVISORS TO SYMPOSIUM COMMITTEE

David J. Allee
Cornell University

H. B. N. Hynes
University of Waterloo

Stuart Neff
University of Louisville

Paul Ruggles
Canada Department of Fisheries

William C. Starrett
Illinois Natural History Survey

Richard Stroud
Sport Fishing Institute

Theodore P. Vande Sande
California Department of Fish and Game

SYMPOSIUM COMMITTEE

Lyndon Bond, Program
Inland Fisheries and Game, Maine, Augusta, Maine

Clarence Carlson, Program Co-chairman and Co-editor
New York Cooperative Fisheries Unit, Ithaca, New York

William Daugherty, Arrangements
Sport Fisheries and Wildlife, Washington, D. C.

Robert Jones, Chairman
Connecticut Board of Fisheries and Game, Milford, Connecticut

James McCann, Program Co-chairman and Co-editor
Massachusetts Cooperative Fisheries Unit, University of Massachusetts,
Amherst, Massachusetts

Ray T. Oglesby, Editor
Department of Natural Resources, Cornell University, Ithaca, New York

Christopher Percy, Publicity
Connecticut River Watershed Council, Greenfield, Massachusetts

Roger Reed, Arrangements
Massachusetts Cooperative Fisheries Unit, Amherst, Massachusetts

Lyle Thorpe, Finance
Northeast Utilities Service Company, Tolland, Connecticut

Dwight A. Webster, Program Coordinator
Cornell University, Ithaca, New York

CONTRIBUTORS

John Bardach, Department of Wildlife and Fisheries, 1006 Natural Resources Building, University of Michigan, Ann Arbor, Michigan

T. Blench, Department of Civil Engineering, The University of Alberta, Edmonton 7, Alberta, Canada

John L. Blum, Department of Botany, The University of Wisconsin, Milwaukee, Wisconsin

William A. Brungs, Fish Toxicology Laboratory, Environmental Protection Agency, Water Quality Office, 3411 Church Street, Newton, Cincinnati, Ohio

John Cairns, Jr., Center for Environmental Studies, Virginia Polytechnic Institute and State University, Blacksburg, Virginia

James Crutchfield, Department of Economics, University of Washington, Seattle, Washington

Kenneth W. Cummins, W. K. Kellogg Biological Station, Michigan State University, Hickory Corners, Michigan

Robert R. Curry, Department of Geology, University of Montana, Missoula, Montana

John D. Dingell, House of Representatives, House Office Building, Washington, D. C.

H. A. Einstein, Department of Civil Engineering, 412 O'Brien Hall, University of California, Berkeley, California

Jack C. Fraser, Premier's Department, Treasury Place, Melbourne, Victoria 3000, Australia

Desmond Hammerton, Clyde River Purification Board, Rivers House, Murray Road, East Kilbride, Glasgow, Scotland

T. E. Langford, Central Electricity Generating Board, Ratcliff-on-Soar, Nottingham, England

xiii

E. D. LeCren, Freshwater Biological Association, River Laboratory, East Stoke, Wareham, Dorset, England

Justin W. Leonard, Department of Resource Planning and Conservation, 2006 Natural Resources Building, University of Michigan, Ann Arbor, Michigan

Reinhard Liepolt, Bundesenstalt für Wasserbiologie und Abwasserforschung, Des bundesministeriums für Land und Forstwirtschaft, 1223 Wien, Kaisermuhlenstrasse 120, Postfach 7, Vienna, Austria

Kenneth H. Mann, Fisheries Research Board of Canada, Marine Ecology Laboratory, Bedford Institute, Dartmouth, Nova Scotia, Canada

Daniel J. Nelson, Ecological Sciences Division, Oak Ridge National Laboratory, Oak Ridge, Tennessee

Ruth Patrick, Academy of Natural Sciences, Philadelphia, Pennsylvania

Walter M. Sanders, III, Nutritional Pollutants Fate Research Program, Environmental Protection Agency, Water Quality Office, Southeast Water Laboratory, College Station Road, Athens, Georgia

William C. Starrett,[*] Illinois Natural History Survey Laboratory, Havana, Illinois

Robert V. Thomann, Civil Engineering Department, Manhattan College, Bronx, New York

Parker Trefethen, National Marine Fisheries Service, Biological Laboratory, 2725 Montlake Boulevard East, Seattle, Washington

Theodore P. Vande Sande, Water Projects Branch, California Department of Fish and Game, 1416 Ninth Street, Sacramento, California

[*]Deceased.

FOREWORD

Informally the notion of a symposium on river ecology began in a motel corridor in Bedford, New Hampshire, site of the 1968 Annual Meeting of the Northeastern Division of the American Fisheries Society. This led to the appointment of an *ad hoc* committee to investigate the feasibility of Divisional sponsorship, and ultimately to the appointment of a River Ecology Symposium Committee in September, 1969. To reach this decision, Division members drew heavily on the counsel and experience of the Southern Division of the American Fisheries Society, sponsors of the symposium on "Reservoir Fishery Resources" at Athens, Georgia in 1967.

A significant early decision of the Symposium Committee was to invite a group of fellow scientists to serve as advisors in determining format and content of the symposium. The advisors, whose names are listed elsewhere, met with the Committee in March, 1970 at the Cornell University Biological Field Station on Oneida Lake. This two-day, freewheeling, "think tank" generated the basic structure of the symposium and the names of individuals qualified to make it a viable and meaningful event. The retreat was surely one of the most stimulating and satisfying activities for the Committee.

Sponsoring an international symposium takes financial backing, and the years 1969-1971 were not the most auspicious ones in which to raise money easily. Donations and grants are also acknowledged elsewhere, but one deserves special comment. At the time the Committee was most pressed for money to defray early operational expenses, an unexpected and unsolicited windfall occurred. The Class of 1970, George School of Bucks County, Pennsylvania, decided to forego their usual senior bash and, in its place, donate the money anteed up by parents to an environmentally oriented project. Their Class Advisor, Alan Sexton, learned of the symposium plans through the American Conservation Foundation and a substantial contribution followed. It was a heartening gesture viewed as an unselfish response of youth to the needs of the times, as well as a response to the needs of the Committee for cash in hand at a critical time in program development.

As funds were being gathered other activities proceeded apace, including the selection of the new Lincoln Conference Center at the University of Massachusetts as the symposium site and June 20-23, 1971 as the date. The program co-chairmen and session chairmen obtained commitments from participants. Two of these are not shown among the authors but both deserve special thanks for their contributions to the program. These were Pete Seeger, the well-known folk singer who volunteered his talents to inspire and challenge us all and Ron Bonn of CBS News who vividly demonstrated the important role of the news media.

Regretfully, as this volume is in final stages of preparation, we must record the death of William Starrett on December 30, 1971. He provided a contagious enthusiasm as an advisor to the Committee and later participated importantly in the symposium program. We gratefully and affectionately acknowledge his efforts in the Division to stage a successful meeting. We are honored to have been associated with Bill Starrett in this substantial contribution to the literature of the aquatic sciences.

At some point (at what now seems the distant past) this undertaking by the Northeastern Division of the American Fisheries Society became officially titled "An International Symposium on River Ecology and the Impact of Man." From time to time many of us associated with the events leading up to the meeting felt that it was presumptious to believe that a topic of such complexity and enormity could be handled by the resources, talent, and time at our disposal. Indeed, it could not have been were it not for a cadre of devoted and diligent Division members. We make no case for having preempted the subject by presumptious- ness, but we trust we have achieved some focus on the state of the art as re- ported by a diverse group of people and that these transactions will serve as a base and departure point for further deliberations on the subject of river ecology.

Dwight A. Webster, Program Coordinator
Department of Natural Resources
Cornell University
Ithaca, New York

ACKNOWLEDGMENTS

The Symposium Commitee thanks the following organizations for financial assistance that helped make this Symposium a reality:

Northeast Utilities Service Co.
Hartford, Connecticut

George School, Class of 1970
Bucks County, Pennsylvania

The Conservation Foundation
Washington, D. C.

American Conservation Association
New York, New York

National Science Foundation
Washington, D. C.

New England Power Service Co.
Westboro, Massachusetts

Connecticut Yankee Atomic Power Co.
Hartford, Connecticut

Sport Fishing Institute
Washington, D. C.

private foundations requesting anonymity.

River Ecology and Man

INTRODUCTION

John Bardach

We know that rivers are the lifelines of nations and threads from which their history is woven. We also know that rivers and their shores are precariously balanced, interacting ecosystems, easily upset by man. The immediate interest of some of us is to find out how these balances and interactions work; but our collective, longer term interest is in rivers as resources in the broadest meaning of the term. This interest inevitably leads us to try to look into the future. Whether we emerge from such attempts at forward glances mildly optimistic or wildly pessimistic depends on whether we look at the world or at a carefully selected slice of it and on what our time perspective happens to be.

Global forecasts are overshadowed by the inexorable constraints encountered in one trying to fit a cowboy economy into a spaceship ecology. The contradiction in these descriptive terms, coined by Kenneth Boulding, further indicates that exponential growth of people and their material aspirations simply cannot be sustained on the Western pattern of resource use of the last two hundred years. Langford, in this symposium, will mention some forecasts of the possible expansion in the power industry in Britain and the U.S.A., the most extreme of which suspects that by the year 2000 the whole of the dry-weather flow in the U.S. could be used if once through cooling was maintained. It is likely that even we in the U.S. will build cooling towers for river-cooled power plants long before then—a necessity that, together with others, might even modify our energy demands.

But the future looks different for different rivers—temperate, tropical, developed, and undeveloped, and if one selects with care one's angle of sight, one's glances might even be tinged by a ray of hope.

Whether we look at the globe or at one river, our guesses, let alone predictions, about the future rely on an examination of past and present. Such an examination of rivers and what man does to them is promised to us in the next few days. The organizers of this symposium, too numerous to mention singly, and through them our host, the University of Massachusetts, recognize the importance of rivers to the welfare of man. They know that an examination of river case histories and present river problems—our fare for the next few days—

1

should give us some guidelines to act with wisdom, rather than panic, when the future becomes the present. Man has modified rivers from the dawn of history although his impact may at first have been elusive.

Nearly fifteen years ago I flew with an FAO team over certain portions of the Mekong River watershed. Some of the areas I had visited by jeep or on foot, but it took the view from a DC-3 to impress on me that man at any stage in culture and at any density is closely associated with and modifies rivers and streams. After this trip plans were begun for the development of tributaries and portions of the main stream of this mighty river, quite possibly to lead within the next two or three decades to a quantum jump in the impact man will have on it. Our knowledge of the ecology of these waters is scant indeed, and therefore predictions of the long-term ecological effects produced by the works of man remain of a general nature—so general, in fact, that Mekong development planners, though cognizant of the warnings even vague preductions contain, cannot find them highly useful. The picture for the West that emerges from reading the papers of this congress is somewhat different. The case histories impress one with the particular problems posed by man's activities, but one can extract some interesting generalities. The first is that, in some cases at least, pollution conditions are being alleviated. Dr. Thomann mentions 1975 as the year after which conditions on the Delaware—a long- and mightily abused river—should improve. He says that "at that time and for subsequent years all look forward to a river system where a balance has been achieved between man and his activities and the ecological structure of his environment." I envy him his optimism even though it seems well founded.

Another river, the Thames in England, shows comparable trends; there man made flow modifications long enough ago that a new ecological balance has been reached. Dr. Mann compares that river and its associated flowages to a kidney, highly taxed by the strains its owner puts on it, but nevertheless working quite well. Incidentally, as a physiologist I note that a high pollution load must not be a very serious problem on the Thames—otherwise the anatomic anology would have to include the liver as a detoxifying organ.

We all know that the Ohio was in far worse shape some twenty years ago, before Orsanco began to do its work, than it is now, and even the treatments of the Illinois and the Wisconsin rivers suggest that although problems exist, attempts are being made to deal with the competitive and multiple demands that industrial man places on his rivers. Integrated river basin planning and administration either is, or can become, a reality.

Conflicts of interest remain. I mention here only one of them, especially applicable to smaller rivers: the more natural a river, the higher the esthetic quality, as affluent societies understand the term, and the more sought after for recreational use. But ten thousand canoeists of the socio-cultural persuasion of the American consumer must change such a river, as is clearly borne out by the fate of the Ausable or Pere Marquette.

I have read the case histories and some of the other papers submitted to the symposium and, believe it or not, I did some outside reading as well. I found that Ralph Nader does not share the optimism of some of the symposium papers. In his report, "Water Wasteland," he castigates the federal government for perennial listlessness, many words, and little action. Fifteen years after the National Water-Pollution Control Act was passed in 1956, after seven laws and after the expenditure of $3\frac{1}{2}$ billion dollars, the level of filth has not been reduced in a single body of water, his report says. Industry's share of pollution is now four to five times that of domestic sources and continues to rise. The report is meticulously documented and in places hard-hitting as it plunges into the politics, legalities, economics, and unsavory facts of water-pollution. It has reinforced my suspicion that we have skirted the economic and political problems—really not surprising since most decision-makers still view the world with the concepts of the first half of the twentieth century, instead of those apt to get us through the second. Perhaps this is the reason for the rise in environmental lawsuits in the U.S., where some individuals see the need to blaze new paths in court, rather than through other forms of public pressure.

Not being able to reconcile the gist of the Nader report with the reports I read on some of the rivers, I turned from the intricacies of an advanced technological civilization to Dr. Desmond Hamilton's remarks on the River Nile. Being interested in the Mekong, I thought that certain experiences of the Nile-modifiers might perhaps be applicable to that Asian river. I was again struck by the differences between tropical and temperate zone rivers. Water fluctuations of the former are usually large, and of the latter, relatively moderate. The number of plant and animal species is far greater in the tropics, where the potential for fast biological regeneration is also high. A comparison of the Mississippi, for instance (even though it is not a true temperate zone river) with the Mekong shows not only more fish families in the latter, but also more genera and species in those families that the rivers have in common; catfish species in the Mississippi number 25, while there are well over 50 in the Mekong. Dr. Bailey tells me there are 95 minnow species in the Mississippi; I suspect their number exceeds 150 in the Mekong.

3

There are differences also in the uses by man of the rivers of the tropics and the temperate zone. The most prominent of these is that the tropical rivers by and large are not yet used to absorb industrial wastes, though in places they too are being loaded with additions of organic materials, with which they can apparently cope because of their naturally high regenerative capacity and productivity.

But back to the River Nile and what the changes on and in it may portend for the many rivers in the tropics where man is now planning, and has in places already instituted, hydraulic controls that dwarf any such previous attempts. On the Volta, the Mekong, the Parana, and the Brahmaputra dams can give power, provide water for irrigation, improve the transportation of goods, provide means to control or curb the incursion of salt water in the delta, and create large lakes in which fish production may be managed. When the economy of the country in question advances, a regulated river may better accept industrial waste than an unregulated one. At that same future time the recreation potential of man-made lakes interposed by sections of river may assume some importance. To this list of potential benefits, add the fact that the power produced is clean—that neither heat nor air pollution is generated in its production.

Dysfunctions, however, can arise from dams and man-made lakes on large tropical rivers. Although not all hydraulic regulations will affect the water regimes of rivers as drastically as the Aswan Dam affected the Nile, we must guard in future dam building against the unexpected side effects associated with some of the dams in Africa. There are changes in the sedimentation patterns of the delta, encroachment of aquatic weeds, losses in fish stocks in the sea that were once associated with the rivers' fluctuating fertility in their effluent cones, exclusion of anadromous fish stocks from spawning grounds, spread of waterborne diseases, loss in floodplain agriculture, damage to power-generating structures through climate-induced meromixis, and the welter of problems associated with the relocation of people displaced by the waters that accumulate behind the new dams.

Some of these resettlement problems threaten to be intractable indeed. The closure of a dam and the resulting benefits of a closed reservoir, that is electricity, perennial water, fishing opportunities, etc., become available only after displacement of people. But it is one thing to say that a resettled person should at least be provided equivalent conditions, and another to execute such plans. If the person has been in an agricultural occupation, it is likely that the good agricultural soils of a region are already occupied.

4

The present estimate of people to be displaced by the proposed Mekong mainstream dam at Pa Mong is around 300,000; by 1990, a possible target date for this development, they will number a half million. I doubt that adequate plans could be executed to move such numbers in the United States without considerable discomfort, if not disaster, to those concerned. Now imagine this to happen after the war in Indochina. Imagine further that you live in a small Laotian town and that rumor reaches you that your house, your school, or your temple may well be covered by water. This rumor is given some substance by men with red and white sticks and some kinds of telescopes appearing from jeeps and crawling all over the countryside. The grapevine has it that you will be subjected to forced uprooting and that you may have to live for a number of years as a refugee. Note also that you are not a typical American used to changing your residence once every two or three years if the company so desires, but a traditional peasant, very much tied to your land, the people in your village or town, your priests, and your temple. Those who govern you have not been efficient in telling you their plans and what these might mean for you. People from across the river come and tell you that if you work with them and join them, social adjustment due to resettlement might be easier for you and you would live in a society where you might have more to say about your own future. Thus social and political boomerangs are added to the ecological ones already mentioned when one attempts large-scale modifications of rivers.

This does not mean that dams should not be built or rivers regulated, but rather that we should be more informed and foresighted.

In the developed world we have come by the political and administrative instruments of river management only with considerable difficulty and after detours and mistakes. One might say we have been on many a creek without a paddle. But surprisingly quickly we have developed river basin commissions, regional planning boards, effluent standards, pollution fines (insufficient though they may be), water resources commissions, and the like. We also have a public with a voice in these matters and we have Nader reports. Tropical rivers do not have those safeguards. Most of them flow through countries where the per capita GNP is but a small fraction of ours and the politicians quite understandably say that they will worry about the ecology later. There are very few riparian watchdogs along those rivers, and those who do exist are usually foreigners to the lands in which the rivers flow and therefore of relatively low persuasive impact. It seems, therefore, likely that our transgressions will be visited on those who now will emulate us. In fact, it is not unlikely that the problems they will face with their rivers, which they plan to modify faster and to a greater degree than we have ours, will even be worse than the ones with which we grappled. It is therefore my conviction that we must use all persuasive means at our disposal

5

through all possible channels to do several things: We must train and educate many riparian river watchers in tropical countries so that the necessary fact-finding can be done; we must assure in so-called development projects the inclusion of adequate environmental forecasting; in doing so, we must try to assess the second- and third-order consequences of river modifications; and even now we must pay particular attention to incipient industrialization along tropical river courses. Enough is now known to engage in predictive modeling of the impacts of man on aspects of river basin ecosystems. Such technical expertise may be applicable to dealing with the human impact problems on tropical rivers, as it was and is on temperate ones. If humble and in the right spirit, one might even apply some of our administrative solutions. It is clear that international and national organizations, as well as private industry, must join in this task. Let the American Fisheries Society be one of them with the dream, perhaps, that the trials and tribulations that once beset the Ohio need not be repeated on the Parana. If a Delaware River Basin Commission can accomplish what it says it can for several *counties* that share a river, perhaps also *countries* can cooperate in developing the liquid lifeline that joins them so as to achieve a balance between the ecological exigencies of a watershed and the activities that man wants to undertake therein. Utopian as this goal may sound, some of the things that will be said here in the next few days will certainly be applicable to it. Will they be applied, though? Your guess is as good as mine. As we hope that they will, let me propose a motto for this conference—one that combines the right amount of levity with the right amount of determination:

IN QUALCUMQUE FLUMINE UT NUNGUAM SINE REMO

which translates freely to:

ON WHATEVER CREEK, STREAM, OR RIVER, MAY WE NEVER
BE WITHOUT A PADDLE

SESSION I

WHAT IS A RIVER?
Chairman: Ruth Patrick

RIVERS – A GEOMORPHIC AND CHEMICAL OVERVIEW

R. R. Curry

The earth scientist views a river as a link in the hydrologic cycle; as a locus of erosion, transportation, and deposition of dissolved, suspended, and tractively carried geologic materials; and as a complete open physical system hydrodynamically balancing and distributing energy and work over the earth's land areas. This view has developed over the last 500 years beginning with Bernard Palissy and Pierre Perrault's development of the role of rivers in the hydrologic cycle in the 16th and 17th centuries. British geologists James Hutton and John Playfair in the late 18th and early 19th centuries set the stage for the development of modern geomorphic concepts by arguing forcefully with geologic evidence that Leonardo da Vinci had indeed been correct in stating earlier that rivers carve their own valleys. Final foundations were set in America in the period 1870 to 1900 by such men as John Wesley Powell and his concepts of base level and geomorphic histories of river systems (1875), Grove Karl Gilbert (1877) and his pioneering work on the hydraulic mechanisms of river erosion and erosional landscape formation, and William Morris Davis (1909) with his once widely accepted ideas about evolutionary stages in the "life" of rivers and concepts of cyclicity and rejuvenation of fluvial landscapes.

After publication of Davis' work in the period 1899-1909, fluvial geomorphology, or the study of the development and dynamics of rivers, was essentially quiescent in America for fifty years. Then J. Hoover Mackin (1948) began to try with difficulty to define accurately the accepted concepts, such as base-level, balance in a river system, and the concept of the graded river, and Kirk Bryan encouraged in his Harvard geomorphology students a serious questioning of the fundamentals of the development of river systems. At present two decades of questioning of earlier concepts are beginning to bear fruit. Gilbert's original ideas of the dynamic response of river systems to changes in climatic and geologic conditions imposed upon them has survived and been strengthened by workers like John Hack (1960) and Luna Leopold (Leopold, Wolman, and Miller, 1964; Leopold and Wolman, 1960; Leopold and Langbein, 1962). This has been coupled with a realization of the importance of basic principles of fluid dynamics, such as put forth by R. A. Bagnold on the physics of windblown sand (1941) and on transport in all fluid media (1966) and has been combined to form the nucleus of modern concepts of the physical development of rivers. Certainly no single body of work has had more fundamental impact on development of current concepts than that of Leopold, as typified by his 1964 text (Leopold, Wolman, and Miller) on fluvial processes in geomorphology. A few of these current geomorphic concepts will be reviewed in this paper, and an attempt will be made to integrate this information with knowledge of the chemical nature of river systems.

Current Concepts in Fluvial Geomorphology

The Geomorphic Limits of a River. Most geomorphologists and geochemists view rivers in terms of their entire watersheds, rather than just in terms of the actual body of water flowing in channels. All are familiar with the hierarchy of channels in a stream network, as carefully defined by R. E. Horton (1945), but some may not realize that the same principles of fluid flow define some processes acting upon the landscape outside the river channel well above the smallest ephemeral rills leading to the smallest tributary of the main river. Emmett (1970) has worked with overland flow on hill slopes to show that the work done by water before it becomes confined to a channel may be greater than that done by the same volume of water in the tributary stream channel and that thus the dissolved and suspended load of the main river, and hence its entire geomorphic character, is in large measure a function of the geologic, biologic, and meteorologic conditions within the entire watershed. The geomorphologist, then, distinguishes between channelized and unchannelized flow and distinguishes different processes and landforms resulting from the action of running water in these two cases; but generally the development and physical character of a given river is considered in terms of the aggregate sum of all kinds of flow contributing water or other materials to that river. Current use, therefore, is to consider a river and its entire watershed as a system.

Rivers as Systems. Beginning with the work of Leopold and Maddock (1953), both rivers and the landscapes upon which they flow have been considered as systems in several senses of the word. For most, the approach of systems theory helps to explain a dynamic interrelatedness of factors affecting flow in streams, their channels, and enclosing canyons (King, 1966/67). In their 1953 paper Leopold and Maddock defined the interrelated hydraulic functions of a river, its channel, and its solid load with the concept of a hydraulic system. These interrelated river-system factors are termed the hydraulic geometry of a river.

By defining and quantifying the variables in a river's hydraulic geometry, one can equate the discharge of a river with its suspended particulate sediment load, its mean velocity, and its channel geometry. Leopold showed empirically that water depth, width, and velocity are functions of the load transported by the river and that one could thus predict, for example, the effects of changes in load

10

supplied by side streams upon the entire geometry of the system. One fundamental effect of this work was to impress upon fluvial geomorphologists the fact
that rivers can change, almost instantly, several factors of their hydraulic geometry in response to changes in other factors imposed upon the main channel by
tributary streams or natural or man-made changes along the main river itself.
For instance, disruption of overload flow in a watershed as the result of
construction may raise the sediment yield from that watershed by two orders of
magnitude (Wolman, 1964). This local load increase imposed by a tributary
stream upon a river would only very slowly produce an increase in gradient of the
trunk river but could immediately begin to alter the width-to-depth ratio of that
river channel in response to the imposed load. To carry a higher suspended load
without increasing its velocity by an increase in grade, the river would probably
become narrower, or the channel bed friction would decrease by siltation, or
both. The net result in this instance would be an increase in available energy to
do the work of transporting the imposed load within a given reach of the river
system. This leads to the second concept of rivers as systems—that of the dynamic
system of energy balance.

The consideration of the energy and work balance of rivers began as an
analogy with thermodynamic systems, as propounded by Leopold and Langbein
in their 1962 paper on the concept of entropy in landscape evolution. The thesis
of this paper was that the distribution of energy in a river tends toward the most
probable state. Prior to consideration of this simple concept, geomorphologists
had been hard pressed to elucidate basic causes of variations in stream patterns
such as meandering, straight, and braided channels and to derive simple postulates
for the factors controlling the longitudinal profiles of rivers. Probability concepts
imply an assumption of randomness for processes of flowing water (Scheidegger
and Langbein, 1966). This assumption has allowed fluvial geomorphologists to
discard the necessity to precisely define, for instance, the exact parameters of
the hydraulic conditions causing the initial formation of meanders in a previously
straight river channel and has opened the way for the use of statistical concepts
to describe the physical state of a river and its dynamic equilibrium.

Current ideas are that river systems proceed toward conditions of a steady-
state balance, wherein the open systems balance a continuous (though not
necessarily constant) supply and removal of water and sediment by adjustment
of the geometry of the system itself. This steady-state open system is only rarely
characterized by exact equilibrium, and generally the river and its landscape tend
toward a mean form, definable only in terms of statistical means and extremes
(Chorley, 1962). The equilibration of river systems is manifest in the tendency
of the river to maximize the efficiency of energy received and utilized, balanced
against its tendency to make the efficiency of this use constant through time.

These opposing tendencies, which Leopold (1970) correctly points out are fundamental to all natural physical systems such as biologic communities, define the conditions of balance in river systems, or ecosystems, and are time independent. This means that William Morris Davis' concept of the geographic cycle, wherein the conditions at one state of a river's history are dependent upon conditions in previous stages of its history, is not necessarily valid because different initial conditions may yield similar results in terms of the dynamic state of a river at any point in time.

The two opposing tendencies are considered by geomorphologists under the theories of minimum variance (Langbein and Leopold, 1966; Scheidegger and Langbein, 1966) and minimization of work (Leopold and Langbein, 1962, p. 6). Biologists may be most familiar with these opposing tendencies through Garrett Hardin's theorem (1963) which states that one cannot simultaneously maximize efficiency and stability in biologic systems. These concepts and their inter-relations define river systems in as fundamental a way as the second law of thermodynamics defines quantifiable closed physical systems. With understanding of these concepts, a biologist, for instance, can better appreciate the dynamics of energy environments in a river as they may affect aquatic organisms; while the hydraulic engineer may better predict the response of a river system to man-caused change.

Let us consider the longitudinal profile of a river. There is first the tendency in the system to operate so as to minimize the work done by each definable quantum of water or sediment moving in the river. The most probable distribution of energy exists when the rate of gain of entropy in each interval along the river is equal. Since the higher the river is above its base level, the greater is its possible number of slopes, the rate of loss of energy at any point is proportional to its elevation above that base level (because entropy represents a ratio between a given state and the number of alternate states). The most probable sequence of energy loss along successively lower reaches of a river corresponds to the conditions of a uniform increase in entropy in each succeeding river segment. When this condition occurs, without restraint on the river's length, the theoretical river profile should be exponential—with a marked concavity.

However, most river profiles merely approach an exponential form and are not quite that concave. This follows from the second balancing restraint, that of minimization of variance. Another way of stating this principle is that a river tends toward a uniformity in rate of doing work in various parts of the system. This is also an expression of Le Chatelier's principle, wherein each variable factor resists change, or the variance of each tends toward zero. A change in a balanced river system, as, for instance, the addition of water at a tributary junction, will produce changes in velocity, depth, width, shear, and bed form, among others, and each of these changes will produce effects in the others that react against the change to maintain the tendency toward stability or balance. Thus, in the case of the longitudinal profile of a river, minimum variance would predict a uniform energy expenditure along the river's length, necessitating a straight profile. Neither straight nor directly exponential profiles are typical due to the opposing effects of the two mentioned tendencies. Actual river profiles are the result of a dynamic balance between the two. The exact equations of balance resist precise definition as yet, but by considering the overall profile as representing a most probable state at a given point in time, one can at least define most of the variables affecting the longitudinal profile and theoretically consider the effects of each as if they acted alone on the river system.

The idea of least work, or minimum energy utilization, as a tendency of river systems is, of course, a statement of efficiency. In rivers, most of the potential energy represented by the height of the water mass above its base level is lost or dissipated into heat by the flow of that water to that base level—analogous to absolute zero in thermodynamics. Bagnold's recent work (1968) has confirmed earlier estimates of the thermal efficiency of rivers with notation that only about 2% of the available energy is used directly for sediment transport. The 98% energy lost may, to a minor extent, be compensated by changes in the dissolved chemical load of the river but, for the most part, is expended as frictional heat energy—of great importance as energy input in biologic communities. In this sense, the aquatic biologist must become thoroughly familiar with the distribution of energy in river systems to understand stability and species diversity in riparian communities. For instance, the tendency toward uniformity of expenditures of energy in river systems in equilibrium predicts a maximization of the total number of potential biologic niches along the length of a river. An imbalance such as a waterfall plunge pool or beaver dam may thus locally decrease species diversity as the variety of energy utilization sites drops. But just as a chain under tension responds by equalization of the work done by each link, so also does the river; and the beaver dam or engineers' dam or waterfall will be sites of compensatory changes in the hydraulic variables of the river which will tend rapidly to remove the cause of the imbalance. The increase in velocity of water falling over a waterfall results in an increase in downcutting at the fall brink and an increase in erosion in the plunge pool, resulting in both backwasting and downwasting of the falls until a condition of equilibrium is reached. Similarly, the decrease in velocity of water where a stream enters a beaver-dammed pond results in a decrease in energy available for transportation of sediment and an increase in sediment deposited both in the pond and progressively upstream from the pond until a continuous gradient is established across the site of the pond for equilization of energy expenditure per unit length of stream course.

In the past it has been popular to question why rivers meander (Callander, 1968). The theory of minimum variance, however, has now focused attention on the more basic problem of why some rivers do not meander. Although it is true that the shortest distance between two points is a straight line, it does not follow that the most probable path of a randomly moving particle between two points is a straight line. It is, in fact, a meandering line (Von Schelling, 1951, 1964; Scheidegger, 1967), precisely as found in natural rivers. Clearly, there is a greater probability that the deviation of particles of water will be in the direction of steepest gradient, but in most rivers the influence of gradient is small. Von Schelling showed that the most probable random walk pattern between a given set of points was a meandering curve of the sine-generated type seen in natural rivers. Langbein and Leopold (1966) have shown that this random walk pattern satisfies the conditions of minimum variance with a reach of river; i.e., that water without constraint will meander. One has merely to observe a very small stream of water flowing across an inclined porcelain tile or clean glass plate, constrained only by surface tension, to note that it flows in a meandering pattern. By regulating the flow of this stream of water and by changing friction and surface tension by adding minute quantities of a wetting agent such as soap, one can see that the meander amplitude and wavelength vary as a function of the ratio between discharge and depth. If too wide a channel is formed, by adding more wetting agent or exceeding a critical discharge, the meandering pattern breaks up into a series of smaller streams of water which coalesce into a braided channel. Langbein and Leopold (1966) showed by field observations that downstream spatial variance in shear and friction were lower in natural meandering channels than in otherwise comparable straight reaches of rivers. I would intuitively suggest that this minimization of variance by meandering holds on a temporal as well as spatial basis; that is, rivers subject to variations in flow at different times are better able to carry their varying discharges and loads with a maximized efficiency of use of energy received both spatially and temporally if they flow in meandering channels. The same holds for all naturally flowing media—from the jet stream, to ocean currents, possibly even to radio and light waves.

A meandering river has greater changes in bed contours than a straight segment of river. However, as depth changes, so also does the slope of the water surface and the energy gradient, with the combined effect of actually reducing the variability in bed shear and friction, thus minimizing work of the system while equalizing the rate of gain of entropy along the river's longitudinal profile. With this understanding, it becomes a relatively simple matter to predict how a river will react to man-caused changes—as, for instance, the straightening of a meander loop. Just as is predicted by Le Chatelier's principle, the straightened river resists change and tends toward the reestablishment of a stable minimum variance regime in both time and space. Specifically, we can envision a situation where straightening a meander would increase the gradient in the straightened reach, resulting in increased velocity, erosion, and transportation and increased depth of scour during seasonal flooding. The net result, in a short time of perhaps one year, would be to reestablish a quasi-equilibrium condition, possibly of alternating pools and riffles, with these bed forms moving slowly downstream until total downstream accumulation of sediment would permit regrading of the whole reach across the cut-off meander. The pools and riffles themselves would effectively dissipate more energy in the altered reach, negating both equalization and minimization of work over the longer reach passing through that altered meander. The eventual result, over a period of years, would be to reestablish a regraded meander pattern developing out of the pool and riffle longitudinal inhomogeneity.

This scenario brings into question the dynamics of gravel bars, riffles, rapids over transported bed materials, and other bed forms in minimizing variance and maximizing efficiency and distribution of work. Langbein and Leopold (1968) have postulated that moving bed-load materials in rivers become kinematically grouped or aggregated in spatially predictable forms as a function of the interaction of the individual particles. This is so because the tractive force that a river can apply to a single cobble of a given size is greatest if that cobble is far from other similar cobbles or holes in the riverbed between such cobbles. Once a group of such cobbles are aggregated into a gravel bar or riffle, they can generally not move as a group in an equilibrium river because equalization of work along the river course precludes concentrations of energy necessary to overcome the resistance to shear of the closely packed particles. Thus a kinematic wave form is produced by all interacting quanta moving as a fluid, from sand and gravel in river, through automobiles moving along a thruway, to radio noise and solar proton bursts in space.

Just as meandering minimizes variance in bed shear and friction, so also will the development of pools and riffles reduce variance by interacting adjustments of depth, velocity, slope, and energy gradient. This, of course, suggests that it is the existence of bed materials or bed load that causes certain rivers to meander, and that it is an outcome of the erosional, transportational, and depositional processes within rivers that leads to the meandering. Most rivers have bed load and most meander; but some meander with no bed load, such as supraglacial meltwater channels and the stream of water on the porcelain plate, due simply to the most probable configuration of particles randomly moving between the two points. Others, such as glacial outwash channels and ephemeral streams in arid regions, may move very large quantities of bed material without meandering. This later condition, noted in braided channels, occurs as a result of the very large imposed bed load. Such rivers achieve a time and space equilibrium through widening their channels and making them shallower to expend energy more evenly and efficiently as shear against the bed. In such braided channels the courses of the principal flow of water change constantly, as do the loci of motion of bed materials; but the total result, averaged over a period of time, is one of statistical minimization of variance for the whole width of the composite channel, as well as maximization of efficiency of transport of the water and its load across its transportational gradient.

Just as an individual river responds rapidly to the constraints of minimization of work and variance, so also do networks of rivers—tributary streams, rivulets, and the feeding overland hill-slope flow. Drainage networks, where unrestricted by differential erosion along varying bedrock units, develop in the well-known dendritic pattern. This pattern has been shown (Leopold and Langbein, 1962) to be the most probable form for individual channels of water moving randomly downslope, constrained only by initial spacing and the condition of absorption when two randomly moving channels happen to come together. Thus river networks may exhibit a pattern of optimum probability in terms of distribution of energy, suggesting that the entire network, as well as the individual river channels, may well represent an expression of a condition of dynamic balance between the tendency to maximize the efficiency of energy received in the watershed and the tendency to make use of this energy constant through time throughout the watershed. However, in the case of a hierarchy of river drainage networks, an additional factor of competition occurs between adjacent watersheds wherein no one watershed can grow except at the expense of the size of its neighbors. This would be expected to create a non-Gaussian size distribution of watersheds but should not otherwise affect applicability of the concepts of spatial and temporal equilibrium within the system.

The Factor of Time in River Systems. William Morris Davis postulated that rivers and their landscapes undergo a very slow "aging" process or a sort of integrationary evolution before reaching an equilibrium in the stages he termed maturity and old age. This concept drew extensively on ideas of Darwinian evolution and organismic growth which were reaching wide public acceptability in Davis' time, around the beginning of the twentieth century. Just as biology has now developed an emphasis on a combination of systems approaches and thermodynamic cellular considerations, so also have our concepts of river systems and hydraulic behavior of individual particles.

In place of Davis' ideas that hundreds to millions of years must elapse before equilibrium is reached in rivers, we now see rivers as adjusting quite rapidly to changes in hydraulic geometry and achieving conditions of dynamic balance in hours to years of time. This is not to say that a waterfall, for instance, may not persist for millennia. Rather, we now see the conditions of balance among the hydraulic variables such as depth, width, hydraulic radius, slope, velocity, shear force, friction factor, sediment transport, intensity of power expenditure, etc., as rapidly equilibrating for minimization of total work and variance in each and every reach of a river under consideration, including all or part of a waterfall. This quasi-equilibrium concept does not attempt to predict the ultimate form a river may take through a sum total of changes through time, but rather suggests a statistical range of immediate dynamic responses that will occur from any given instantaneous change in the system. Rivers are recognized as not being static entities, and we now understand the river channel as a form representing the most efficient—in terms of energy utilization—geometry capable of accommodating the sum total of the means and extremes of variability of flow that have occurred in that channel throughout its history. Thus a hundred-year-old irrigation ditch can reflect only the history of forces working upon it for the hundred-year period and will not reflect the extremes or means of flow to be expected from any randomly occurring hundred-year sequence of local meteorologic and runoff conditions. This difference, although perhaps subtle, is of utmost importance in considering the effects of man on river systems and his ability to predict extremes of natural variation in flow regimes.

There is a marked deficiency in the use of present-day hydrologic data by fluvial geomorphologists to attempt to explain the gross morphology of river systems. Past climatic variations, affecting the geometry of actual river channels, are not accurately known. Further, the effects of human interference in runoff and precipitation variations cannot be eliminated from present records. Discharge records may not include the largest floods that have ever occurred in a given site, yet the effects upon the landscape of such floods may have been so catastrophic as essentially to define the gross morphology of the entire river channel. The Guil floods recorded by Tricart (1961) are an example of such a geologic event which accomplished more geomorphological change in a week than in the whole of the previous ten to fourteen thousand years of post-glacial time.

The idea of "grade" in the Davisian sense of ultimate balance in a river system is really very different from the idea of dynamic equilibrium in individual reaches of a river. Although the river responds rapidly to imposed changes, it merely does so within certain limits. Thus a river can change its width-to-depth ratio in a matter of days in response to a change in imposed bed load, for instance; whereas it may take millennia to adjust its total energy gradient or longitudinal profile to the optimum for that new load with the range of discharge available to move it through the difference in elevation existing along the river course. Some variables can adjust rapidly, while others take long periods of time; however, the total system remains in quasi-equilibrium as one variable slowly

17

compensates for the other. The ultimate condition achieved will be the one with the most efficient utilization of energy received and the maximum constancy of this use through time. Therefore, the total history of the river will affect its geometry as a function of the magnitude of variance of hydraulic variables occurring during that time.

The channel of the Connecticut River, for example, represents an adjustment to the range of means and extremes of discharge and load imposed upon it in both deglacial and non-glacial times. In such condition of high prehistoric variability, a most efficient transport system is one with temporary storage facilities for the large volumes of bed and suspended load that can be moved only during the times of maximum discharge, as at times of melting of major ice sheets and draining of proglacial lakes. Such a storage facility is manifest in the fluvial terrace sequence, which we tend to think of as a relic of past conditions, rather than a measure of the range of conditions, that the river is adjusted to move at any appropriate time.

Likewise, the present floodplain deposits, the gravel bars, and all the alluvial materials in a river channel represent, in their depositional forms, a transient state that results from the maximization of the efficiency of energy use through time. Thus the description of the physical state of these materials at a point in time can only represent a transient condition and cannot define the hydraulics of the system any more than the recording of the genetic character of a single man can define the biologic character of mankind. For example, it has been demonstrated that the observed downstream reduction in particle size in a riverbed cannot always be the result of hydraulic sorting or intergranular abrasion of the bed-load material, and hence cannot represent a short-term product of the river hydraulics (Schumm and Lichty, 1965). We are forced to assume in some cases that the observed phenomenon is the result of weathering of the particles as they reside in floodplains and other storage features for long periods of time between periods of downstream transport (Bradley, 1970; Clayton, 1968).

Thus time is of import in consideration of the geomorphic development of rivers from the standpoint that the river channel rapidly assumes a shape that most efficiently accommodates the variations in flow and load occurring over time. The adjustment occurs rapidly, but the variations to which the channel may be adjusted may occur over very long periods of geologic time. For example, many present-day river channels, including their floodplains, are adjusted to maximum flow conditions occurring during periods of deglaciation. The lower portion of the channel, commonly called the active floodplain, reflects adjustment to that subset of conditions occurring through post-glacial times, whereas one or more higher terraces represent a subset of conditions occurring under higher discharge and load regimes.

Another important aspect of the time factor in the geomorphic development of river systems is that of the frequency of events. Although it is probably true that events of intermediate frequency of recurrence largely dictate the short-term character of river-channel geometry (Wolman and Miller, 1960; Piest, 1963), an appreciation of the rapid balancing of the hydraulic variables in rivers forces acceptance of the fact that the gross morphology of the floodplain and terrace sequences within river valleys may exist in response to very infrequent events of large magnitude. Catastrophic events of great geomorphic significance are merely events of exceedingly infrequent recurrence intervals, such as once every ten thousand years. If we had long enough records of the flow of some major rivers of the world, we might indeed note that the 50,000-year flood of the Colorado River may be of a very different magnitude from the 10,000-year flood because of wholly different causal mechanisms—as, for instance, during periods of deglaciation with hydrologic regimes of high flow augmented by periodic release of glacier-dammed lakes.

If rivers can adjust rapidly to equilibrium conditons in response to a wide variety of flow conditions occurring over a long period of time, it follows that in alluvial materials there should be subsets of nesting river channels, with each younger set reflecting optimum adjustment within the time period allowed for such adjustment. Thus in some areas the frequency and magnitude of flow events over the last 100 years in a given drainage basin may determine the character of the present channel and its floodplain, whereas the range of conditions occurring over the last thousand years and 10,000 years might be reflected in larger terrace remnants or sizes of bed-load material found within and around the present channel. This is not to negate the concepts of tectonic and base-level changes occurring through long periods of geologic time as being of primary import in determining the differential elevations of, for instance, the Rhine Terraces. But certainly many have puzzled over the geometry of many terrace-sequence situations wherein successively younger terrace remnants are of progressively lesser lateral and aerial extent. We may attempt to explain this situation by saying that stable flow conditions occurred for longer times when forming the older terrace surfaces, or that flood magnitudes have been decreasing with time as the drainage system develops, but in the case of many depositional terrace sequences, these arguments are often without much foundation.

The Factor of Man in River Systems. Man's influence on the load of sediment carried by rivers is well known and documented. Wohlman (1964) has shown that impact of urban construction in eastern United States watersheds may increase sediment yield 100 times, from an average of 1,000 tons per square mile per year to 100,000 or more tons per square mile per year. Data from sparsely populated watersheds in western United States (Wahrhaftig and Curry, 1967) show that similar results can occur from man's mere interference with vegetative cover and land stability when conducting logging and road-building operations. Much more difficult to assess is the actual impact of man on the geometry of the river channels. We know that urbanization, logging, land clearing, and many agricultural practices all cause an increase in the magnitude of a flood occurring as a result of a given intensity of precipitation or snowmelt; and we know that the time of peaking of this flood is shortened over what it would have been if the surface water had to drain through a mature climax vegetative cover before reaching watercourses (Leopold, 1968). However, we cannot go to the Ganges River, for example, and establish its flow regime before the advent of mankind. The present river channel represents a condition of equilibrium established after clearing of trees from both plains and slopes and after considerable change in both the character and magnitude of suspended and bed-load materials carried by the river as the result of erosion in its watershed. Open slopes in an area where a dry season is followed by a monsoon are subject to rapid erosion of materials that may then be easily transported by a river with flood discharges of 3 million ft^3 per second. Since the interrelationships implicit in the hydraulic geometry of a river suggest that an imposed increase in suspended load may result in a decrease in energy available for bank erosion or bed-load transport, certainly one would predict substantial changes in the Ganges River channel shape, energy distribution, biologic populations, and hydraulic flow regime as a result of man's occupation of its watershed. We have not yet developed the ability to predict the precise adjustment that could be expected from such a change but can say that the new equilibrium conditions must be based upon a different distribution of energy use if maximum efficiency of such use is to be maintained.

Similarly, we can predict a range of change that would occur after channelization of streams that are either ephemeral or subject to frequent changes in course during flooding. Current practice in much of North America and Europe is to straighten and reinforce river channels, with materials varying from discarded automobile bodies to reinforced concrete, where flooding becomes a nuisance to man's use of adjacent lands. For instance, where cities have been located along river courses for long periods of time, it can often be demonstrated that the frequency of damaging floods of a particular magnitude is increasing and causing increased hazard to the city dwellers. Such increase in frequency of floods of a given magnitude can usually be attributed to decreases in infiltration in the upstream watershed and changes in upstream sediment yield and thus downstream channel geometry—all caused by mankind's alteration of the land and its vegetation. A frequent response to this flood hazard is to encase the river in a concrete channel or at least to build concrete walls along the river course through the city. Such walls often decrease the roughness or friction along the river banks, upsetting energy expenditure balance for the reach of river through the city.

Such decrease in friction may increase velocity locally, or it may increase depth of scouring during flooding, or both, and would increase rate of transport of bed materials through that reach. Such disequilibrium would then readjust down-stream where the channelization ended, often with resulting increases in bank and bed erosion due to the increase in velocity of the river exiting from the channelized reach. In southern California, for instance, this downstream effect of upstream change has led to the necessity, in some cases, of building a complete concrete channel for many streams all the way from the cities to the ocean, since not to do so would increase the flood hazard and damage suffered by each succeeding city downstream. Where such concrete channels fail, as on occasions of large magnitude floods, results are often much more disastrous than they might have been had the channel not been altered in the first place. Straightened channels, with their steeper gradients and higher velocities will, upon abandon-ment during flood, often reassume a braided or meandering pattern much to the detriment of the city established along its banks. Even liability law in the United States is beginning to recognize these hydrodynamic realities (Davis, 1968).

Such man-caused changes in energy balance in rivers have the utmost impor-tance upon the river's ecology. If one accepts the premise that organisms occupy niches in such a manner as to achieve a maximization of efficiency of energy utilization, both in space and time, then it follows that human effects upon the distribution and non-biological use of that energy must affect the biologic systems.

River Water Chemistry

Mechanisms Controlling River Water Chemistry. The chemical load of a river, generally considered its dissolved load, is derived from atmospheric precipitation, from mineral matter dissolved by the surface and ground waters as they flow over and through the land, and from atmospheric gases dissolved into the surface water. The mechanisms controlling surface water chemistry have been recently reviewed by Gibbs (1970), who distinguishes three basic origins for loads of dis-solved salts in surface waters. These are atmospheric precipitation, rock weather-ing, and the evaporation-crystallization process. We may add to these the abiogenic and biogenic contributions of dissolved atmospheric gases (Holl, 1955; Lindroth, 1957) to broadly define the overall origins of the chemical load of rivers under natural conditions.

Figure 1 illustrates some of the controlling mechanisms for water chemistry for some of the major rivers of the world. One can note that highly saline rivers, with large dissolved loads, have relatively higher weight ratios of sodium to sodium plus calcium. These rivers, such as the Rio Grande and the Pecos, often show marked increases in salinity downstream as the river water evaporates.

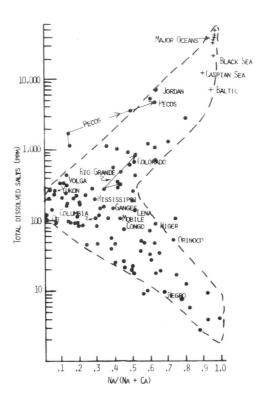

Figure 1. Weight ratio of Na/(Na + Ca) as a function of total dissolved salts for rivers of the world. (Modified from Gibbs, 1970, p. 1088)

Gibbs postulates that these relative enrichments in sodium as evaporation and crystallization progresses in rivers in hot, arid regions is due to precipitation of $CaCO_3$ from solution and thus a transfer from dissolved to suspended load or a precipitation of this material along the river channel. Such a downstream increase in suspended load may help explain why rivers such as the Rio Grande build such extensive overbank silt deposits along their lower reaches. In the vicinity of Brownsville, Texas, the Rio Grande flows along the crest of a $CaCO_3$-rich silt berm 3-6 m above the surrounding coastal plain. This partly chemical aggradation resulting in rivers located along the highest, rather than lowest, regional topographic features, produces peculiar map anomalies and is yet another example of equilibrium reaction in rivers. The large heat of formation of $CaCO_3$ (288,450 cal/mole) results in an increase in heat energy within the evaporating river. Although the river loses heat energy through evaporation, it takes more energy to evaporate saline than fresh water, and the net result will be an imbalance in the rate of heat lost by the system. This loss of heat energy in a

22

given reach of river will increase as the temperature of the river increases. This means that the rate of heat transfer per unit time would increase downstream as evaporation increased, and thus the entropy would increase. Since the most probable longitudinal river profile is one in which the downstream rate of production of entropy per unit mass is constant, and yet in these peculiar rivers mass is decreasing while entropy increases, clearly the profile must be in disequilibrium. In the case of the lower Rio Grande, aggradation decreases the gradient, thus decreasing the rate of hydrodynamic entropy production for a constant volume of water for a net constant rate of entropy production for the river as a whole.

Most rivers of moderate salinity (about 100 mg/l total dissolved salts) have a river water chemistry dominated by the available soluble components of the mineral materials through which the river flows. In Figure 1, for instance, we see that major cation dominance in mid-temperate to high latitude rivers vary from a 10 to 1 or better dominance of Ca over Na in the Ob and Yukon to an approximately equal concentration of Na and Ca in the Mobile, Colorado, and Mississippi. In a general fashion, this portion of the figure suggests that the warmer the river and the greater the weathering in the watershed, the less the dominance by calcium in the rock chemistry-dominated rivers. Relief and vegetation are utterly immaterial in determining the chemistry of these rock-weathering dominated river systems except as they affect the degree of weathering of the soil and rock of the watersheds.

Humid equatorial rivers, with deeply weathered watersheds, are in the class of surface waters that are precipitation chemistry dominated. In the case of the Negro, Orinoco, and Niger, for instance, the low total salinities of these rivers show that little contribution of cations is being made directly from watershed minerals. Gibbs (op. cit.) has shown that these humid equatorial river chemistries closely match the dissolved load components of watershed precipitation. About the only cation notably present in the rivers that is in very low concentrations in the precipitation is silica. The approximately 1 mg/l silicon in these rivers is to be expected since SiO_2 may make up a half or more of the total surface colloidal mineral matter in deeply weathered equatorial watersheds, and it is about the only readily soluble constituent to be found there (solubility = about 100 mg/l SiO_2 at 25°C, pH 4-8; Krauskoph, 1967, p. 167). The proportion of Na to Ca generally increases as total dissolved salts decrease, irrespective of distance from the oceans, suggesting that in areas of humid equatorial atmospheric upwelling, Na is the dominant cation in natural precipitation. The plot of the major anion content of world rivers (Figure 2) shows that chlorine dominates over bicarbonate in these equatorial regions also, demonstrating that the oceans are the most probable source of these constituents.

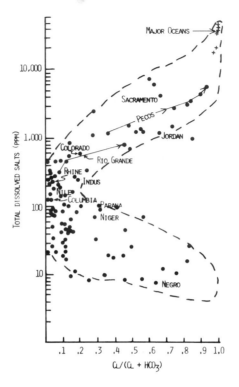

Figure 2. Weight ratio of Cl/(Cl + HCO₃) as a function of total dissolved salts for rivers of the world. (Modified from Gibbs, 1970, p. 1089)

In temperate latitudes, however, the river water chemistry becomes in large measure a function of continentality with respect to the chemical components added through precipitation. Although this precipitation-added component usually does not dominate, in the United States, for example, absolute concentrations of Na^+ and Cl^- in precipitation vary quite directly as a function of distance from the ocean (Junge and Werby, 1968). Calcium and HCO_3 dominate precipitation chemistry in the more arid and more continental regions of the continent, and most workers presume that these are derived from wind-borne mineral matter picked up from the western two-thirds of the United States. This mineral matter is more important in determining river water chemistry than is generally thought. Whitehead and Feth (1964) have shown that bulk precipitation, a mixture of dry fallout and meteoric precipitation, is 4 to 10 times higher in mineral concentration than mere rainwater. Thus by sampling precipitation alone to determine the chemical input into a river system, one may be noting only a small fraction of the total potentially soluble chemical components. The dry fallout addition is usually of much more local origin than the rainfall

24

component, but since the particles are small and may represent partly weathered minerals and clays in large measure, their exchange capacities are high and thus their ability to transport substances like Ca^{++}, Mg^{++}, and SO_4^{--} to rivers are significant.

Langbein and Dawdy (1964), working with rivers in the United States only, have shown that total dissolved loads of rivers increase directly as a function of mean annual runoff up to maximum concentrations of 100 to 200 mg/l at run-offs of 15 cm or more. Above 15-cm runoff dissolved loads of rivers do not increase significantly. This can be explained in the United States by noting that the record stations used in this study with less than 15 cm of runoff all lie west of the Mississippi, where, in general, evaporation exceeds precipitation and fresh mineral matter is generally available along the river courses. Here we would expect total dissolved loads to increase as a function of runoff, as the more water that flows over the land, the more that can be dissolved from it. In the eastern part of the country, however, where surface soils can be more leached of readily soluble components and clay minerals forming at the surface are largely kaolinite, rather than montmorillonite, we would expect many fewer soluble cations, except where fresh, bare mineral matter is exposed as through human activity.

Rock-weathering-dominated eastern U.S. rivers, with undisturbed watersheds wherein precipitation exceeds evaporation and runoff exceeds 15 cm, would be expected to be dominated by 3 to 6 mg/l SiO_2, about 0.5 to 4.0 mg/l Na and Cl (depending on distance from the coast), and 3 to 8 mg/l SO_4 (depending upon air pollution in the air shed). In the west, where evaporation exceeds precipitation, Ca, Mg, and K would make up a larger, although highly variable, component of the river water, with Al and SiO_2 being in relatively lower concentration than in similar eastern rivers.

The Effects of Man on River Water Chemistry.   Direct effects of water pollution on river water chemistry are obvious and becoming well documented in the literature. Of interest also are the indirect effects of man's alteration of watersheds upon the resultant river water chemistry. Physical disturbance in the watershed and pH and anion changes in precipitation brought about by air pollution can have dramatic effects on river water chemistry. Among the many interesting examples is that of the work of Likens, Bormann, and others in the Hubbard Brook watershed in the Connecticut River drainage (Likens, et al., 1970); Bormann, et al., 1969; Fisher, et al., 1968). Several years of careful net chemical balance studies in a series of matched watersheds, one of which was subject to complete deforestation, have helped to reveal the dynamic complexity of the effects of man upon river water chemistry. Sulfate enters the watershed from both sea salt sources and from industrial pollution. Junge and Werby's (1968) map of $SO_4^=$ concentrations in rainwater show concentrations of 3 or more mg/l in the industrialized lower Great Lakes region west of the Hubbard Brook site, corresponding to the average of 3 mg/l noted in the annual precipitation at Hubbard Brook (Fisher, et al., 1968, p. 1119). Values go to 7 mg/l in the summer months, as rainfall seems more effective than snowfall in washing the $SO_4^=$ from the atmosphere. The pH of Hubbard Brook undisturbed watersheds is about 5.1, with a dominance by a weak sulfuric acid. This is typical of areas of eastern United States, with undisturbed forested watershed and podsolic weathering, where precipitation exceeds evaporation. This contrasts with dominance by weak carbonic acid in precipitation and surface waters of the central and western states where evaporation exceeds precipitation. The lower pH sulfurous and sulfuric acid-dominated eastern waters are much better able to leach cations from clays and partly decomposed silicate minerals—which explains the prevalence of lateritic and podsolic weathering and the concomitant effects upon surface water chemistry where precipitation exceeds evaporation. Generally, wherever sulfurous or sulfuric acid is the dominant acid in rainfall, leaching of minerals can proceed more rapidly than where it is HCO_3. Thus in the West, where a source of SO_3 or SO_4 air pollution is introduced, as around smelters or refineries, into an area previously dominated by HCO_3 precipitation, disastrous soil nutrient leaching takes place, which at rainfall pH's of 5 to 6 can leach 10 to 15 thousand years of accumulated exchangeable soil nutrients in periods as short as 30 to 100 years. These soil nutrients, dominantly monovalent and divalent cations, move out through the rivers and should theoretically increase total dissolved loads in rivers draining small intermontane polluted watersheds by an order of magnitude or more. Unfortunately, few accurate before-and-after water chemistry data are available for western U.S., and where they are, as in Los Angeles, the increased ionic loads are most often attributed directly to water pollution rather than soil leaching.

However, many a western mining town in Colorado, Idaho, Montana, Nevada, California, and Utah can show deforested and virtually sterilized hill slopes where soil leaching associated with recovery of sulfide ore minerals within the air sheds has rendered the headwaters of major rivers destitute of all vegetation except species adapted to bare nutrient-poor soils. Where such air pollution has ceased with depletion of the sulfide minerals, the hills remain bare awaiting thousands of years of slow nutrient build-up; but where leaching begun in the recent past is still going on today, surface waters are so enriched in nutrients that algal mats and other signs of accelerated eutrophication occur in local rivers and lakes.

At Hubbard Brook, Bormann and Likens have found that after deforestation of watersheds, local streams may show fivefold increases in hydrogen ion contents and change from a sulfuric to a nitric acid dominance (Likens, et al., 1970),

In undisturbed systems yearly input of nitrate nitrogen, some of it from air pollution (Junge, 1958), exceeds output from rivers. However, in deforested regions in the northeastern U.S., where soil microflora populations are not greatly reduced, as through forest fire, the soil organisms will apparently oxidize ammonia to nitrite and eventually nitrate. This nitrate is readily soluble and flushed from the ecosystem to the rivers, which in Hubbard Brook showed 41-fold and 56-fold increases in stream water nitrate concentration in the first and second years after deforestation. It is then this HNO_3 nitrification that rapidly leaches metallic ions and silica from the weathered mantle into the rivers, even in the humid summer rainfall areas of the east coast where such exchangeable ions within the soils are in less abundance per unit of soil volume than in the less weathered soils of the arid west. This biotic nitrification in the upper Connecticut River watershed has been shown to alter river water chemistry so as to favor rapid eutrophication of surface waters and, in the case of Hubbard Brook, to upset water chemistry from the deforested watershed to the extent that forest streams may be rendered unpotable by U.S. Public Health Service standards due to nitrate content (60-80 mg/l NO_3^-).

The nitrogen cycle and its upset through the activities of man may account for some of the indirect effects of man upon river water chemistry. Aside from direct dumping of chemicals into rivers, man's use of nitrate and ammonium fertilizers and urbanization and other removal of vegetation to break the nitrogen cycle probably have had the most direct effects on river water chemistry. Hans Jenny has shown in conifer forests of central California that forest fires do not always upset the nitrogen balance, but rather apparently temporarily "freeze" the nitrogen cycle with most of the available nitrogen remaining in the soil (Jenny, 1962). With logging and vegetation removal, however, even on flat ground without surface drainage, soil nitrogen losses of 30% were recorded after a single winter season (Jenny, 1967). Since no surficial erosion had taken place, we can only conclude that this nitrogen loss must have been by water soluble nitrate and should result in cation leaching, decreased soil productivity, and increased surface water ionic loads.

Many more data must be collected regarding man's indirect effects on surface water chemistry. Good water quality data with multielement cation and anion analyses are available for rivers in heavily populated watersheds with much diverse human activity affecting the water chemistry. However, almost nothing is known about the effects of single imbalancing activities upon the chemistry of otherwise undisturbed river waters, and it is this writer's opinion that we have much vital information to collect before the waters are so muddied by population and resource expansion that the effects of one imbalance cannot be separated from those of others.

Literature Cited

Bagnold, R. A. 1941. Physics of blown sand. William Morrow, New York. Reprinted 1954, Methuen, London. 265 p.

————————. 1966. An approach to the sediment transport problem from general physics. U.S. Geol. Surv. Prof. Paper 422-1. 37 p.

————————. 1968. Deposition in the process of hydraulic transport. Sedimentology 10:45-56.

Bormann, F. H., et al. 1969. Biotic regulation of particulate and solution losses from a forest ecosystem. Bio. Sci. 19(7):600-610.

Bradley, W. C. 1970. Effect of weathering on abrasion of granitic gravel, Colorado River (Texas). Bull. Geol. Soc. Amer. 81:61-80.

Callander, R. A. 1968. Instability and river channels. J. Fluid Mech. 36(3): 465-480.

Chorley, R. J. 1962. Geomorphology and general systems theory. U.S. Geol. Surv. Prof. Paper 500-B. 10 p.

Clayton, K. M. 1968. The evolution of river systems. Proc. Inst. Civil Eng. (London) 41:397-399.

Davis, R. J. 1968. Special problems of liability and water resource law, p. 103-140. *In* H. J. Taubenfeld [ed.], Weather modification and the law. Oceana Pub., Dobbs Ferry, New York.

Davis, W. M. 1909. Geographical essays. Reprinted 1954, Dover Pub., New York, N. Y. 777 p.

Emmett, W. W. 1970. The hydraulics of flow on hillslopes. U.S. Geol. Survey Prof. Paper 662-A. 68 p.

Fisher, D. W., et al. 1968. Atmospheric contributions to water quality of streams in the Hubbard Brook Experimental Forest, New Hampshire. Water Resources Res. 4(5):1115-1126.

Gibbs, R. J. 1970. Mechanisms controlling world water chemistry. Science 170: 1088-1090.

Gilbert, G. K. 1877. Report on the geology of the Henry Mountains. Dep. Interior: U.S. Geographical and Geological Survey of the Rocky Mountain Region, Washington, D.C. 160 p.

Hack, J. T. 1960. Interpretation of erosional topography in humid temperate regions. Amer. J. Sci. 258A:80-97.

—————. 1965. Geomorphology of the Shenandoah Valley, Virginia and West Virginia and the origin of the residual ore deposits. U.S. Geol. Surv. Prof. Paper 484. 84 p.

Hardin, Garrett. 1963. The cybernetics of competition: A biologist's view of society. Perspect. Biol. Med. 7(1):58-84.

Holl, K. 1955. Chemische Untersuchungen an Fliesgewassern. Verh. Int. Ver. Limnol. 9:360-372.

Horton, R. E. 1945. Erosional development of streams and their drainage basins: hydrophysical approach to quantitative morphology. Bull. Geol. Soc. Amer. 56:275-370.

Jenny, Hans. 1967. Abstracted statement, p. 31-32. In Section 1—Man's effects on the California watersheds. Subcommittee on Forest Practices and Watershed Management, Assembly Committee on Natural Resources, Planning and Public Works, vol. 25, Rep. 8. Assembly of the State of California, Sacramento.

Junge, C. E. 1958. The distribution of ammonia and nitrate in rain water over the United States. Trans. Amer. Geophys. Union. 39(2):241-248.

—————, and R. T. Werby. 1968. The concentration of chloride, sodium, potassium, calcium, and sulfate in rain water over the United States. J. Meteorol. 15(5):417-425.

King, R. H. 1966/67. The concept of general systems theory as applied to geomorphology. The Albertan Geographer 3:29-34.

Krauskopf, Konrad. 1967. Introduction to geochemistry. McGraw-Hill, New York, N. Y. 721 p.

Langbein, W. B., and D. R. Dawdy. 1964. Occurrence of dissolved solids in waters in the United States. U.S. Geol. Surv. Prof. Paper 501-D, p. 115-117.

Langbein, W. B., and L. B. Leopold. 1966. River meanders—Theory of minimum variance. U.S. Geol. Surv. Prof. Paper 422-H. 15 p.

——————, and L. B. Leopold. 1968. River channel bars and dunes—Theory of kinematic waves. U.S. Geol. Surv. Prof. Paper 422-L. 20 p.

Leopold, L. B. 1968. Hydrology for urban land planning—a guidebook on the hydrologic effects of urban land use. U.S. Geol. Surv. Circ. 554. 18 p.

——————. 1970. A tentative and incomplete hypothesis on energy utilization in some physical and biologic systems. Unpub. written communication of Oct. 7, 1970. 6 p.

——————, and W. B. Langbein. 1962. The concept of entropy in landscape evolution. U.S. Geol. Surv. Prof. Paper 500-A. 20 p.

——————, and T. Maddock, Jr. 1953. The hydraulic geometry of stream channels and some physiographic implications. U.S. Geol. Surv. Prof. Paper 252. 57 p.

——————, and M. G. Wolman. 1960. River meanders, Bull. Geol. Soc. Amer. 71:769-794.

——————, M. G. Wolman, and J. P. Miller. 1964. Fluvial processes in geomorphology. W. H. Freeman, San Francisco, Calif. 522 p.

Likens, Gene E., et al. 1970. Effects of forest cutting and herbicide treatment on nutrient budgets in the Hubbard Brook watershed-ecosystem. Ecol. Monogr., 40:23-47.

Lindroth, Arne. 1957. Abiogenic gas supersaturation of river water. Arch. Hydrobiol. 53(4):589-597.

Mackin, J. H. 1948. Concept of the graded river. Bull. Geol. Soc. Amer., 59: 463-512.

Piest, R. F. 1963. The role of the large storm as a sediment contributor. Proc. Fed. Inter-Agency Sediment. Conf., U.S. Dep. Agr. Misc. Publ. 970.

Powell, J. W. 1875. Exploration of the Colorado River and of the west and its tributaries, explored in 1869, 1870, 1871, and 1872, under the direction of the secretary of the Smithsonian Institution, Washington, D.C. 241 p.

Scheidegger, A. E. 1967. A thermodynamic analogy for meander systems. Water Resources Research 3(4):1041-1046.

—————, and W. B. Langbein. 1966. Probability concepts in geomorphology. U.S. Geol. Surv. Prof. Paper 500-C. 14 p.

Schumm, S. A., and R. W. Lichty. 1965. Time, space and causality in geomorphology. Amer. J. Sci. 263(2):110-119.

Tricart, J., et al. 1961. Méchanismes normaux et phénomenes catastrophiques dans l'évolution des versants du bassin du Guil. Zeit. Geomorphol. 5(4):277-301.

Von Schelling, H. 1951. Most frequent particle paths in a plane. Trans. Am. Geophys. Union 32:222-226.

—————. 1964. Most frequent random walks. General Electric Advanced Tech. Lab. Rept. No. 64GL92, General Electric Co. 60 p.

Wahrhaftig, Clyde, and R. R. Curry. 1967. Geologic implications sediment discharge records from the northern Coast Ranges, California. In Man's Effect on California Watersheds, Part III, Rept. of the Inst. of Ecology, University of California, Davis, Calif., p. 35-58.

Whithead, H. C., and J. H. Feth. 1964. Chemical composition of rain, dry fallout, and bulk precipitation at Menlo Park, California, 1957-1959. Jour. Geophys. Res., 69(16):3319-3333.

Wolman, M. G. 1964. Problems posed by sediment derived from construction activities in Maryland. Report to the Maryland Water Pollution Control Commission, Annapolis, Md. 125 p.

—————, and J. P. Miller. 1960. Magnitude and frequence of forces in geomorphic processes. Jour. Geol., 68:54-74.

WHAT IS A RIVER? – ZOOLOGICAL DESCRIPTION*

Kenneth W. Cummins

A survey of the ecological literature clearly illustrates that running waters have been investigated predominantly by zoologists. For example, well over half the references cited in Hynes' (1970) excellent book deal with zoological aspects of rivers and streams. This division of lotic ecology into zoological, botanical, and microbial sections has been extremely counterproductive. Fortunately, in the last few years the producers and microbial (micro-) consumers, of lotic habitats have begun to receive additional attention. However, there is still an overemphasis on animal (macro-) consumers, and essentially nothing has been done to treat these three components as the inseparable whole that they are.

In theory, the earth's natural running waters form a continuum, from spring brooks to massive rivers. However, due to the activities of engineers in concert with power companies and agronomists, most of the large European and American rivers are now only a series of impoundments, and their inclusion in a general theoretical framework for running waters is questionable. A zoological description of these larger rivers is greatly simplified because of their highly degenerate state resulting from decades of gross pollution. In the West, with the East pressing to catch up, the majority of our large rivers are now, or soon will be, of little biological interest except as sources of human toxicants and pathogens. Therefore, to avoid discussing only a few protozoans, nematodes, and oligochaetes, smaller bodies of running water will be stressed. Also, as Hynes (1970) has documented, among the lotic habitats it is only the smaller stony streams that have a truly distinctive fauna the world over.

---

*Contribution No. 197 of the Kellogg Biological Station. The ideas developed here and original data reported have been possible over the years only because of the research support supplied by the U.S. Atomic Energy Commission, Environmental Science Branch, Contract AT-(11-1)-2002, the Department of the Interior, Office of Water Resources Research, Grant B-008-Mich and a National Science Foundation Coherent Area Research Project in Freshwater Ecosystems, Grant GB-15665.

Since any discussion of running waters should be directly relatable to an understanding of their structure and function as entire ecosystems, some attention must be directed toward the structural component represented by animal consumers in relation to overall running water compartmentalization. Figure 1 reflects the basic functions inherent in lotic ecosystems. The initial substructuring recognizes the origin of organic substance from within or outside the system. The second level of partitioning indicates the fundamental biochemical difference between producers and consumers. The third-order compartmentalization is based on turnover or replacement times of the lotic organisms, which is related to size. A final pattern reflects differences in organism structure and function in relation to physical differences in running water habitats; there are organisms in the water column (plankton and nekton) and associated with the sediments or structures fixed to the sediments (benthos). In addition, as originally suggested by Shelford (1914), Shelford and Eddy (1929), and Moon (1939), running waters can be viewed as erosional and depositional habitats. Erosional areas are dominated by rapid flow and removal of finer sediments, depositional regions are characterized by slow currents and the deposit of finer sediments. Actually, adaptations of animal benthos reflect another type, intermediate flow, dominated by sand and gravel sediments (Cummins, 1966).

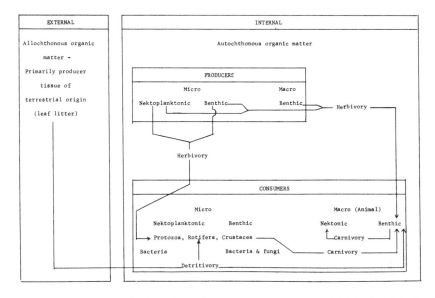

Figure 1. Compartmentalization of running water ecosystems. The initial substructuring is based on the origin of organic substance, the second level recognizes photosynthetic capability, the third is based on size, which is related to turnover times, and the fourth partitions the organisms of the water mass from those of the sediments. Arrows indicate trophic interactions between levels of organization

34

In the following discussion of the zoological component of running waters, the animal adaptations to the three habitat types—erosional, intermediate, and depositional—are considered primarily energetic, with problems of concealment from predators and competition secondary. That is, the exploitation of the various microhabitats, which comprise the matrix of each of the broad categories, concerns the primary function of the majority of the animals—feeding and growth—and, to a lesser extent, reproduction. Most of the insects which normally dominate the benthos of healthy streams and rivers meet the reproductive problems in the terrestrial environment. Within the energy-expenditure patterns the most costly portion is respiration, with growth (and reproduction) being accomplished only after that cost is assigned. Since highest respiration costs are closely linked to locomotion, that activity can be viewed as being in direct "competition" with growth. These relationships are summarized in Figure 2. The general strategy is to minimize locomotion and maximize feeding, thereby maximizing growth. Although considerable attention will be paid to trophic characterizations of the animals, it is difficult to generalize at the species level since quite often there is an age-, rather than a species-, specific pattern (Coffman, Cummins, and Wuycheck, 1971).

Lotic Nekton

Strictly defined, the nekton comprises forms independent of water movements; that is, capable of significant locomotion on their own (Hutchinson, 1967). Although large rivers, particularly because of the extensive interference with unrestricted flow, often contain a biota that might be described as planktonic, these organisms are simply in transport and at least the animal portion can best be described as drift (Waters, 1965). However, most of the animals are capable of significant locomotion, especially benthic forms in the drift. The term nekton has therefore been employed to cover all animals in the water flowing over the substrate. Obviously, the nekton is much less restricted to the erosional-depositional compartmentalization than is the benthos. In addition to organisms in transport, the larger nekton—namely fish—often feed, hide, and spawn in different regions.

Erosional Assemblages. Almost all components of the benthos are encountered in the drift, with the erosional areas representing the primary region of origin. Therefore the drift and the benthos are dominated by insects, annelids, crustaceans (especially amphipods Gammaridae and Talitridae), and in the "microdrift," protozoans and rotifers. A great deal has been written concerning the streamlined shape of the fishes inhabiting fast-running waters (e.g., Hubbs, 1941), primarily species feeding in or just below the erosional areas. Actually, the fish typical of erosional lotic habitats spend a large proportion of their time

35

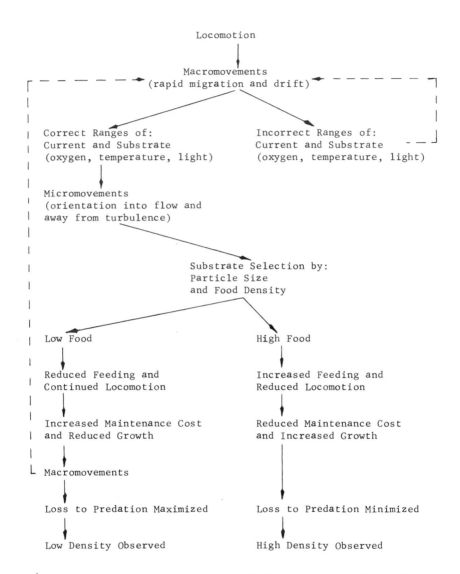

Figure 2. The relationship between locomotion, feeding, and growth, indicating the strategy of reduced locomotion and increased feeding

in micro-depositional areas; for example, behind large boulders or logs. The salmonids (*Salmo* and *Salvelinus* spp.), introduced the world over, are a good example of erosional inhabitants. At certain times the regions are populated with spawning fishes or lampreys, more strictly part of the benthos during such activity because of the intimate association with the sediments (see below). In addition to the salmonids, certain whitefish species and daces are common inhabitants of fast-water sections, particularly smaller rivers and streams in North America and Europe. A wider variety of genera is encountered in larger rivers where the flow is rapid over coarse substrates.

Depositional Assemblages. As indicated above, regions of slack water, characterized by the deposition of finer sediment particles, are often indistinguishable from truly lentic habitats of the same region. The same genera, and often the same species, of protozoans, rotifers, and micro-crustaceans as those encountered in lakes and ponds can be collected from slow-water areas, particularly large rivers. All of these organisms compensate for weak swimming capabilities with high reproductive rates and short generation times, adequate to replace any displacement from depositional regions to erosional sections further along the watercourse. The food base for engulfing (e.g., sarcodin protozoans) and filter-feeding organisms (e.g., ciliates, free-swimming monogonont rotifers, cyclopoid and diaptomid copepods, and cladocerans) is mainly algae, as it is in standing waters. However, because lotic systems have a higher density of particulate organic matter than do standing waters, a greater proportion of detrital feeding is frequently encountered. This density is caused by the physical factors keeping particles in suspension and the extensive interface with the terrestrial system.

The fish typical of depositional sections, primarily cyprinids (minnows), centrarchids (sunfishes), ameiurids (bullheads), and catostomids (suckers), have a laterally compressed and less rounded and streamlined body form. The food base for these species is slow-water benthos, organic detritus, and some "planktonic" forms, supplemented by terrestrial invertebrates. For example, when cyprinids and catostomids collected from depositional sections (pools) of a small Pennsylvania stream were examined for gut contents, all these faunal elements were present, but chironomid midge larvae, typical of the benthos of the region, were overwhelmingly dominant in the food.

The presence of vascular hydrophytes in depositional areas further transforms the habitat into a direct homologue of shallow ponds and the lotic margins of lakes. In addition to the benthos associated with such weed beds, numerous small fishes and planktonic and semi-planktonic invertebrates are found among the plants.

37

Although not strictly according to definition, the surface-dwelling inverte-brates common in slack-water areas of streams and rivers are included here as depositional zone nekton. The hemipteran water striders (Gerridae) and the aggregating whirligig beetles (Gyrinidae) are easily found wherever reverse-flow backwaters develop, on the surface of pools, or in river sections where the flow rate is generally reduced. These species are isolated from the remainder of the flowing water system, and their food is primarily of terrestrial origin.

Intermediate Zone Assemblages. The faunal elements of both erosional and depositional nekton are found in intermediate transitional areas of moderate flow over sand-gravel substrates. In small rivers and streams, where such regions form an easily recognizable compartment, the nekton is represented by intermit-tent incursions of benthic micro-crustaceans, e.g., chydorids, and a persistence of "planktonic" forms of depositional zone origin. In small streams of midwestern and northeastern United States, the cladocerans *Simocephalus* and *Chydorus* are often encountered in intermediate regions.

If these zones have a typical fish fauna, it is dominated by cyprinid minnows. In rivers and streams examined in the regions mentioned above, the common shiner (*Notropus cornutus*) was a constant faunal representative of intermediate zones.

Lotic Benthos

The benthos is defined as those animals living on or in the sediments or structures fixed to the substrate, such as vascular hydrophytes. These surface and interstitial organisms are usually dominated, both in numerical and biomass density, by the insects, although in some large rivers high densities of pelecypods may be encountered. Also, in pollution-altered, running waters annelids are often the dominant forms, tubificids sometimes reaching amazing numerical densities.

Erosional Assemblages. As mentioned above, the erosional benthic assemblages of running waters constitute the only strictly lotic benthic fauna. Their only counterpart is found on coarse sediments in the severely erosional, wave-swept shoreline areas of large oligotrophic lakes. The adaptations of the benthos of fast-flowing areas have already been extensively reviewed (Wessenberg-Lund, 1943; Bardach, 1964; Hynes, 1970), and the methods of permanent and tempo-rary attachment to the substrate and dorso-ventral flattening are well known.

Poriferans (Demospongiae, Spongillidae) permanently attached to cobbles, boulders, and submerged logs and branches are abundant in small running-water systems. Temporary attachment organs are more common among invertebrates

38

of the erosional habitats. A ring of hooks allows the filter-feeding simuliid dipteran larvae to remain firmly attached to smooth surfaces overridden by laminar flow, even in torrential currents. The actual attachment and movement is over a mat of silk spun from the salivary glands and laid down by the larvae over the substrate. Hooks on the posterior proleg hold the larva in place while it feeds by entrapping small particles in a pair of expandable, fan-like structures on the head. Movement along the substrate is accomplished by alternate attachment of the ventral anterior and posterior prolegs.

True suckers are limited to leeches and the curious dipteran family *Blepharoceridae* inhabiting rapidly flowing waters, and the former are very seldom observed on the exposed upper surfaces of sediments. The foot-shell structure of gastropods and the adhesion pad of *Hydra* species inhabiting erosional areas provide sucker-like attachment devices. The most familiar gastropods are the freshwater limpets (Ancylidae, e.g., *Ferrissia rivularis*) whose dome-shaped shells, slightly recurved, are common in most streams and rivers. The very heavily shelled, operculate snails, such as the genus *Goniobasis*, are also widely distributed in North America. Interference with suction attachment caused by the operculum is compensated for by the protection that structure affords and by the heavy shell. When *Goniobasis* is disturbed, it releases its hold on the substrate (usually large cobbles and boulders) and, due to its heavy shell structure, drops quickly to the bottom behind the attachment site.

Sucker-like structures are achieved by the peripheral arrangement of bristles, spines, and hairs in the coleopterans *Psephenus* and *Ectopria*, known as water pennies, and the well-known mayfly example *Ephemerlla doddsi*. The enlargement of the anterior gills of species in the ephemeropteran genera *Rithrogena* and *Epeorus* also provides peripheral contact with the substrate (Needham, Traver, and Hsu, 1935). All of the above structural arrangements function primarily to prevent water from flowing under the larvae and nymphs.

Hooks for attachment are often the primary device for maintaining animals in position, for example, the curved tarsal claws of elmid and dryopid beetle larvae and adults. Both the tarsi and anal prolegs are equipped with hooked claws in the free-living rhyocophilid caddis larvae and megalopteran corydalids, such as *Nigronia* and *Corydalus*. The dipterans *Deuterophlebia* (Deuterophlebiidae) and *Limnophora* (Anthomyiidae) *have prolegs equipped with* recurved hooks that are used for attachment. Hooked claws are combined with the construction of fixed retreats in the hydropsychid, philopotamid, polycentropid, and psychomyiid caddisflies, orthoclad midges, such as *Rheotanyarsus* sp., and lepidopteran larvae in the genus *Cataclysta*. In large rivers, piers, channel markers, and other man-made "fixtures" in rapidly flowing water are often

39

heavily colonized by net-spinning hydropsychid caddis larvae. High densities are also common below impoundments, where the larvae filter out large quantities of lentic plankton.

Perhaps the most common adaptation among the rapid-water invertebrates is doro-ventral flattening. Compressed body form allows the animal to avoid the major thrust of the current by moving through crevices among the coarse sediment particles of the erosional habitat and along the boundary layer at the water-sediment interface. The boundary or dead layer (a zone of greatly reduced flow, about 1 mm thick, above the surface of the substrate) has been demonstrated elegantly by Ambuhl (1959, 1961). Small size, enhanced by dorso-ventral flattening, allows many species to move over the sediments below high velocity currents. The heptagenid and some of the leptophlebeid mayflies, perlid stoneflies, and fast-water-dwelling flatworms (*Dugesis* and *Phagocata*) are examples of organisms that move primarily in the boundary layer.

A number of fishes are properly considered part of the benthos in erosional habitats. The best known of these in North America are the catostomid hog sucker (*Hypentelium*), the cyprinid stone roller (*Campostoma*), the darters (*Etheostoma*, Percidae) and the sculpins (*Cottus*, Cottidae). *Hypentelium* species have ventrally-placed suckers used to attach to the substrate and to move pebbles and cobbles, dislodging the associated food organisms. When these species are feeding, the drift in the immediate vicinity is visibly increased. The darters and sculpins rest on the bottom with their pectoral fins extended as braces. Both genera frequent microhabitats with reduced current within the general flow regime of the erosional habitat. For example, the sculpins are found in the type of habitat where crayfish (especially *Orconectes*) hide—in crevices under cobbles, boulders, and logs in riffle sections. The sculpins are particularly suited for bottom dwelling since they lack the buoyant swim-bladder.

The microfauna of erosional habitats is less distinctive than the macrofauna. However, the increased general flow in rapidly running sections results in a microfauna associated with a predominently aerobic sedimentary system which differs from that of the anaerobic sediments that often characterizes depositional areas and occasionally develops in intermediate zones. Accumulations of particulate organic matter—at the interstices of the sediments, in the micro-depositional "shadows" that develop behind cobbles and boulders, among the periphytal stalks on substrate surfaces, and in packs of large organic particles, such as terrestrial leaves trapped in front of larger sediment units—harbor a diverse protozoan fauna, as well as crawling bdelloid and monogonont rotifers, crustaceans, typically harpactacoid copepods, chydorid cladocerans, and mites (Hydracarina). Bacteria, actinomycetes, aquatic hyphomycetes, and certain phycomycetes also abound in such habitats, functioning both in the breakdown of organic matter and as food for the micro- and macrofauna of the zone (Figure 1).

The leaf packs and other accumulations of large organic particles, primarily of terrestrial origin, constitute special habitats within the erosional areas. A specific fauna develops in these accumulations, including some large animals such as craneflies (*Tipula*; Vannote, 1969), stoneflies (e.g., *Pteronarcys* and *Peltoperla*; Wallace, Woodall, and Sherberger, 1970, and McDiffett, 1970), and caddisflies (e.g., *Pycnopsyche* and *Hydatophylax*), that act as "shredders" in the initial fragmentation process. Many species which feed on finer organic particles can be found, notably ephemerellid, baetid, heptageneid, leptophlebeid and siphlonurid mayflies, and orthoclad midges. Also common in organic pockets are predators such as perlid stoneflies (e.g., *Acroneuria* and *Phasganophora*) and the megalopteran *Nigronia*.

Some additional faunal elements of the erosional benthos are the heavy-shelled pelecypods (e.g., the unionid *Fusconaia*) and heavy case-bearing trichopterans such as the snail-shaped case-builder *Helicopsyche*, and *Neophylax* and *Gorea*, with heavy lateral ballast stones that keep the animals in place and inhibit rolling. In certain mayflies, species of Baetidae, Ephemerellidae, and Leptophlebidae, the primary adaptation is in modifications of the caudal cerci and a streamlined body shape, resulting in minnow-like swimming capability which allows the nymphs to dart from one protected spot to another. Not only does the siphlonurid mayfly genus *Isonychia* have a streamlined body and caudal filaments fitted with overlapping hairs forming a swimming "tail," but also the inner surfaces of its forelegs have long setae which can be erected to form a basket used in straining food particles from rapid flowing water.

Depositional Assemblages. Although annelids, nematodes, and midges in the tribe chironomini are encountered among both the erosional and intermediate assemblages, they are much more typical of depositional zones, where they burrow into the upper layer of the sediments. These fine sediments represent not only the inhabited matrix, but also the food supply for such organisms. The high organic content of such a substrate attracts microconsumers (decomposers and microfauna) similar, if not identical, to those typical of organic accumulations in the fast and intermediate flow sections. Whereas the lumbriculid annelids are more common in erosional and intermediate zones, the families Naididae and Tubificidae are found particularly in fine sediments. Among the more interesting burrowing forms are the flattened predaceous gomphid dragonfly nymphs, which lie with only the eyes and tip of the abdomen exposed. Burrowing mayflies (especially *Hexagenia*) which maintain broad U-shaped tubes in the fine sediments are common in large rivers and frequently represented in small streams.

The thinner shelled pelecypods (e.g., *Anodonta*) are common in depositional habitats, particularly in large rivers. The incurrent and excurrent siphons of these clams extend above the water-sediment interface. Siphon currents, along with the regular undulations of the burrowers, are responses required in an environment of continual sedimentary "rain." A similar function is performed by the operculate-like first gills of the caenid and certain ephemerellid mayfly nymphs. These flaps protect the delicate underlying gills, preventing the respiratory surfaces from becoming fouled with silt particles. When observed in their natural habitat, sprawled on the surface of the soft sediments, the nymphs can be seen periodically to flutter the operculate flaps.

The damselfly *Calopteryx* is extremely common in depositional habitats formed along stream margins, where it preys upon midges, amphipods, and other silt-dwelling forms. The predaceous hermipterans *Belastoma, Ranatra,* and *Lethocerus* usually inhabit the same areas as Calopteryx, undoubtedly with some overlap in food habits. Detrital—and animal-feeding stoneflies, such as *Taeniopteryx*, are often abundant in cold, small streams. Where vascular plant stands develop along such stream margins, or in pools or slack-water depositional regions of large rivers, a distinctly lentic fauna can be found. The main exception may be the *Potomogeton,* or *Elodea,* patches that develop in depositional margins and pools of small streams, where extremely high densities of amphipods (*Gammarus, Hyallela,* and *Crangonyx*) thrive. Where small springs enter streams, watercress beds are a common feature, and the habitat is best described as depositional, even though the watercress beds often are only islands in generally erosional or intermediate sections. The most common inhabitants of these watercress beds are amphipods.

As in the lentic environment, lotic plant beds abound with climbing damselflies (e.g., *Ischnura* and *Agrion*), dragonflies (e.g., *Anax* and *Libellula*), and organic case-bearing caddisflies (e.g., species of Phryganeidae, Leptoceridae and Limnephilidae).

Intermediate Zone Assemblages. The benthos of intermediate areas exhibits representatives from both erosional and depositional habitats; however, there are some faunal elements maximally abundant in sand-gravel sections of moderate to low flow. Two very common gravel-burrowing genera of streams and rivers are the *Ephemera* mayflies and some species of *Pycnopsyche* caddisflies. The fingernail clams *Spaerium* and *Pisidium* are also gravel-dwelling forms, the former tending toward coarser sediments, the latter to finer ones. A similar gastropod pair is represented by the genera *Limnaea* and *Physa*.

The lighter mineral-cased caddis larvae and those with cases constructed partially of organic fragments are consistent representatives of intermediate sections; the leptocerid *Mystacides* is a good example. Some species of the small, predaceous leptocerid *Oecetis*, which constructs a slightly curved fine mineral case, are also common over gravel-sand substrates.

Dominant predators of intermediate assemblages are species of Sialis (Megaloptera) and *Atherix variegata* (Diptera, Rhagionidae), which feed on the diamesin midge fauna of these areas.

General Ecological Relationships Between
Faunal Elements of Running Water

Trophic Relationships. As stated earlier (Figure 1), food gathering, and therefore the morphological-behavioral adaptations that form its basis, is considered the paramount animal consumer function in lotic ecosystem operation; this includes the general strategy of animals to restrict energy-costing locomotion in favor of feeding (Figure 2).

In a detailed study of the trophic relations of the benthos (including sculpins) of a woodland stream (Cummins, Coffman, and Roff, 1966; Coffman, et al., 1971), several basic features became apparent. Most organisms were predominantly herbivore, detritivore, or carnivore, but within these categories the opportunism rule applied: The material in the digestive tract faithfully reflected the relative environmental densities of the food items falling within the ingestible size range. Food habits were more closely related to size than to species; for example, the early instars of most insect species (Figure 3), even those destined to be predators later in their growth cycle, were detritivores, feeding on particulate and amorphous organic material and the associated microbial assemblages. As this indicates, the early stages of the majority of the insects are associated with organic accumulation in macro-depositional zones (the margins of the running watercourse, backwaters, pools, or long, slow-flowing sections) or in micro-depositional pockets (the interstices of the sediments or "shadows" behind large substrate units), which thus serve as the "nurseries" for many of the benthic species.

Herbivory is more common in the central channel portions of erosional and intermediate sections, where the dominant food source is epilithic diatoms (Figure 4) and certain filamentous green algae. In depositional areas, where more blue-green algae are found, the vascular hydrophytes constitute a food base for herbivores, primarily leaf-mining dipterans and lepidopterans and a few leaf-eating caddisflies, such as certain species of Phryganeidae, Leptoceridae, and

43

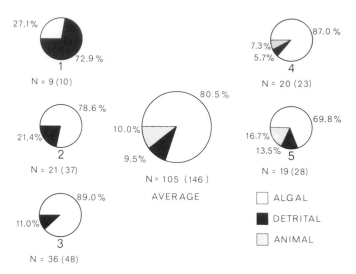

Figure 3. Trophic relations of the herbivore-detritivore mayfly <u>Stenonema fuscium</u> Clemens from Linesville Creek, Pennsylvania. The relative amounts of algal, detrital, and animal ingestion are shown for five age classes (each consisting of two or three instars) and as an average of all age classes

Limnephilidae. The vascular plant tissues are seldom eaten while growing, more often entering the food cycle as detritus, particularly after autumnal die-backs.

The primary interaction between the nekton and the benthos, other than the resettling of drift organisms in new areas, is through feeding (Figure 1): removal of fine particles from the water by benthic animals through a variety of filtering devices and predation by nektonic fishes on benthic species. The benthic system has many more self-contained predator-prey cycles than does the nekton system. The only predation by benthic species on nektonic organisms is on animals in the drift of benthic origin.

Just as some predaceous benthic species are detritivores in their very young stages, many large detritivore-herbivores ingest small animals as they reach the end of their growth cycle. For example, in a Pennsylvania stream the food of the mayfly Stenonema fuscum included small midges when the nymphs reached their terminal instars (Figure 3). Undoubtedly, these organisms were swept into the mouth along with other large particles during the scraping-brushing feeding procedure typical of these nymphs. There is every reason to believe that because of its generally higher caloric content, such animal ingestion is of nutritious significance in the growth cycle of such species.

Similarly, most benthic predators contain small amounts of detritus and algae (Figure 5), representing either the gut contents of prey or materials swept into the mouth during capture of the prey. Again, such material may be of

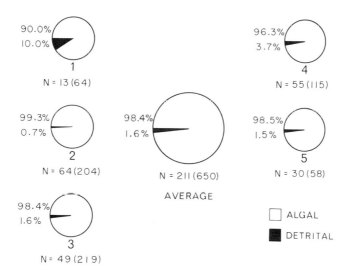

Figure 4. Trophic relations of the herbivorous caddisfly *Glossosoma nigrior* Banks from Linesville Creek, Pennsylvania. The relative caloric ingestion of algae and detritus is shown by larval instar and as an average for all instars. **N** is the number of individuals examined out of those collected (parentheses) in quantitative samples

nutritional significance to the predators. Although the detritus and algae as the gut contents of prey have a reduced calorie-per-gram value (Cummins and Wuycheck, 1971), such material is a mechanically disrupted, partially digested, pre-packaged food supply which may well be assimilated with considerable efficiency. Certainly the early instars of benthic insects before they have begun any significant fat deposition, are an undigestible chitinous shell surrounding an enormous, almost continuously packed digestive tract. Benthic predators of lotic systems may, therefore, feed significantly on the digestive tracts of prey— convenient instruments of collecting and packaging.

Animal Production

The long-standing interest in the estimation of animal production in running waters has recently been renewed by Ricker (1968), Mann (1969), and Waters (1969), and the interest in benthic invertebrates, by Hynes and Coleman (1968), Hamilton (1969), and Fager (1969). Actually, very little progress has been made beyond the original statements made about fish by Ricker (1946) and Allen

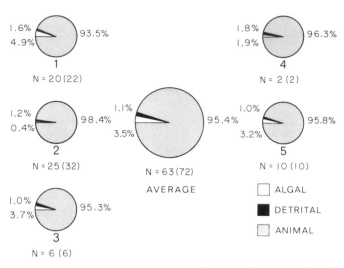

Figure 5. Trophic relations of the carnivorous megalopteran <u>Nigronia</u> <u>serricornis</u> Say from Linesville Creek, Pennsylvania. The relative amounts of algal, detrital, and animal ingestion are shown for five age classes (each consisting of one to two instars) and as an average of all age classes

(1951), although Waters (1969) modified the statements about invertebrates (primarily lotic). Whereas there are many estimates of standing crop numbers per unit area of lotic habitat, there are few values for biomass of nekton and benthos in running waters and even fewer estimates of production, which are usually simply the difference in standing crop over some unreasonably long time interval. There is reason to believe that the animal standing crop biomass is relatively constant over time in any one system (Coffman, et al., 1971), barring human-engendered perturbations. Therefore the turnover rates of the various components are critical, especially in manipulating the animal consumers as processors of organic matter in rivers and streams and maximizing (or minimizing) the yield of selected species populations.

Although much is known about the growth rate of fishes from lotic systems, an almost insignificant fund of data exists for benthic invertebrates. The initial determination of the generation time has been made for very few running-water species. Some completed studies make it clear that a simple temperature food-density relationship, as applied to fish populations, will not serve for benthic invertebrates. For example, many lotic species show cold temperature compensation; that is, their metabolism is optimized at low temperatures. At higher latitudes the autumnal community is dominated by species that grow mostly between 0 and 5°C. Also, population growth rates are probably often adjusted, not by increase or decrease in the growth of individuals, but by changes in density through adjustment of survivorship. The implication is that for many benthic invertebrates the survivors are always surrounded with a food supply adequate for normal growth and that such growth is characteristic of the species under a given temperature regime.

We need a great deal more information on the generation times, growth rates, and factors determining both at least in key species of representative systems.

Summary

Some of the relationships discussed above have been summarized in the table on the following pages. The organization of the material follows the habitat partitioning discussed in the introduction: the vertical distinction between nekton and benthos (between the water mass and sediments) and the horizontal distinction between erosional, depositional, and intermediate regions. This compartmentalization, together with trophic specialization, has been employed as a framework to view the structure and function of running-water animals, primarily through their adaptations to various habitats of lotic ecosystems.

Examples of some taxa typical of North American running waters receiving only minor organic and inorganic enrichment from human perturbations. Faunal elements are organized according to the compartmentalization discussed in the text

Vertical Compartmen- talization	Faunal Elements	Horizontal Compartmentalization		
		Erosional	Intermediate	Depositional
Nekton	Invertebrates	Drift: all components of the benthos during catastrophic drift; during normal (diurnal) drift especially Ephemeroptera, e.g., Baetidae; Diptera, e.g., Chironomidae; Simuliidae; and Amphipoda, e.g., *Gammarus*; also "planktonic" forms displaced from depositional areas	Drift: similar to erosional sections but reduced Occasional incursions by certain benthic forms, e.g., Chydoridae	Drift: settling out, especially smaller classes and lighter forms not removed or recolonized in other sections "Planktonic" Protozoa (esp. Ciliata), Rotifera (free-swimming Monogononta), and Microcrustacea (esp. cyclopoid and diaptomid copepods and daphnid and chydorid Cladocera) .
	Fish	Salmonidae (esp. *Salmo* and *Salvelinus*); Cyprinidae (*Rhinichthys*) (in large rivers *Esox, Micropterus dolomieui, Catostomus commersoni*)	*Notropus cornutus* (and elements from both erosional and depositional areas	Cyprinidae (e.g., *Notropus*), Centrarchidae, Ameiuridae
Benthos	Invertebrates	Sessile: Porifera (Spongillidae) "Suckers": Hirudinea, Diptera (Blepharoceridae), Gastropoda (Ancylidae), *Hydra*	Elements from both erosional and depositional zones but typical gravel	Burrowers: Annelida (Naididae, Tubificidae), Nematoda, Diptera (Chironomidae), Odonata (Gomphidae), Ephemeroptera —*cont.*

Table — (cont.)

Vertical Compartmentalization	Faunal Elements	Horizontal Compartmentalization		
		Erosional	Intermediate	Depositional
Benthos (cont.)	Invertebrates (cont.)	Marginal contact: Coleoptera (Psephenus, Ectopria), Ephemeroptera (Rithrogena, Epeorus, Ephemerella doddsi) Fixed retreats: Trichoptera (Hydropsychidae, Philopotamidae, Psychomyiidae, Polycentropidae), Diptera (Orthocladiinae, e.g. Rheotanytarsus), Lepidoptera (Cataclysta) Hooks and claws: Diptera (Simuliidae, Deuterophlebiidae, Anthomyiidae—Limnophora), Megaloptera (Nigronia, Corydalus), Trichoptera (Rhyacophilidae), Coleoptera (Elmidae) Dorso-ventral flattening: Turbellaria (Dugesia, Phagocata), Ephemeroptera (Heptageneidae, Leptophlebiidae), Plecoptera (Perlidae)	fauna of: Ephemeroptera (the burrowers, Ephemera); Trichoptera (Limnephilidae Pycnopsyche; Leptoceridae Mystacides, Oecetis) Diptera (Atherix, Orthocladiirae, Diamesinae); Megaloptera (Sialis); Gastropoda (Limnaea, Physa), Pelecypoda (Sphaerium, Pisidium)	(Hexagenia), Pelecypoda (thin-shelled Anodonta) Sprawlers: Ephemeroptera (Caenidae, Ephemerellidae with gill flaps), Amphipoda Climbers: along stream margins in rapid and intermediate flow areas: Plecoptera (Taeniopteryx), Odonata (Calopteryx), Hemiptera (Belastoma, Ranatra) Climbers: in vascular plant beds: an esentially lentic fauna, e.g., Odonata (Ischnura, Enallagma, Anax, Libellula), Trichoptera (Phryganeidae, Leptoceridae, Limnephilidae)

—cont.

Table — (cont.)

Vertical Compartmen- talization	Faunal Elements	Horizontal Compartmentalization		
		Erosional	Intermediate	Depositional
Benthos	Invertebrates (cont.)	"Crevice seekers": Decapoda (e.g., *Orconectes*), Diptera (*Tipulidae*) Cases and heavy shells: Trichoptera (esp. *Helicopsyche, Neophylax, Goera*), Pelecypoda (e.g., *Fusconaia*) Swimmers: Ephemeroptera (Ephemerellidae, Baetidae, Leptophlebiidae, Siphlonuridae) Leaf packs—entrapment against obstructions: A special fauna with coarse-particle feeders "shredders", e.g., *Tipula, Pycnopsyche, Pteronarcys, Peltoperla, Lepidostoma*; fine-particle feeders, e.g., *Gammarus, Stenomena, Baetis*; and predators, e.g., *Nigronia Acroneuria*		Leaf packs at stream margins: similar fauna to packs in erosional sections

Literature Cited

Allen, K. R. 1951. The Horokiwi Stream. Bull. New Zealand Marine Dept. Fisheries 10:1-231.

Ambuhl, H. 1959. Die Bedeutung der Stromung als okologischer Faktor. Schweiz. Z. Hydrol. 21:133-264.

——————. 1961. Die Stromung als physiologischer und okologischer Faktor. Experimentelle Untersuchungen an Bachtieren. Verh. int. Verein. Limnol. 14:390-395.

Bardach, J. 1964. Downstream. A natural history of the river. Harper and Row, New York.

Coffman, W. P., K. W. Cummins, and J. C. Wuycheck. 1971. Energy flow in a woodland stream ecosystem: I. Tissue support trophic structure of the autumnal community. Arch. Hydrobiol. 68:232-276.

Cummins, K. W. 1966. A review of stream ecology with special emphasis on organism-substrate relationships, p. 2-51. In K. W. Cummins, C. A. Tryon, and R. T. Hartman [eds], Organism-substrate relationships in streams. Spec. Publ. No. 4, Pymatuning Symposia in Ecology. Edwards Bros., Ann Arbor, Mich.

——————, W. P. Coffman, and P. A. Roff. 1966. Trophic relationships in a small woodland stream. Verh. int. Verein. Limnol. 16:627-638.

——————, and J. C. Wuycheck. 1970. Calorific equivalents for investigations in ecological energetics. Mitt. int. Verein. Limnol. 18:1-158.

Fager, E. W. 1969. Production of stream benthos: a critique of the method of assessment proposed by Hynes and Coleman (1968). Limnol. Oceanogr. 14:766-770.

Hamilton, A. L. 1969. On estimating annual production. Limnol. Oceanogr. 14:771-782.

Hubbs, C. L. 1941. The relation of hydrological conditions to speciation in fishes, p. 182-195. In A symposium on hydrobiology. Univ. Wisconsin Press.

Hutchinson, G. E. 1967. A treatise on limnology. Vol. II. Introduction to lake biology and the limnoplankton. John Wiley & Sons, New York. 1115 p.

Hynes, H. B. N. 1970. The ecology of running waters. Univ. Toronto Press. 555 p.

——————, and M. J. Coleman. 1968. A simple method of assessing the annual production of stream benthos. Limnol. Oceanogr. 13:569-573.

Mann, K. H. 1969. The dynamics of aquatic ecosystems. Adv. Ecol. Research 6:1-81.

McDiffett, W. F. 1970. The transformation of energy by a stream detritivore, *Pteronarcys scotti* (Plecoptera). Ecology 51:975-988.

Moon, H. P. 1939. Aspects of the ecology of aquatic insects. Trans. Brit. Ent. Soc. 6:39-49.

Needham, J. G., J. R. Traver, and Y. Hsu. 1935. The biology of mayflies. Comstock Publ. Co., Ithaca, N. Y. 759 p.

Ricker, W. E. 1946. Production and utilization of fish populations. Ecol. Monogr. 16:373-391.

——————[ed.]. 1969. Methods for assessment of fish production in fresh waters. IBP Handbook No. 3, Blackwell Sci. Publ., Oxford. 313 p.

Shelford, V. E. 1914. An experimental study of the behavior agreement among animals of an animal community. Biol. Bull. 26:294-315.

——————, and S. Eddy. 1929. Methods for the study of stream communities. Ecology. 10:383-392.

Vannote, R. L. 1969. Detrital consumers in natural systems. p. 20-23. *In* K. W. Cummins [ed.], The stream ecosystem. Mich. State Univ. Inst. Water Research, Tech. Rept. No. 7, 42 p.

Wallace, J. B., W. R. Woodall, and F. F. Sherberger. 1970. Breakdown of leaves by feeding of *Peltoperla maria* nymphs (Plecoptera: Peloperlidae). Ann. Amer. Entomol. Soc. 63:562-567.

Waters, T. F. 1965. Interpretation of invertebrate drift in streams. Ecology 46:327-334.

——————. 1969. The turnover ratio in production ecology of freshwater invertebrates.

Wessenberg-Lund, C. 1943. Biologie der Susswasserinsekten. Gyldendalske Boghandel, Copenhagen and Springer, Berlin. 682 p.

PLANT ECOLOGY IN FLOWING WATER

John L. Blum

To all those who prefer the synthetic or integrated approach to ecology, the plant ecologist's view of streams is unwarrantedly narrow. Since no other contribution will deal exclusively with plants, I feel justified in presenting this view of the stream from the standpoint of the plant ecologist. But this paper has no pretension other than to present a few salient facts and opinions of recent coinage about river plants. Plant ecology and limnology are both relatively enormous fields, and the mass of available information about autotrophs and heterotrophs growing in flowing water is derived from both. Discoveries pertaining to aquatic plants generally have implications for river plants, but because of the specific nature of the habitat designated, I have drawn the line at such discoveries so in general they are omitted, and my paper will be limited to remarks about a group of relatively recent contributions which have particular urgency for river plants. Greater emphasis has been given to the benthic algae than to the vascular macrophytes, and no attempt has been made to survey work on potamoplankton or on river bryophytes; the latter has been done by Hynes.

To answer our agenda question "What is a river?" is no simple matter for a botanist. Plants like animals find themselves cast at an infantile stage into clement as well as inclement environments. Unlike animals, they generally have relatively little ability to move away to greener pastures. A majority of aquatic plants—cryptogams and phanerogams alike—accept the current environment with apparent indifference and are nearly devoid of means for coping with it. In other words, they are general practitioners of the aquatic habitat, nonspecialists in the river habitat. There are surprisingly few plants which have become such partisans of the lotic habitat as to have acquired structural adaptations to it, or to exclude facultative growth in still water. Such specialist lotic plants as we have are mostly algae and constitute a part of the periphyton or "Aufwuchs" community, but the remarkably algoid angiosperm family Podostemonaceae, characteristic of warm tropical streams, deserves prominent mention.

Few plants have become specialists to a point where they can be called lotic. We have neither a typical geographic succession of plant communities such as the faunistic succession described by Huet (1946) for European streams nor, although there is no shortage of described families of algae and angiosperms,

53

anything to compare with the zonation of insect families in streams as described by Illies and Botosaneanu (1963). Indeed, many of the plants in a given stream, and especially in lower reaches, seem to be principally nonspecialists, so that giving a botanical description of a river based on floristics is often difficult or impossible. Hence our attempt toward a definition leads to an examination of works on river communities, and to the autecology of river plants as elucidated by physical and theoretical models.

New Approaches to Stream Ecology

Model Streams; Physical Models. The complexity of the stream environment, with its many variables, has tended to foil attempts to explain and predict. Nevertheless, recent specific successes have been realized. The number of variables has been reduced by the use of artificial streams or by suitable enclosures in which stream conditions can be maintained with overall fidelity and in which one of the environmental variables is entered at a series of values representing at least a sizable portion of its range in the natural community. In the United States some of the more elaborate efforts in this direction have included those of McIntire and his co-workers (McIntire, 1966a, b, 1968; McIntire, et al., 1964; McIntire and Phinney, 1965). This work was partly carried out with indoor streams consisting of a pair of parallel wooden troughs 3 m long, open at the ends on adjacent sides, thus permitting complete circulation. Water and substrate were brought in from a nearby Oregon stream. Light was provided by white fluorescent tubes. Under such conditions variables are present much as in a natural stream; however, the control of light, temperature and substrate type are important stabilizing factors and have made it possible to demonstrate the effect of varying the intensity of illumination on primary production, in benthic organisms which consisted largely of diatoms (*Synedra ulna*) and *Oedogonium* spp. The algae in this system grew on rubble in submerged trays which could be removed for determination of photosynthetic and respiratory activity. Two communities, one light adapted at 550 fc and the other shade adapted at 225 fc were exposed to different intensities of illumination and their production compared. The curves for both communities which differed somewhat in species composition approached a plateau between 1000 and 2000 fc with, however, measurable differences in the slope of the two curves. The light adapted community showed higher potential productivity than did the shade adapted one (McIntire and Phinney, 1965). The shade-adapted community performed better at lower light intensity, and the light-adapted community performed better at relatively high light intensity. Similar studies were made with CO_2 as variable, with light and current rate (McIntire, 1968) and temperature and current rate (McIntire, 1966b) as variables.

54

Of the numerous experiments and theoretical contributions of McIntire and his co-workers, only a fraction are cited here. It is inadequate indeed to give so summary a treatment to work which has provided such a wealth of quantitative data; the interested researcher is necessarily referred to the principal papers listed.

Another series of artificial stream studies has been pursued in Switzerland. Zimmermann (1961) demonstrated significant quantitative effects due to current; so massive was the *Tribonema* developed in response to current and sewage additive that the algae had to be removed manually at intervals from the experimental troughs. The studies of Eichenberger and Wuhrmann (Eichenberger, 1967*a*, *b*; Eichenberger and Wuhrmann, 1966; Wuhrmann, 1964) were also carried out partially to relate changes in the vegetation to varying amounts of sewage loading. This work was done utilizing a series of outdoor, concrete troughs with rubble substrate. The subsurface water used to fill the troughs received domestic sewage from the City of Zürich, in four different concentrations, as additive. Like the Oregon studies, these Swiss models were colonized by algae naturally appearing from influent sources, which included the Limmat; the dominant plants, however, were cryptogams of more conspicuous diversity. In the channel with heaviest sewage loading *Sphaerotilus* maintained dominance. *Cladophora, Melosira, Vaucheria* sp., *Batrachospermum* and the moss *Amblystegium* were prevalent or dominant under less concentrated loading.

The Eichenberger-Wuhrmann studies are of special interest in many ways. A significant feature of the streams studied was their substantial length (200 m) which in consequence of the relatively small volume of water carried, permitted sufficient metabolic interchange to affect the water significantly in its course. Noteworthy was the change in vegetation in the course of the stream, a change which pointed to self-purification and which supported the assumption that organisms on the stream bed compete for specific nutrients in the water or in the sewage additive. Growth at given levels of sewage admixture declined and stopped once competing upstream organisms had removed the presumed critical amounts of important factors, hence the sharp differences which frequently were observed between upper and lower portions of any one stream. This phenomenon as it relates to biomass depends on the initial loading of the stream. At a relatively heavy load (10% sewage) biomass tended to increase downstream.

Another interesting feature of the Swiss studies was the temperature regime which, due to the subsurface water employed, followed the natural temperature changes of surface streams by 2 to 3 months. Thus, the coldest temperatures occurred in April and May, the warmest in September and October. This made it possible to test the relative significance of temperature and light schedules in the growth of certain river organisms. To *Sphaerotilus natans* Eichenberger (1967*a*)

55

ascribes best development in winter months; in the artificial stream, the best development of *Sphaerotilus* was achieved in the period November to April, but this occurred between temperatures of 15° and $6^{o}C$, a sufficiently wide range to discredit winter temperatures as the decisive factor in determining *Sphaerotilus* growth at this period. *Sphaerotilus* die-off occurred in April at $6^{o}C$, a temperature typical for winter. Eichenberger's conclusions, stemming from this and related observations, are that such die-off is due to effects of light conditions on the autotrophic organisms which compete for space and perhaps for metabolites with the *Sphaerotilus.*

For the growth maxima of the larger algae growing in the model streams Eichenberger concluded that temperature plays a major role. The best development of *Cladophora* and *Vaucheria* was achieved in fall and winter. He states that these algae are seldom observed in natural waters in late fall—a conclusion which is not supported by my own observations of fresh waters in North America.

Another of the artificial systems is that devised by Thomas and O'Connell (1966). This consisted of closed chambers of 10-liter capacity, of Plexiglas and polyethylene in which algal mats or tufts were exposed to circulating water for measurement of dissolved oxygen production and consumption and determination of the P/R ratio. This work was done *in situ* in the Truckee River, a Nevada stream polluted by the City of Reno. The method incorporated a D.O. probe in the chambers, which permitted frequent recording of D.O. in light as well as dark chambers.

More recently Sperling and Grunewald (1969) have utilized a small artificial indoor two-channel stream to demonstrate current effect on biomass and on ^{32}P uptake rates by algae. In earlier work Whitford and Schumacher (1961, 1964) brought out sharp differences in respiratory levels and ^{32}P uptake between freshwater algae in still water and in current and especially by typically lotic algae, some of which were shown to have higher respiratory rates than species characteristic of still water. While the differences in rates demonstrated by Sperling and Grunewald are relatively small, they nevertheless clearly show current influence on uptake rates in the thermal alga *Mastigocladus* when it is cultured at flow rates of 45 cm/sec as opposed to 2 cm/sec. Features of the Sperling-Grunewald model include the maintenance of constant temperature, essentially constant medium at constant pH, and the use of styrofoam as a substrate. The styrofoam permits great ease and frequency of sampling of the *Mastigocladus*, which grows as a tomentum of periphyton on the styrofoam surface.

Of the systems discussed above, both the McIntire-Phinney and Eichenberger-Wuhrmann systems utilize natural water sources of presumably somewhat variable chemical composition and the latter system was further influenced by the addition of sewage of fluctuating composition and BOD. Neither system introduced a known biotic community whose composition was fully controllable. The experiments of Whitford and Schumacher, and of Thomas and O'Connell were performed at small scale on known algal communities, and important characteristics of the ambient fluid could also be monitored and controlled. Perhaps the best refinement respecting the community make-up was achieved by Sperling and Grunewald who worked with unialgal cultures of *Mastigocladus laminosus*. This is not so artificial a society as it might seem, since *Mastigocladus* is known to grow in unialgal stands in hot springs and streams where it is commonly found. Their apparatus permitted significant controls of the ambient medium, the only obvious variables giving interference being those related to and dependent on the variable current rate they were studying.

None of the above streams provides perfect control of all potential variables. Indeed, streams are so variable in time that an experimental setup in which all physical and chemical factors are stabilized would be extraordinary indeed and it may be observed that information on stream ecology can still be obtained in the least sophisticated of all stream systems: a natural stream which in one of two comparable reaches experiences a single determinable, measurable and sustained environmental aberration.

The papers mentioned, and numerous others involving means of obtaining similar basic data, suggest one way of defining a stream: as pointed out by Odum (1956) primary productivity levels are generally high by comparison with standing water productivity.

Theoretical Models. Among the theoretical models relating to the life and growth of stream plants is that of Westlake (1966) for vascular plants, principally *Potamogeton pectinatus* growing in streams of southern England. Westlake established that light at the surface saturates the photosynthetic mechanism for much of the day, but only in surface layers. When saturation occurs at the surface, the compensation point is at about 0.5 m; at lower light intensities this occurs closer to the surface. It is noteworthy that the growth of this weed is massive just below the surface, very little of the weed biomass occurring below the compensation point for bright sunlight. As is common in situations of optimum growth of green plants the net oxygen balance increases, in fact doubles, with population density during the early stages of growth; later the weed begins to fill available space and the branches and leaves near the surface utilize more and more of the light. Finally the lower branches no longer function in assimilation, but their respiration proceeds and the oxygen balance falls off.

Westlake's model uses graphic techniques to obtain values for photosynthesis and respiration and to predict the effect of changing various factors. The weed biomass, its photosynthetic characteristics, and certain data about the stream are the principal base from which the overall hourly or daily net production or consumption of oxygen can be calculated, per unit area of stream surface. Checks of the Westlake model against experimental observations showed no gross discrepancies. As growth proceeds, maximum biomass is attained and at high population densities the respiration of the shaded weed becomes greater than the photosynthesis of the surface, illuminated weed. When weed is not cut, the population is likely to exceed optimum value before the end of the growing season. Plants or parts of plants can adapt to the average light intensity of their habitat. Within a single plant it is possible that surface leaves are sun-adapted and deeper leaves shade-adapted. Westlake points out the usefulness of the ratio between net and gross oxygen production. In his basic example, this was 80% at noon on a fine day; if the full 24 hours are considered, 54% was expected in fine weather, and 29% on an average day, but on very dull days respiration always exceeded photosynthesis.

Periodic Phenomena. Among recent investigations on algal growth in streams are European papers identifying a series of parameters which, at least in relatively small streams, fluctuate diurnally. Some years ago an observation made on southern Michigan streams (Blum, 1954) led to the identification of a daily pulse in phytoplankton drift. This was demonstrable only for the diatom *Nitzschia palea* and it appeared at the time that the phenomenon might be limited to this diatom and this stream. In recent years Müller-Haeckel (1966, 1970) has demonstrated a similar diatom periodicity in streams in Central Europe and in Sweden; the phenomenon clearly has wide prevalence. Serra-Tosio (1969) has shown the existence of regular diurnal variations in the chemistry of mountain streams; O_2, SiO_2, NO_3, PO_4, dissolved solids and alkalinity all showed fairly clear diurnal trends, although their changes, particularly their respective periods of increase and decrease are not synchronous. Serra-Tosio states that plants are responsible for these changes, but does not indicate which plants these are.

Müller-Haeckel's work has shown interesting seasonal trends in the drift phenomenon: In the Swedish stream studied a striking annual periodicity was recorded in the *Synedra minuscula* drift and the *Ceratoneis arcus* drift. These fluctuations are in agreement with known periodicity trends in these species. Superimposed on this annual fluctuation is a diurnal periodicity in the drift of both these species. The diatoms studied have a daily rhythm whose maximum in the drift or phytoplankton coincides with the light maximum at midday, as in the diatoms I observed in Michigan. Certain green algae studied by Müller-Haeckel enter the drift during the night and have their maximum representation before and just after dawn. When colonization by these drift organisms is related to time, the period of principal colonization correlates in each case with the period of principal representation in the phytoplankton drift. Müller-Haeckel related these facts to the general tendency of diatoms to divide by day, whereas green algae generally divide by night. It can be supposed that division is followed by the freeing of one or both daughter cells from the substrate, thus by their dislodgment and temporary renunciation of benthic existence. But it is clear that they do not go far. Sampling for drift at various downstream points on the same day gave no indication of an overall increase in the drift population with distance; thus the organisms settle out, and the drift populations are for a few hours each day in a state of equilibrium, gaining along one reach of the stream approximately as many new drift organisms as are lost by colonization or other attrition. Plastic slides placed in this stream were colonized after only one-half hour. Thus the drift maxima which seem to be so striking a feature of many small streams may be related not directly to photosynthesis and oxygen buoyancy, as is the case for filamentous algae when they are torn from the stream bed and enter the current (Blum, 1957), but indirectly, and apparently through the diurnal timing of cell division which may differ from group to group.

Classification and Structure of Stream Communities. Stream ecologists would probably be gratified to have and to work with well drawn maps representing the extent of lotic plant communities. Although this has been attempted and in part achieved for animal communities (and for benthic marine algae), we still have nothing of this type to show for streams, although such maps could probably be based on intensive studies of specific geographic areas, such as the studies of Whitford in North Carolina, Foged in Denmark, Kann in Austria, and Backhaus in Germany. There is relatively solid knowledge of the vegetation of many somewhat extreme lotic habitats, hot springs, acid streams, saline streams, streams with heavy sewage loads, and of many rather ordinary mesotrophic streams and reaches of streams, but so few of the total number have been investigated, and the vegetation is so complex that the plant communities of "ordinary" streams in much of the world remain uncharted. The dictum that an ecological grouping of watercourses is unfeasible, that only reaches of watercourses can be thus defined (Berg, 1948) has particular significance for botanists since the periphyton and phytoplankton with which they work are small and sometimes poorly understood at the species level, and tend to join with other species in complex and rather confusing communities.

Some of these difficulties are clearly brought out by a detailed study of the biota of the Savannah River, U.S.A. Patrick, Cairns and Roback (1967) were unable to identify a constant association between any two individual species either within one group or between major systematic categories. A few species did indeed always occur together, but they were among the infrequent species and thus their association was open to question as a mere chance grouping. A constant association, resulting in high correlation between individual species and specific ranges of chemical or physical conditions, was likewise difficult to demonstrate and few such correlations were possible, partly because most species were collected infrequently—in spite of the team of trained biologists who collected this stream with thoroughness. The choice of river basin may have a strong influence here. It is noteworthy that the Savannah team did not set out to demonstrate such correlations, but only to study and characterize a given river. Streams with a wider range of chemical or physical conditions might give a different picture, but the fact remains that stream vegetation is often discouragingly difficult to characterize and classify ecologically, or to use in formulating the definition of a river.

In the Savannah study, in which a broad spectrum of animal and plant species were collected and catalogued, the number of species in any major group was found to remain from season to season and from year to year with relatively little change. An interesting aspect of community composition was brought out by "overlap" studies of species in geographically separate communities in the Savannah River. On a given survey (same time but different stations) the overlap in diatom species, i.e., in species found in both stations of a given pair, was nearly always greater than 60%, with diatoms tending to overlap more than algal species. There was, however, no definite correlation found between the amount of overlap and the proximity of stations. When a station was studied at different times, a much smaller overlap was found: Again diatom species showed higher overlap values, about 40 to 55% as against about 23% for nondiatomaceous algae.

In a recent summary of factors affecting the structure of benthic communities in streams, Patrick (1969) has expressed a salient fact of life for river communities, namely that their nutrients—dissolved, suspended and organismal—are continually entering a given area from upstream or from the watershed either by direct runoff or through ground water. Unlike the conditions in lakes, where there is characteristically a recycling of nutrients, streams enjoy a continual renewal of dissolved nutrients. For the primary producers this presumably signifies a need for rapid uptake of such nutrients as are available. Inevitably, in a growing community, the downstream plants find themselves sooner or later losers in a competition for factors which the upstream plants are in a better position to obtain; thus the survival value of rapid uptake ability, a large surface-to-volume ratio, as pointed out by Cushing (1967) and a body plan which promotes abundant contact with the moving water. Other advantages of the current (which may again become disadvantageous in downstream areas) include the possibilities it offers for dissemination of reproductive bodies and for local riddance of excretions and excreta (an advantage which human communities seized upon long ago). The basic facts of stream loading apply here: Like the reduction in nutrients characteristic of a crowded stream bed population, the increment of excreta presumably becomes deleterious, depending on volume, flow rate, distance traversed, and the size and condition of the upstream populations.

Fjerdingstad has done much significant work on benthic communities, largely in relation to water pollution. His larger works of the past decade (Fjerdingstad, 1964, 1965) are based on the premise that optimal conditions for growth of a species are much narrower than the conditions of its possible occurrence, thus that the estimation of pollution levels should be based rather on communities of organisms present (reasoning that community formation must depend on optimal conditions for its species) than on isolated indicator species.

Fjerdingstad's work and that of Patrick and co-workers are probably the most detailed studies to date on the autoecology of individual species of aquatic microphytes. Both studies have provided environmental data for an impressive list of organisms. Fjerdingstad's work stresses "saprobic valency" and is limited to plant organisms, but it covers a relatively broad spectrum, giving for example, extensive consideration of bacteria, and it represents a synthesis of information from a literature of more than 600 titles. The very multiplicity of observers quoted is something of a drawback here. The work on the Savannah River (Patrick, Cairns and Roback, 1967) is limited to the results of work on that stream but it is of particular value since the chemical data are presumably comparable throughout, for the results with protozoan, insect and other animal families as for those with benthic plants.

While the above workers have given us precise information about many stream species when they are exposed to polluted water, other species, as information becomes more abundant and more reliable, show such broad tolerance that their usefulness as indicator organisms is nil. With the years, the indicator concept has undergone much modification. It has become clear that many algae, particularly the blue-greens, exhibit a kind of genetic diversity essentially unlike that of other eukaryotic algae. It is not unusual to encounter five or six collections of a blue-green which, through environmental influence (Drouet, 1963) or genetic foibles (Golubić, 1969; Eberly, 1967), are distinguishable one from another, but are nevertheless felt to be referable to a single taxonomic species. Under such conditions the structure of the binomial nomenclature begins to totter. More apropos of our present topic, the raison d'être of an indicator is clearly a biochemical one—a growth response, positive or negative, to nutrients or toxic compounds in the water. Depending on the presence or prevalence of biochemical variability as between different strains, the performance of any one indicator could be lamentably spotty. Some years ago I attempted to demonstrate this fact for strains of *Scenedesmus quadricauda*, thinking that strains derived from clean water might be more susceptible in culture to common industrial pollutants than others from polluted water, but results were inconclusive.

Several well-known plants appear to be reliable indicators. Unfortunately, they are not reliable in the sense that they can be counted on to turn up in every stream system where we need an estimation of water quality. Other methods and systems of assessing quality on the basis of plant populations have been reviewed elsewhere (Fjerdingstad, 1964; Patrick, 1949, 1963).

One of the difficulties which has hindered the successful cataloguing of stream algae is in part removed by the work of Backhaus (1968) on Danube headwaters. For the first time we have in his work an extensive iconograph of the small and frequently nondescript stream algae. Although the work has strict geographic limits, it should prove very useful to stream ecologists generally.

The botanical picture of a river, the view from the mesophyll—or the eyespot—which emerges from the above remarks is of a nutrient medium in unidirectional motion, less characterized by recycling phenomena than a lake, with complex and diverse chemical and physical parameters which are likely to be fluctuating at different frequencies and in many different ways but generally in such a way as to permit a high level of productivity. It is a niche offering suitable rewards to those autotrophic and heterotrophic specialists which can cope with it, or with any geographically limited part of it, a favorable but erratic medium for many plants which find the current a fact of life but have no reliable means of attachment. Regrettably this description has nothing about it of the dictionary definition. I hope it expresses to a small degree certain of the conclusions to be drawn from the labors of the scientists I have cited.

Literature Cited

Backhaus, D. 1968. Ökologische Untersuchungen an den Aufwuchsalgen der obersten Donau und ihrer Quellflüsse IV. Systematisch-autökologischer Teil. Arch. Hydrobiol./Suppl. 34 (Donauforschung 3)(4):251-320.

Berg, Kaj. 1948. Biological studies on the River Susaa. Folia Limnologica Scandinavica 4:318 p. and plates, etc.

Blum, John L. 1954. Evidence for a diurnal pulse in stream phytoplankton. Science 119:732-734.

—————————. 1957. An ecological study of the algae of the Saline River, Michigan. Hydrobiologia 9(4):361-408.

Butcher, R. W., F. T. K. Pentelow et J. W. A. Woodley. 1930. Variations in the composition of river waters. Int. Rev. Hydrobiol. 24:47-80.

Cushing, C. E. 1967. Periphyton productivity and radionuclide accumulation in the Columbia River, Washington, U.S.A. Hydrobiol. 29:125-139.

Drouet, F. 1963. Ecophenes of *Schizothrix calcidcola* (Oscillatoriaceae). Proc. Acad. Nat. Sci. Philadelphia 115(9):261-281.

Eberly, W. R. 1967. Problems in the laboratory culture of planktonic blue-green algae. *In* Environmental Requirements of Blue-Green Algae, Univ. Wash., U.S. Water Pollut. Contr. Adm. Sympos. (1966) v, 111 p. Corvallis, Oregon.

Edwards, R. W., and M. Owens. 1962. The effects of plants on river conditions— IV. The oxygen balance of a chalk stream. J. Ecol. 50:207-220.

Eichenberger, E. 1967a. Ökologische Untersuchungen an Modellfliessgewässern 1. Die jahreszeitliche Verteilung der bestandesbildenden pflanzlichen Organismen bei verschiedener Abwasserbelastung. Schw. Zeitschr. Hydrol. 29(1):1-31.

—————————. 1967b. Ökologische Untersuchungen an Modellfliessgewässern 2. Jahreszeitliche Veränderungen der Biomassebildung bei verschiedenen Abwasserbelastungen. Schw. Zeitschr. Hydrol. 29(1):32-52.

—————————, et K. Wuhrmann. 1966. Über jahreszeitliche Veränderungen der Besiedlungsdichte in Modellfliessgewässern mit verschiedener Abwasserbelastung. Verh. Internat. Vereining. Limnol. 16:886-896.

Fjerdingstad, E. 1964. Pollution of streams estimated by benthal phytomicroorganisms. Int. Rev. Ges. Hydrobiol. 49(1):63-131.

—————————. 1965. Taxonomy and saprobic valency of benthic phytomicroorganisms. Int. Rev. Ges. Hydrobiol. 50(4):475-604.

Golubić, S. 1969. Tradition and revision in the system of the Cyanophyta. Verh. Internat. Verein. Limnol. 17:752-756.

Huet, M. 1946. Note préliminaire sur les relations entre la pente et les populations piscicoles des eaux courantes. Règle des pentes. Dodonaea 13:232-243.

Hynes, H. B. N. 1970. The Ecology of Running Waters. xxiv, 55 p. Univ. Toronto.

Illies, J., et L. Botosaneanu. 1963. Problèmes et méthodes de la classification et de la zonation écologique des eaux courantes, considérées surtout du point de vue faunistique. Mitt. Int. Ver. Theor. Angew. Limnol. 12:1-57.

McIntire, C. D. 1966a. Some effects of current velocity on periphyton communities in laboratory streams. Hydrobiologia 27:559-570.

—————————. 1966b. Some factors affecting respiration of periphyton communities in lotic environments. Ecol. 47:918-930.

—————————. 1968. Structural characteristics of benthic algal communities in laboratory streams. Ecology 49:520-537.

—————————, R. L. Garrison, H. K. Phinney, et C. E. Warren. 1964. Primary production in laboratory streams. Limnol. and Oceanogr. 9:92-102.

—————————, et H. K. Phinney. 1965. Laboratory studies of periphyton production and community metabolism in lotic environments. Ecol. Monogr. 35: 237-258.

Müller-Haeckel, Agnes. 1966. Diatomeendrift in Fliessgewässern. Hydrobiologia 28(1):73-87.

—————————. 1970. Tages und Jahresperiodik einzelliger Algen in fliessenden Gewässern. Österreichs Fischerei 5/6:97-101.

Odum, H. 1956. Primary production in flowing waters. Limnology and Oceanogr. 1:102-117.

Patrick, Ruth. 1949. A proposed biological measure of stream conditions based on a survey of the Conestoga Basin, Lancaster County, Pennsylvania. Proc. Acad. Nat. Sci. Philadelphia 101:277-341.

—————————. 1969. Summary of factors affecting the structure of benthic communities in streams, p. 38, 39. In The Stream Ecosystem, AAAS Symposium, Tech. Rep. No. 7. (3), 42 p.

Patrick, Ruth, J. Cairns, and S. S. Roback. 1967. An ecosystematic study of the fauna and flora of the Savannah River. Proc. Acad. Nat. Sci. Philadelphia 118(5):109-407.

Schmassmann, H. J. 1951. Untersuchungen über den Stoffhaushalt fliessender Gewässer. Schweiz. Zeitschr. Hydrol. 13:300-335.

Serra-Tosio, M. B. 1969. Mise én evidence, dans les cours d'eau de montagne de variations nycdthémérales de certains facteurs chimiques sous l'influence des organismcs benthiques. C. R. Acad. Sci. Paris 269:2431-2434.

Sperling, J. A., and R. Grunewald. 1969. Batch culturing of thermophilic benthic algae and phosphorus uptake in a laboratory stream model. Limnology and Oceanogr. 14:944-949.

Thomas, N. A., and R. L. O'Connell. 1966. A method for measuring primary production by stream benthos. Limnology and Oceanogr. 11:386-392.

Westlake, D. F. 1965. Some basic data for investigations of the productivity of aquatic macrophytes. Mem. Ist. Ital., Idrobiol. 18 Suppl:229-248. (Also) Proc. of the I. B. P. Symposium on Prim. Productivity in Aq. Environ. Pallanza, Ital.

—————————. 1966. A model for quantitative studies of photosynthesis by higher plants in streams. Air, Water Pollut. Int. J. 10:883-896.

Whitford, L. A., et G. J. Schumacher. 1961. Effect of current on mineral uptake and respiration by a fresh-water alga. Limnology and Oceanogr. 6:423-425.

—————————. 1964. Effect of a current on respiration and mineral uptake in Spirogyra and Oedogonium. Ecology 45:168-170.

Wuhrmann, L. 1964. River bacteriology and the role of bacteria in self-purification of rivers, p. 167-192. In Heukelekian, H., et N. C. Dondero [ed.]. Principles and Applications in Aquatic Microbiology, xiii, 452 p. New York.

Zimmermann, P. 1961. Experimentelle Untersuchungen über die ökologische Wirkung der Strömungsgeschwindigkeit auf die Lebensgemeinschaften des fliessenden Wassers. Schweiz. Zeitschr. f. Hydrol. 23:1-81.

A COMMENTARY ON "WHAT IS A RIVER"

Ruth Patrick, Session Chairman

According to the Oxford dictionary, the word "stream" is an all-inclusive term for flowing water. A river, therefore, is a type of stream. In America we think of a river as a fairly large, flowing body of water whose characteristics are largely determined by the geology, topography, soils, and land use of the watershed. Recent studies show that rivers may also be affected by pollutants in the atmosphere and by activities in other watersheds through contaminated rain.

Rivers in the United States may be classified in many ways. One method is by the concentration of nutrient chemicals. As Hutchinson (1969) states, we must consider the total potential concentration of nutrients in the aquatic system. Oligotrophic streams are those low in dissolved nutrients and relatively cool. These kinds of streams are often found in mountainous regions in a drainage basin largely composed of igneous rocks. Mesotrophic streams are those which have a moderate amount of nutrients. Eutrophic streams are those rich in a well-balanced assortment of dissolved nutrients. The term "eutrophic" has been greatly misused in recent times to refer to streams with excessive and often imbalanced nutrients, for example, streams polluted from organic wastes of farms and cities. A fourth type of stream is one that is rich in humates and typically has a pH below 7. Often these streams are quite acid. These are the black- or brown-water streams that are commonly found in the coastal plains of the eastern United States.

Another way to classify rivers is according to the kinds of cations that are present. For example, in the West we have rivers rich in alkali metals—that is, sodium and potassium or a combination of sodium, potassium, calcium, and magnesium, with sodium and potassium the dominant ions. A second class is hard-water rivers which are commonly found in midwestern United States and are rich in calcium and magnesium. A third class is moderately soft-water rivers which have relatively small amounts of calcium, magnesium, sodium, and potassium. This is the type of river most commonly found in the eastern United States. A fourth class is streams which have no measurable calcium and are characterized by calciphobe floras such as diatoms of the genera *Eunotia* and *Actinella.* These rivers may be found in the northern part of the United States and in our coastal plains.

67

Structure of Rivers

The main stem or trunk of a river is formed by the confluence of many smaller streams. As Dr. Curry (in these *Proceedings*) points out, the profile of a large river approaches an exponential shape with a marked concavity, when the elevation is plotted against miles from the mouth. An example of such a profile is that of the Savannah River (Figure 1).

We have found in our studies of rivers in eastern United States that from the shape of the profile one can determine many characteristics of the watershed and the river. In the region of steep gradient the watershed is hilly or mountainous. The streams are small and shallow and, as a result, the immediate banks are low. If their flow is mainly derived from springs, the water temperature in the upper reaches of the headwater streams is fairly similar throughout the year, being warmer in winter and cooler in summer than streams which are mainly derived from surface drainage. Because the watershed is typically forested in this area, erosion is minimized and the water is clear—as a result, the photosynthetic zone extends across the bed of the river.

The amount of primary production is dependent on the chemical composition of the water and the amount of shading of the forest. Typically, predator pressure is high because of the large number of invertebrates that live in these headwater streams, and, as a result, the standing crop of algae may give a false impression of the productivity.

The bed of the typical stream of the eastern United States is rock, rubble, and sand, although a small amount of mud and silt may be present. In contrast, in the Midwest the beds of headwater streams are sometimes hard mud. The stream structure is typically a riffle-pool sequence. This diversity of current flow and habitats provides many niches for species occupancy, and the species carrying capacity of the stream is high.

Man's use of the watershed in this region may produce severe perturbation in these streams. For example, the work of Wert and Keller (1953) in the Potomac River Basin showed that when the forest cover was reduced from 80% to 20%, the sediment yield increased eight times. Bormann and Likens in their studies in Hubbard Brook have shown that cutting the forest increased the annual runoff 28% to 39%; however, between June and September the increased runoff became much greater—414% in 1966-1967 and 380% in 1967-1968—because the loss of water by transpiration had greatly decreased. There was also a great increase in dissolved and particulate solids entering the stream. Due to the loss of binding action of the tree roots and the protective effect of leaves, the erosion of the stream channel increased.

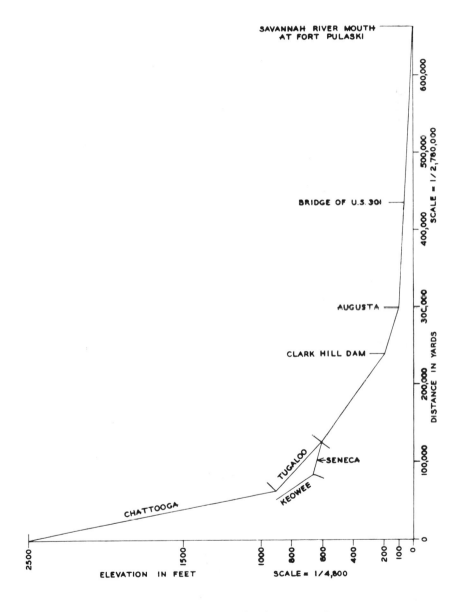

Figure 1. Longitudinal profile of surface gradient

The main trunk of the river is formed in the region of intermediate gradient. The flow greatly increases and the meanders become larger. Depending on the geology of the area, the river may be broad and shallow with a large surface-to-depth ratio or it may be deep with a smaller surface-to-volume ratio.

Typically, the silt load becomes much heavier in this region. The photosynthetic zone is thus restricted to the shoaling areas of meanders, to floating debris, and to oxbows and sloughs. In the latter two areas the current is greatly reduced, the suspended solids settle out, and the clearer water permits extensive algal growths. It is here that one often finds the greatest concentration of fish, algae, and invertebrates.

The floodplain of the river is often large, and small ponds are numerous. These are often connected with the river during high flow and, as a result, add considerable numbers of various kinds of aquatic life to the river. If it were not for these ponds, oxbows, and sloughs, the carrying capacity of the river for aquatic life in this reach would be much less.

This intermediate gradient area is the section of the river that has been used by man heavily over time for farming, urban development, and construction of highways. The large flow and turbulence of the river provide excellent mixing and dilution. The river is also deep enough in many instances to allow navigation if dredged channels are maintained. For these reasons in many rivers the aquatic life has often been subjected to a great amount of perturbation. These effects could have been minimized if floodplains had been protected, cities and highways designed to conform to the constraints of the watershed, and waste water recycled or given maximum treatment before being discharged.

The natural flow of the river in this section is often altered by the creation of dams which would not have been necessary or could have been greatly reduced in size if man had planned the use of the watershed so that the rain penetrated the soil rather than producing erratic runoff. The effects of dams on the river are observed both upstream and downstream in chemical and physical changes in water quality and changes in the structure of the river channel, depositional and erosional rates of sediments, and the abundance and diversity of aquatic life.

The estuary is the area of little or no gradient. It is here that the banks of the river merge into open swampland and the river channel ceases to have definite boundaries. These swamplands are extremely good buffering areas for floods. They are also traps for the silt load being carried downstream. It is in these estuaries that a considerable amount of aquatic life breeds, and the productivity due to algae is relatively great. Beds of rooted aquatics are extensive in these areas.

They, as well as the algae, improve the water quality of the estuary by the removal of nitrogen and phosphorus and the production of oxygen (Grant and Patrick, 1970). The rooted aquatics are also a source of food for the emergent fauna, and their detritus is a valuable food source for aquatic life.

The flow of the river, instead of being unidirectional, is two-directional because of tidal action. Typically, a salt tongue extends up the stream along the bed of the river. The extent of this tongue varies with the force of the fresh-water flow and the structure of the estuary. Therefore in the summer the saltwater tongue usually extends upstream a greater distance than it does in the winter. Because the water is salty, we often have considerable stratification in this area, with the fresh water flowing on top of the salt water. Depending upon the shape and structure of the estuary, the degree of stratification may vary. Because this is the area of interface of fresh and salt water, the colloids of the suspended solid load are precipitated. As a result, the beds of estuaries are typically muddy, and this mud is often fairly rich in organic material. Thus during the summer stratified "lake-like" conditions may develop, and the estuary may have little or no oxygen near the bed of the river.

Pollutants placed into an estuary may not be subject to the same degree of mixing that occurs when they are discharged into the region of intermediate gradient. This is because the density of the waste may be such that it is less than that of salt water but greater than that of fresh water. Any organism that tries to move from the bottom to the surface of the estuary or vice versa may therefore be killed by a thin toxic layer of effluent. Concentrations of discharges of waste may build up in this area due to two-way tidal flow and the resultant hold-up time. Thus in calculating the effect of a pollutant on a river one must take these facts into consideration.

Aquatic Life

The aquatic life in streams is classified by Cummins (in these *Proceedings*) as lotic nekton and lotic benthos. Each of these two main divisions he divides into erosional assemblages, intermediate assemblages, and depositional assemblages. He also discusses the type of species, particularly animals, which may be present. Dr. Blum (in these *Proceedings*) discusses the algal flora.

When one examines the kinds of plant species composing the aquatic flora, he finds little evidence of one type of flora succeeding a previous type, as in land communities. It is true, as pointed out by Hutchinson and others, that there is seasonal succession of species and succession due to shifts in nutrients, but not an orderly succession over time as a result of the maturing of an ecosystem.

71

This is no doubt due to the fact that the communities of streams, as set forth below, are young or in a condition of steady state. It is apparent that the occurrence of an algal species is more dependent on the chemical and physical characteristics of the water and river basin than on the previous sequence of occurrence of other species.

The communities from headwaters to mouth of the river have certain basic similarities and differences. One might say that the aquatic life in most of the river forms a loosely interacting community of species because of the phenomenon of downstream drift and the migration upstream of fish and invertebrates; this is particularly true of fish which migrate to the headwaters to spawn. There are also many small communities such as one finds in pools, sloughs, or riffle areas.

These communities consist of four or five stages of energy transfer: decomposers, primary producers, herbivores, and carnivores. The decomposers are mainly fungi and bacteria but also include some species of invertebrates. The primary producers are algae and rooted aquatics. The algae are a much more important food source than the rooted aquatics for most aquatic organisms. The herbivores may include a great variety of organisms ranging from protozoa to fish. We think of herbivores as organisms that feed on living plants, but there are many organisms that feed on dead plant materials. Indeed, in forest streams plant detritus may be the major food supply. Likewise, carnivores may include many different phyla of animals. Omnivores are very prevalent in streams and often form a major component of the community. Many species are involved in each of these stages of energy transfer. The species represent diverse systematic groups, and there are many pathways in the food web.

The life spans of these species are variable; some reproduce twice a day, others once a week, others once a year, and others at less frequent intervals. Compared with many species in the terrestrial community, all of them have relatively short life spans. They typically produce many progeny and are the kinds of species often found in a relatively young or immature community. Because of their short life span and rapid reproduction rates, the sizes of populations may vary greatly in a short period of time and reflect frequent environmental changes.

These communities are continuous over time and are often referred to as being in a condition of steady state. Dissolved nutrients, detritus, and living organisms which are food sources are continually entering and leaving the community at highly variable and unpredictable rates. Likewise, allelo-chemicals, as described by Whittaker and Funz (1971) are entering and leaving the community.

The species composition of these communities varies greatly over time, apparently because of the high invasion rate and variability of the environment. For example, in our studies of diatom communities when the invasion rate was greatly reduced, the diversity of the diatom communities dropped from 129 to 100 species (Patrick, 1967). The species pool capable of invading an area is much greater than the number of niches for species occupancy in the area. Thus small shifts in the environment result in a shift in species composition. If the species pool for an area is reduced, as happens when a river is polluted, the change in species composition over time is much less.

Although the kinds of species have a high rate of change in a given reach of a stream such as a riffle-pool sequence, the number of species stays quite similar over time. Furthermore, in equivalent ecological areas in similar natural streams the number of species is quite similar (Patrick, 1961).

Equability or evenness of distribution of specimens among species is usually quite low, probably because of the unpredictable and variable ecological characteristics of the stream and the continual renewal of nutrients, which tends to minimize the competition between species for food.

In the fresh-water reaches of a river the same systematic groups of species carry out a given function over time although individual species may change. In an estuary, however, conditions are quite different. In the winter, when the surface waters are fresh, the communities of organisms living in shallow surface waters are mainly fresh-water species and species which can tolerate fresh and brackish waters. For example, when we studied the Escambia River several years ago, we found that fresh-water aquatic insects with short life cycles invaded the estuary in the winter but were absent most of the year when the water was brackish. Under such conditions the surface and shallow waters contain fresh-water species and species which can withstand fresh and brackish water conditions, while the deeper waters, which are always brackish, support a brackish to marine fauna.

The estuary communities also differ in that they have a typical plankton community, as well as nekton and benthic faunas and floras. They have many more invaders from other ecosystems, such as crabs, shrimp, and fish which spend part of their lives in the sea and part in the estuary of the river. From this brief account we see that the communities of aquatic life in a river are affected by the watershed and the ocean, and through drift and migration the smaller communities are integrated into a large and looser community which is the river ecosystem.

73

Literature Cited

Grant, R. R., Jr., and R. Patrick. 1970. Tinicum marsh as a water purifier, p. 105-123. *In* Two studies of Tinicum marsh. Conserv. Found., Washington, D.C.

Hutchinson, G. E. 1969. Eutrophication, past and present. Eutrophication—Proceedings of a Symposium. Nat. Acad. Sci., Washington, D.C. 661 p.

Patrick, R. 1961. A study of the numbers and kinds of species found in rivers in eastern United States. Proc. Acad. Nat. Sci. Philadelphia 113:215-258.

—————. 1967. The effect of invasion rate, species pool, and size of area on the structure of the diatom community. Proc. Nat. Acad. Sci. 58:1335-1342.

Wark, J. W., and F. J. Keller. 1963. Preliminary study of sediment sources and transport in the Potomac River Basin. Interstate Commission on Potomac River Basin, Washington, D.C., Tech. Bull. 1963-11, 28 p.

Whittaker, R. H., and P. P. Funz. 1971. Allelochemics: chemical interaction between species. Science 171:757-770.

SESSION II

USES OF A RIVER: PAST AND PRESENT
Chairman: E. D. LeCren

MAN'S IMPACT ON THE COLUMBIA RIVER

Parker Trefethen

The Columbia River, discovered in 1792, is one of the most highly developed rivers in the United States. In early colonial history American Indians and white settlers depended on the river for transportation and for food and trading material, on the huge runs of salmon that ascended the river each year on their spawning migration. As the northwestern United States was further settled, the river was used more extensively for transportation and commercial fishing, and water was diverted for many purposes. To keep pace with the growing demands on this water resource and to foster commerce and industry in the Columbia River Basin, the river and its tributaries were further harnessed for power, irrigation, and navigation. These developments, along with the increasing use of the river for disposal of various waste materials, have profoundly affected the valuable fishery resource. Because of this impact, ecological studies have emphasized Pacific salmon (*Oncorhynchus* spp.) and steelhead trout (*Salmo gairdneri*). It is our purpose here to review changes in the environment that threaten salmonids and activities directed toward alleviating conditions critical to their survival during the freshwater phases of their life cycle.

The Columbia River and Basin

The Columbia River (Figure 1) is the longest North American river, 1,950 km long, flowing into the Pacific Ocean. In volume of flow it is the second largest in the United States. Annual flows average 5,700 cms (cubic meters per second) and range normally from about 1,400 cms in winter months to 18,400 cms during spring freshets; the largest flow of record—33,100 cms—occurred in 1894. The Columbia River flows from its headwaters in the Canadian Rocky Mountains of British Columbia, Canada, about 730 km in the United States boundary. After crossing into north-central Washington, it flows 1,218 km through the semi-arid Columbia River plateau, the Cascade Mountain range, and the coastal plain to the Pacific Ocean.

The Columbia River Basin drains an area of about 671,000 sq km, including 100,000 sq km in Canada, most of Washington and Oregon, and parts of Idaho, Montana, Utah, Wyoming, and Nevada. Principal tributaries of the Columbia include the Kootenai, which lies mostly in Canada, and the Pend Oreille, Okanugan, Wenatchee, Yakima, Snake, Willamette and Cowlitz rivers in the U.S.

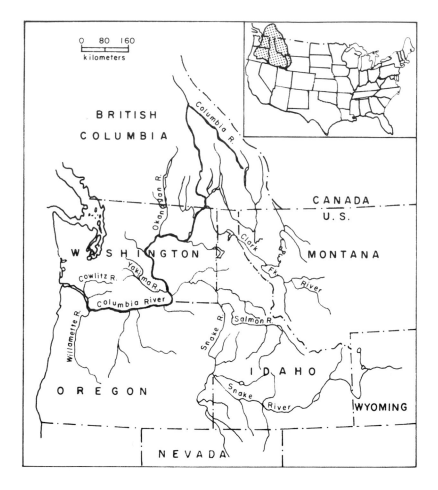

Figure 1. The Columbia River and Basin. The drainage area includes six northwestern states of the United States, and British Columbia, Canada

Transformation of the River

The population of the area which is now the states of Washington, Oregon, and Idaho was estimated at 100,000 in 1850, 58 years after the Columbia River was first explored. By 1900 the number of inhabitants had increased to 868,737; in 1950 it was 3,297,131; and in 1970 it was over 6,500,000. This influx created demands that could be met only by developing the major water resource.

Indians were the first to irrigate in the Columbia River Basin, and by the mid-1800's individual settlers were diverting small streams for agriculture. This development continued slowly, but as the demand for domestic food supplies grew, more land was reclaimed for growing crops. Irrigation projects, which included dams and large diversion canals, were first built in the upper tributary areas, in 1889, 162,000 ha were under irrigation, of which 112,000 were in the mid-Columbia region; by 1933 about 1,619,000 ha were under irrigation. Today about 2,700,000 ha are being irrigated in the Columbia Basin. The earliest projects probably had only a limited effect on the rivers and streams, but as dams and reservoirs increased in size, river flow patterns were severely altered and water quality was adversely affected. Fish passage facilities were rarely provided at these early dams. Hence they effectively blocked anadromous fish from their spawning areas, notably in the Yakima and Okanogan rivers and tributaries of the Snake. Irrigation diversions remained unscreened for many years, and thousands of young fish migrating seaward were lost.

The development of hydroelectric power in the Columbia Basin was first centered at natural waterfalls, such as Spokane (1885) and Willamette (1889). With technical advances in the utilization of water power and in the efficiency of hydro turbines, dams were built on feeder streams to the Willamette, Spokane, and other rivers tributary to the Columbia. These dams affected water quality and flows, and although fish facilities were provided at some dams, many were ineffective in passing fish.

Comprehensive planning to develop the main Columbia River for electric power and other uses began in 1927 with a major survey by the Department of the Army. In 1933 the first dam on the river, Rock Island, was completed by private interests. Bonneville Dam, the second and lowermost dam, was completed in 1938 by the U.S. Army Corps of Engineers. Construction of the uppermost dam in the United States—Grand Coulee—was started in 1933 and completed in 1941 by the Bureau of Reclamation. In the next 26 years eight more dams were completed. During the same period a number of major dams were also constructed on main tributaries to the Columbia River (Figure 2).

The dams have brought about dramatic changes in the character of the Columbia River. The once-free-flowing river was transformed into a series of pools; flow patterns and temperature regimes were altered. The dams themselves created temporary barriers to migrating fish, and the reservoirs behind them flooded spawning areas, modified predator-competitor relations, and affected food organisms. Changes that have already taken place may be amplified by the continuing development of the Columbia River in Canada and of the Snake River, a major tributary in the United States. In about ten years virtually

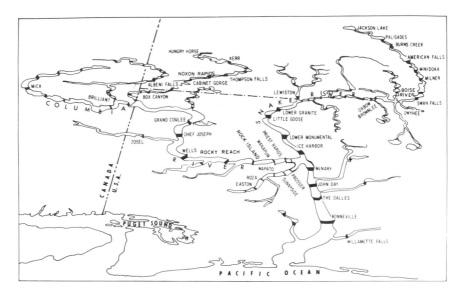

Figure 2. Location of major dams in the Columbia River system

all of the hydroelectric sites suitable for development will have been utilized; river flow will be more completely controlled; and there will be less spilling. When thermal electric plants in the Northwestern states assume base power production, hydroelectric plants of the Columbia River system will operate on peak demand, further changing downstream flows.

Early Investigations

Over the last half century efforts at water resources investigations on the Columbia River were minimal. The U.S. Geological Survey established a gauging station in the upper Columbia to record flows, water levels, and temperatures in 1894. In 1910 the Columbia Basin was included in a study of the quality of surface waters of Washington, and assessments were made of mineral content, color, and alkalinity of the water in relation to projected uses for industrial, municipal, and agricultural developments. After that period only limited water quality data were obtained until 1949, when scientists in Canada commenced studies on water quality of the Columbia River in Canada. This research was followed in 1954-56 by studies in the United States which indicated a general increase in water constituents examined in 1910, although the quality was still high and acceptable for use in public or industrial water supply, irrigation, or propagation of fish. It was noted, however, that there was a need for further study to determine the influence of additional dams and reservoirs on water temperature and its effect on fishery resources. In anticipation of the projected developments in the Columbia River Basin, federal, state, and privately endowed agencies began studies on topography, hydrology, hydraulics, engineering, erosion, sedimentation, and the effect of irrigation on the land and of return flows on the river.

Early investigations of the fishery resource centered around the taxonomy, biology, and physiology of Pacific salmon and other species of resident and anadromous fish. As stresses on salmon populations increased, more attention was directed toward timing of migrations, delineation of spawning areas, estimates of abundance, and status of individual runs. The commercial fishery reached a peak production in 1911 of 22,440,000 kg, but has since declined to an annual average of about 6,800,000 kg. The loss of production can in large measure be attributed to dams and other water use developments. In 1938 a federal program of research was initiated in cooperation with state agencies to investigate fundamental problems of migration and survival of Pacific salmon in the Columbia River. Fish passage facilities were required at most dams, but none were included in plans for Grand Coulee, which blocked about 1,600 km of spawning area. Here provisions were made for artificial propagation of part of the run, and the balance was relocated in other drainages. Extensive surveys were made of all streams in the Basin to classify their current potential use. Fishways were improved and fish counts made at dams to obtain an index of the magnitude of runs and movement to tributary streams.

The management of the fishery resource is the responsibility of state fishery agencies established in the late 1800's and early 1900's. Through their investigations, and with data from research by cooperating agencies, fishing regulations were established and hatchery operations increased rapidly. Laws were passed requiring fish facilities at all dams, and work was advanced on screening water diversions, providing passage facilities at impassable falls, and improving conditions in spawning streams. In 1947 Washington and Oregon proposed that the federal government finance a cooperative program for rehabilitating the Columbia River fisheries. Funds authorized in 1949 for the lower Columbia River Development Program below McNary Dam were justified on the basis of fish and fish habitats lost due to federal water use projects. The program was expanded in 1957 to include areas in the upper Columbia Basin.

Recent Activities

The Columbia River Development Program, concentrating activities on increasing the abundance of salmon, constructed or remodeled 21 hatcheries. Production of juvenile salmonids was increased from 10 million fish in 1949 to

100 million in 1970. In 1945, 5.8% of fall[1] chinook (*O. tshawytscha*) over Bonneville Dam were hatchery origin; now about 33% of the chinook and 85% of the coho (*O. kisutch*) are estimated to be hatchery fish; production of spring chinook salmon and steelhead trout was also improved. A major evaluation program, started in 1962, indicates that about 50% of fall chinook and coho taken in the commercial and sport fishery in the Columbia River coastal areas originated from hatcheries.

Field surveys were increased to locate obstructions that stopped or delayed migration of adult fish to spawning areas, and many of these were removed or bypassed with fishways. To help prevent losses of young fish, screens were designed and installed in water diversions used for power, irrigation, and municipal and industrial supply. The Development Program was also responsible for reviewing proposed water development projects and recommending measures to safeguard fish resources such as water storage, maintenance of flow, seasonal scheduling of construction, protection devices, and fish passage facilities.

Coincident with the goal of increasing abundance, research concentrated on defining sources of high mortality of salmonids in various stretches of the river and on devising methods to increase survival. Behavior and orientation of fish related to various stimuli and environments were investigated, and experiments were conducted on attracting adult fish into fishways, on passage through fishways and reservoirs, and on the use of stimuli and screening devices to prevent juvenile fish from entering hazardous areas.

Present Research

As major development on the Columbia River neared an end, attention was directed to the Snake River and its potential. Several high dams were proposed which could severely alter flows and water quality, thereby affecting major fish runs. To minimize effects on fish of developing the Snake River, an extensive investigation of high and low dams and of reservoirs of the Snake and Columbia rivers began in 1961, planned and conducted jointly by state and federal agencies.

Environmental Studies

Increasing emphasis is being placed on environmental changes related to survival of anadromous fish and on devising methods of increasing survival. Of

---

[1] Several races of chinook salmon in the Columbia River system are classified as spring, summer, or fall chinook, according to the time of migration upriver to spawn.

special concern is the effect of water temperature, nitrogen gas, competitors, river flows, and turbines on the survival of migrating anadromous fish.

Water Temperature. It is apparent from past trends that water temperature in the Columbia River is near levels critical for salmon (Figure 3) during certain periods of the year and that the high temperature cycle is being delayed. At its confluence with the Snake River, a temperature differential blocked movement of adult salmon into the Snake, where ambient temperature was 26°C or 3.9°C higher than the Columbia. Farther upstream in Brownlee Reservoir, a 92-km lake-like impoundment, environmental conditions were critical during the summer (Figure 4). Surface temperature reached 27°C, and at depths where more favorable temperatures prevailed, dissolved oxygen ranged from only 0 to 4 ppm. Conditions in the reservoir affected the river below by prolonging high temperature and by delaying increases in oxygen and other parameters into the fall months.

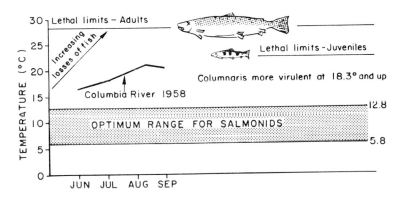

Figure 3. Water temperatures in some areas of the Columbia River are above optimum for salmon and trout

Investigations on the effects of water temperatures higher than those now prevailing began well before construction of thermal electric plants proposed for the Columbia River. The problems of migration and survival of anadromous fish could be compounded if these thermal plants are expanded on a large scale and waste heat is released indiscriminately into the river. Fish could be subjected to sudden increases of 9 to 17°C in thermal plumes at the point of discharge in the river and to lesser increases in the entire river cross section. Sudden exposure of juveniles in a laboratory to maximum temperature of 32°C—a predicted temperature of a thermal plume at peak ambient temperatures—resulted in death within seconds; at smaller increments fish died more slowly. Indirectly, increasing temperature delays migration of adult fish, fosters the development of fungus and disease organisms, affects food supplies, and favors some species of fish which compete with or prey upon salmon and trout (Figure 5).

83

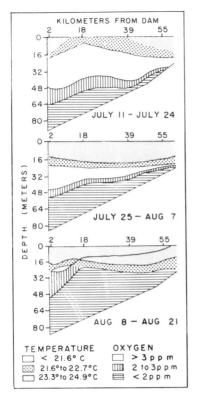

KILOMETERS FROM DAM

JULY 11 - JULY 24

JULY 25 - AUG 7

AUG 8 - AUG 21

TEMPERATURE
- □ < 21.6° C
- ▨ 21.6°to 22.7°C
- □ 23.3°to 24.9°C

OXYGEN
- □ > 3 p p m
- ▥ 2 to 3 p p m
- ▤ < 2 p p m

Figure 4. Temperature increases at surface and oxygen decreases at depths in Brownlee Reservoir decreases area (unshaded) suitable for survival of juvenile salmon and trout

Federal and state agencies continue to cooperate in investigating the effects of elevated temperature on fishery resources. A water temperature laboratory was established in the lower Columbia River near Prescott, Oregon, to examine further the direct, indirect, and synergistic effects of high temperature on anadromous fish. Model studies are in progress to aid in predicting temperatures in the Columbia and Snake rivers, and the possibility of using heated effluents beneficially for fish culture and agriculture is being investigated. Combined efforts by federal and state agencies have discouraged direct use of the river for cooling of thermal electric facilities, and plans for locating thermal plants in the Columbia River Basin and elsewhere are being critically reviewed by many agencies concerned with the ecology of the river.

Dissolved Nitrogen. Concentrations of dissolved nitrogen gas in the Columbia River are higher than would be expected with the warming of the river in summer. Extremely high concentrations below dams in the spillway area, but not in the turbine area, indicate that supersaturation is related to spilling characteristics of dams (Figure 6). Some decrease in nitrogen was noted in unimpounded sections of the river, but saturation remained at a dangerous level for salmon and trout in most impounded areas during periods of high spill discharge.

High nitrogen saturation is a problem where fish are exposed for prolonged periods during migration and in propagation facilities supplied with river water. Symptoms of gas bubble disease (Figure 7) were observed in fish in spawning channels and in juveniles migrating downstream in the Columbia River; survival in the river of juveniles migrating from upriver areas decreased from about 40% before spilling began at dams to 15% after spilling.

Concentrations of dissolved nitrogen gas did not become serious in the Snake River until water was impounded at lower Monumental Dam in 1969 and Little Goose Dam in 1970. In those years nitrogen saturation was as high as 146%.

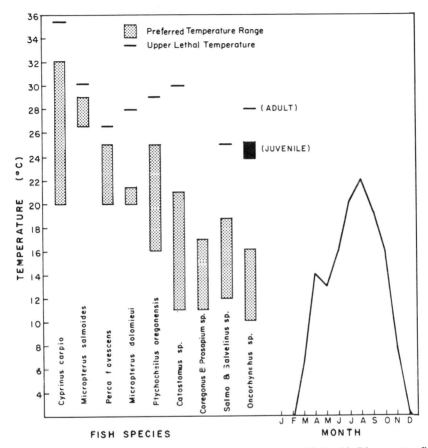

Figure 5. Preferred and lethal temperatures of some species of Columbia River system fish compared with water temperature of lower Snake River

The effect of this high supersaturation of nitrogen gas on fish migrants became apparent in 1969 and was especially critical in 1970, when approximately 70% of the estimated migration of young chinook salmon were lost en route to Ice Harbor Dam from upstream areas (Figure 8). During 1970 about 30% of the juvenile chinook migrants and 40% of young steelhead trout observed at Ice Harbor Dam showed symptoms of gas bubble disease. Also, about 30% of adult fish arriving at a hatchery in the upper Snake River drainage showed effects of the disease.

Recognizing that serious nitrogen gas levels occur during critical periods of the salmon migrations, the U.S. Army Corps of Engineers is vigorously attacking the problem on several fronts. A simulation model has been developed to predict nitrogen saturation levels under such physical parameters as river flow, spillway

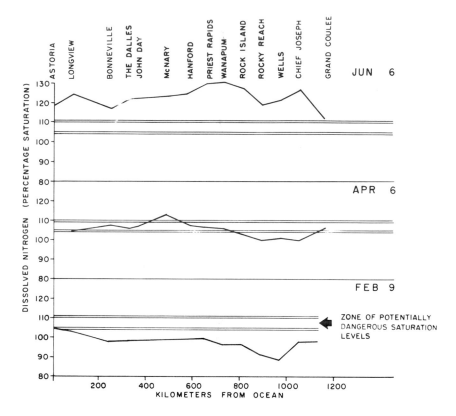

Figure 6. Extensive spilling in June increased saturated nitrogen levels above potentially dangerous levels in Columbia River

design, and stilling basin characteristics. Means of controlling nitrogen gas levels through adjustment of spillway flows or control of major upstream storage releases are under study. Spilling basin design is also being examined, although it is conceded that major changes at existing facilities are unlikely because of prohibitive costs. In the interim other means of protecting the migrants by collection and transport around the hazardous areas are being supported by the Corps.

River Flows. Development of the Columbia River and Basin has depressed the peak flows of the past, and further effect is anticipated (Figure 9) when dams are completed in the upper river and tributaries. So far, dams and impoundments have not appeared to seriously affect the rate of migration of adult chinook and sockeye salmon (*O. nerka*). For example, over a 24-year period the time required for fish to migrate from Bonneville to Rock Island Dam was about the same (Figure 10), although during this period the river was almost totally

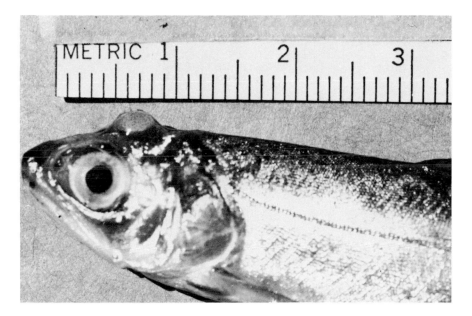

Figure 7. Gas bubble in eye tissue of young coho salmon

impounded. Sockeye salmon now require only two days more to reach Rock Island Dam than they did before intermediate barriers existed.

In the Snake River adult chinook salmon released in 92-km Brownlee Reservoir passed through the lake-like impoundment and spawned in upriver areas as successfully as fish transported and released in the river above the impoundment. However, it appears that dams and reservoirs have seriously affected the survival of adult salmon and trout during their spawning migration. Discrepancies in fish counts at successive dams on the Columbia River have indicated cumulative losses of trout and salmon as high as 40% on some runs. These losses are attributable neither to counting error nor to utilization of tributary streams by the adult migrants. Low survival may be due to failure of fish to locate fishways and pass over a dam, fall back over the spillway during high flows, and critical environmental conditions in the reservoirs.

Impoundments have reduced the velocity of flow of the Columbia, thus affecting downstream migration of juvenile salmon and trout. Juvenile migrants required almost three times as long to pass from McNary Dam to John Day Dam after John Day was completed in 1968. Similar patterns are evident in the Snake River (Table 1). Passage of juvenile chinook salmon through Brownlee Reservoir on the Snake River was related to volume of flow and size of the reservoir at the

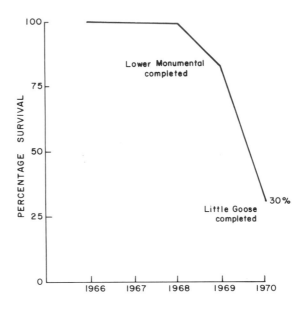

Figure 8. Survival of juvenile chinook salmon migrating down Snake River to Ice Harbor Dam

time of migration. Escapement past the dam was best when a substantial draw-down, high inflows, and a rapid spring runoff resulted in large discharges. Delays in migration at dams and reservoirs prolong the exposure of salmonids to high water temperature, high concentrations of nitrogen gas, predation, and competition with other species. As a result, mortalities during migration of juveniles from the upper basin to the estuary have been estimated as high as 90% in recent years.

Studies are in progress on methods of controlling flow of the Columbia River system to alleviate critical conditions. Mathematical modeling may result in new procedures for regulating reservoirs that will better accommodate all aspects of river use, including fishery resources. At dams it may be possible to discharge cool water from reservoir depths, easing high water temperature, and to control flows in other ways, minimizing losses of downstream migrating juvenile fish. Although more complete control of flows will be achieved in the future, even the high storage dams now projected will lack the capability of full control in major flood years.

Increasing Survival

Along with environmental studies major emphasis has been placed on methods of increasing survival of salmon and trout. Special effort has been directed toward passage of adults and juvenile fish at dams and collecting and transporting juveniles.

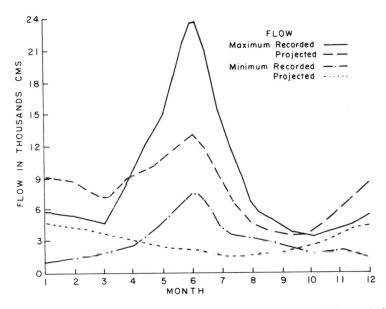

Figure 9. Average monthly flow of Columbia River at The Dalles Dam before and after completion of dams in the upper river

Fish Passage at Dams. Experiments in a fisheries engineering research facility at Bonneville Dam have been aimed at determining abilities, behavior, and requirements of adult migrants during their upstream passage over dams. Among the studies were tests on fishway dimensions and slope, effect of flow on rate of passage, response of fish to orifices of different size and configuration, and passage through pipes. Based on resulting data, more economical and efficient fishways were designed and installed at newer dams (Figure 11). Fishway operational procedures and facilities at existing dams are also being modified where feasible.

Methods were investigated to evaluate and reduce losses of downstream migrants at dams, as it was anticipated that more young fish would be forced through turbines when river flows became more controlled and less spilling over dams occurred. Direct losses of juveniles in turbines were estimated at 10 to 20%, and indirect mortality from predation on stunned fish which were carried laterally into slack water areas increased total mortality to as high as 35%. To avoid lethal agents within turbines, such as differential pressures, hydraulic shear planes, and turbine blades, a method was devised to divert young fish concentrated near ceilings of turbine intakes into a gatewell upstream of the turbine. From here they could be passed through an orifice into a bypass that would deposit them safely downstream of the dam. This system installed at a dam

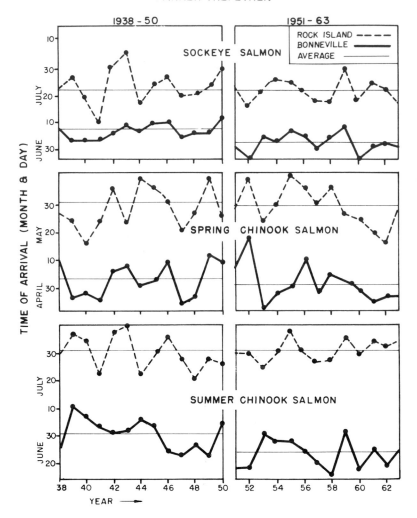

Figure 10. Arrival of adult salmon at Rock Island Dam over a 24-year period did not change appreciably, even with formation of impoundments after 1951

could alleviate immediate mortalities at the dam, but would not protect fish from the harsh environment between dams.

Collection and Transportation. As studies on the survival of downstream migrants progressed, it became obvious that the severe losses of young fish caused by widespread deterioration of the environment could jeopardize the entire salmon and trout resource of the Columbia River Basin. The method of bypassing fish around individual dams, although practical in the short term, would have limitations in that the fish would still be subject to successive

Table 1. Effects of impounding the Snake and Columbia rivers on the rate of migration of juvenile chinook salmon

Section of river	Distance (km)	Elapsed time (average days)	
		Pre-impoundment	Post-impoundment
Salmon River to Ice Harbor Dam	370	15	25
Ice Harbor Dam to McNary Dam	68	3*	9
McNary Dam to John Day Dam	122	5	13
Peak of migration John Day Dam		May 2	June 3

*Estimate based on rate of migration between McNary Dam and John Day Dam site before river was impounded.

stresses during passage through downstream impoundments. Experiments were designed to determine if young fish collected at an upstream dam could be safely transported around the entire hazardous area and returned to the river far downstream to continue their migration to the sea. Basic questions to be answered were: (1) Would survival be increased, and (2) Would transport of young fish affect their homing ability and attainment of natal spawning areas when they returned from the sea as adults.

Experiments on collecting and transporting juvenile salmon were based at Ice Harbor Dam, which is the lowermost dam on the Snake River. Young fish were diverted by a traveling screen into a collection area, where they were marked with magnetic wire tags, branded, and either released as controls at the dam or transported and released about 300 km downstream before and during periods of spilling at dams. Survival of young fish of both groups was determined by sample recovery at downstream areas. Survival of these fish when they returned as adults was determined from recoveries in the commercial and sport fisheries, from the fishway at Ice Harbor Dam, at hatcheries, and on spawning grounds.

Figure 11. Fishway at Ice Harbor Dam completed in 1961 included features to ease passage of adult salmon and steelhead trout

An increase in survival of 64% was measured for transported fish before major spilling began at dams. However, during periods of spill when nitrogen levels were high, the increase was as high as 350%. These transported fish successfully returned to Ice Harbor Dam as adults in 1970 and migrated upstream to spawning areas, indicating that transport did not affect homing ability. Adult returns of nontransported fish also demonstrated the effect of nitrogen and other lethal factors on survival; 90% of these returns were from marked juveniles released before spilling began, as opposed to only 10% from fish released during spill. The results suggest that collecting and transporting fish around dangerous areas is a practical method for increasing survival.

Because of the marked increase in mortalities of juvenile salmonids in the Snake River in 1969 and 1970, threatening the existence of major salmon and steelhead trout populations, investigation on effects of collection and transportation are being moved upstream to Little Goose Dam. Here, in a cooperative effort, the National Marine Fisheries Service and the U.S. Army Corps of Engineers, on approval of the state fishery agencies, are examining the effects on survival and on homing ability of transporting juveniles about 480 km downriver. Traveling screens will divert fish into turbine intake gatewells (Figure 12), from which they will be bypassed to large tank trucks for transport. Of those collected, about 100,000 are to be marked with wire tags and thermal brands; one-half of these will be released at the dam as controls, and the rest will be released below Bonneville Dam, the lowermost power facility on the Columbia River. Survival estimates will be made at downstream dams and in the Columbia River estuary. Returning adults will be examined at various points on their migration through the fishery and to the spawning grounds.

Forseeable Developments

Requirements for electrical energy in the Pacific Northwest are predicted to increase about threefold in the next three decades. Some of this increase will be met by new dams in the Columbia River system and by expanding the existing hydroelectric facilities. These developments may beneficially affect the ecology of the river by reducing spilling and attendant concentrations of dissolved nitrogen and by providing a source of cool water from the deeper reservoirs. These colder flows could be released to counteract high water temperatures in downstream areas. Since virtually all suitable hydroelectric sites have been developed, however, it is anticipated that many thermal power plants will be required to meet growing demand for electricity.

In the next two decades it is predicted that twenty thermal electrical plants will be operating in the northwestern United States; sites for 16 of these have been studied (Figure 13). Although not all of these plants will be in the Columbia Basin, they will gradually assume the base power needs of the region, and by around 1985 hydroelectric plants will be used primarily for peak power

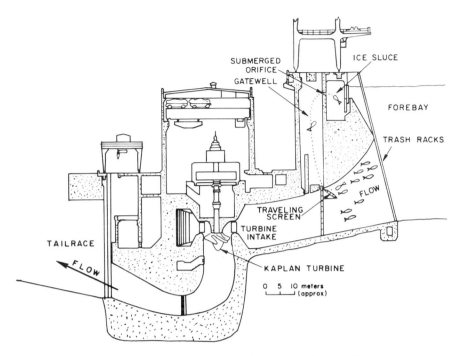

Figure 12. Section through typical turbine unit showing traveling screen for diverting fish into sluiceway

demand. Widely fluctuating river flows will result, affecting the migration of fish, further altering temperature regimes, especially in hot, dry years, and decreasing the ability of the river to dilute pollutants such as industrial waste, insecticides, fungicides, chemicals, and heavy metals during low flow periods. Furthermore, reduced flows in downstream areas will allow tidal flows to extend farther upstream for longer periods of time, thus prolonging exposure of some water masses to potential pollutants from industrial plants located in the estuary and lower Columbia River.

Other demands on the Columbia River will grow as the population in the United States continues to expand. Industries associated with the production of aluminum and paper are being proposed or expanded along the lower river; there are plans for locating new industries in other river areas. Additional water will be withdrawn from the river for irrigating agricultural lands and for domestic and other consumptive uses. Accompanying this industrial development will be an increase in river traffic with the potential for oil and chemical spills which

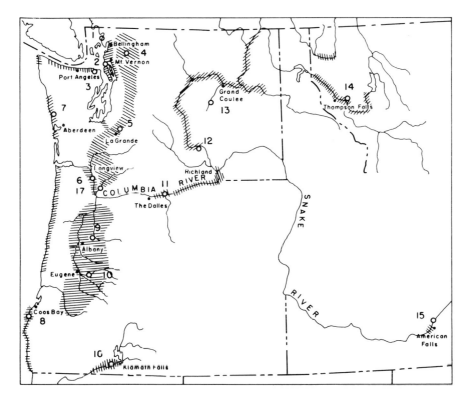

Figure 13. Sites of proposed thermonuclear electric plants in the Pacific Northwest.
Cross-hatching represents preferred areas

can adversely affect aquatic organisms. Coupled with population growth will be the need for more housing, schools, shopping centers, and highways. As these are developed, great areas of land will be paved, surface runoff will be sharply increased, and water quality will be affected by organic and inorganic materials.

It has been proposed that water from the Columbia Basin be diverted to the southwestern United States for agriculture, industry, and domestic uses. Technologically, it would appear that such a diversion is feasible, but the massive transport of water could profoundly affect the ecology of the Columbia River through reduction of flow. This threat has been postponed, but if water shortages become more critical, further attempts may be made to obtain water from the Columbia River system.

The Future

The outlook for the Columbia River is not as bleak as it was ten or twenty years ago. It is evident that ecological change will continue, but it is also evident that the rate of change will diminish. Continuing scientific and technological developments, plus improved methods of data processing and mathematical modeling, will make it possible for scientists and engineers to predict changes more quickly and accurately than in the past. These forecasts will enable management agencies to devise actions to counteract conditions that adversely affect fishery resources and other uses of the river.

Most of the commercial and industrial expansion is expected to occur near established population centers along the mid- and lower Columbia River. It seems unlikely, however, that the developments will be allowed to affect adversely the aquatic environment. In the past few years citizens in the Columbia River Basin and elsewhere in the region have become increasingly aware of the effects of water pollution on recreation, esthetics, and fishery resources, and they have demanded action to prevent further deterioration of the environment. As a result, federal and state laws have been strengthened, state agencies have been formed or reorganized to attack problems, and federal-state and local governments have increased coordination of their research and management activities in in the Columbia and other rivers.

Water from headwaters of the Columbia River and some of its tributary streams will contribute toward obtaining and maintaining a stabilized environment. Many sources of the river lie in forested areas protected by federal and state parks and forests. Since little change is expected in these areas, water quality will remain about the same for years to come, and judicious releases from dams in those areas can enhance the quality of the water downstream.

It is, therefore, quite possible that man's adverse impact on the Columbia River will decrease as his ability to predict changes and control developments increases. With continued vigilance, it is likely that the river ecology can be stabilized within the next two or three decades and that it will not deteriorate as much as some rivers in the eastern United States or in other areas of the world.

Reading References

Calderwood, J. J. 1928. Salmon fisheries and hydro-electricity. Trans. Amer. Fish. Soc. 58:152-158.

Clark, S. M., and G. R. Snyder. 1970. Limnological study of lower Columbia River, 1967-68. U.S. Fish Wildl. Serv., Spec. Sci. Rep. Fish. 610, 14 p.

Collins, G. B., and C. H. Elling. 1960. Fishway research at the Fisheries-Engineering Research Laboratory. U.S. Fish. Wildl. Serv., Circ. 98, 17 p.

Corps of Engineers, U.S. Army. 1951-52. Columbia River and tributaries, northwestern United States. House Doc. 531, 81st Congr., 2d Sess. U.S. Govt. Print. Office, Washington, D.C., 8 v.

Craig, J. A., and R. L. Hacker. 1950. The history and development of the fisheries of the Columbia River. U.S; Bur. Fish., Fish. Bull. 49:133-216.

Davidson, F. A. 1953. Historical notes on development of Yakima River basin. Unpublished manuscript, 22 p.

Ebel, W. J. 1969. Supersaturation of nitrogen in the Columbia River and its effect on salmon and steelhead trout. U.S. Fish Wildl. Serv., Fish Bull. 68(1):1-11.

—————. 1970. Summary report 1970: Dissolved nitrogen surveys of the Columbia and Snake rivers. Bur. Commer. Fish., Biol. Lab., Seattle, Wash. 7 p + 10 p. App. table.

—————, and C. H. Koski. 1968. Physical and chemical limnology of Brownlee Reservoir, 1962-64. U.S. Fish Wildl. Serv., Fish Bull. 67(2).295-335.

Hutchinson, S. J., and L. Perry. 1964. Columbia River Fishery Program—1963. U.S. Fish Wildl. Serv., Circ. 192, 20 p.

Jaske, R. T., and M. O. Synoground. 1970. Effect of Hanford Plant operations on the temperature of the Columbia River 1964 to present. Battelle Mem. Inst., Pac. Northwest Lab., Richland, Washington, BNWL—1345, US-70. Unpublished manuscript.

Jones, F. O. 1947. Grand Coulee from "Hell to Breakfast." Binfords and Mort, Portland, Oregon, 64 p.

Long, C. W., and R. F. Krcma. 1969. Research on a system for bypassing juvenile salmon and trout around low-head dams. Commer. Fish. Rev. 31(6):27-29.

—————, and W. M. Marquette. 1967. Research on fingerling mortality in Kaplan turbines. Proc. Bienn. Hydraulics Conf., Moscow, Idaho. 6:11-36. Wash. State Univ., Pullman.

National Resources Committee. 1937. Drainage basin problems and programs. U.S. Govt. Print. Office, Washington, D.C., 540 p.

Raymond, H. L. 1968. Migration rates of juvenile chinook salmon in relation to flows and impoundments in Columbia and Snake rivers. Trans. Amer. Fish. Soc. 97(4):356-359.

——————. 1970. A summary of the 1969 and 1970 outmigration of juvenile chinook salmon and steelhead trout from the Snake River. Nat. Mar. Fish. Serv., Biol. Lab., Seattle, Wash. Unpublished manuscript, 14 p.

Snyder, G. R. 1968. Research on thermal pollution: report on the Columbia River and estuary. Bur. Commer. Fish., Biol. Lab., Seattle, Wash. Unpublished manuscript, 9 p., 7 tables, 9 figures.

——————. 1969. Nuclear power and anadromous fish:current problem in the Pacific Northwest. Bur. Commer. Fish., Biol. Lab., Seattle, Wash. Unpublished manuscript, 16 p.

——————. 1970. A summary of the thermal pollution problem. Bur. Commer. Fish., Biol. Lab., Seattle, Wash. Unpublished manuscript, 5 p.

Sylvester, R. O. 1958. Water quality studies in the Columbia River Basin. U.S. Fish Wildl. Serv., Spec. Sci. Rep. Fish. 239, 134 p.

——————. 1959. Water quality study of Wenatchee and Middle Columbia rivers before dam construction. U.S. Fish Wildl. Serv., Spec. Sci. Rep. Fish. 290, 116 p.

Trefethen, P. S. 1968. Fish-passage research—review of progress 1961-66. U.S. Fish Wildl. Serv., Circ. 254, 24 p.

——————, and D. F. Sutherland. 1968. Passage of adult chinook salmon through Brownlee Reservoir, 1960-62. U.S. Fish Wildl. Serv., Fish. Bull. 67(1):35-45.

Tyrell, J. B. [ed]. 1916. David Thompson's narrative of his exploration in Western America, 1784-1812. The Champlain Society, Toronto, 582 p.

Van Winkle, W. 1914. Quality of the surface waters of Washington. Dep. Int., U.S. Geol. Surv., Water-Supply Paper 339, 105 p.

Worlund, D. D., R. G. Wahle, and P. D. Zimmer. 1969. Contribution of Columbia River hatcheries to harvest of fall chinook salmon (*Oncorhynchus tshawytscha*). U.S. Fish. Wildl. Serv., Fish. Bull. 67(2):361-391.

THE DELAWARE RIVER — A STUDY
IN WATER QUALITY MANAGEMENT

Robert V. Thomann

"I love any discourse of rivers, and fish and fishing" — I. Walton, The Compleat Angler

Background History

Rich in history and encompassing both outstanding beauty and devastating ugliness, the Delaware River represents a microcosm of all of America's aspirations, dreams, and failures. As a river system, the Delaware is small. Yet it embodies the entire range of river ecology from the clear mountain headwaters 1900 ft above sea level, through the famous Delaware Water Gap, on past historic Washington Crossing into the lower tidal river, past the metropolitan and industrial complex of Philadelphia, through the increasingly brackish water, broadening into Delaware Bay and finally discharging into the Atlantic Ocean after a distance of more than 400 miles (see Figures 1 and 2).

Henry Hudson, that erstwhile traveler, was the first white man to enter the Delaware River in 1609. The region at that time harbored about 1000 aborigines. About six million people now live along the Delaware River, and over 20 million people including those in nearby New York City look to the Delaware Basin for water supply, recreation, and industrial development.

From its earliest beginnings the quality of the waters of the Delaware has largely been taken for granted. Early settlers were overwhelmed with the pleasant tasting water and reaped a bountiful harvest of fish and shellfish. However, as population centers became more dense, especially in the Philadelphia area, outbreaks of waterborne disease began to occur with increasing frequency. As it became evident that these outbreaks were directly associated with the discharge of human wastes, substantial changes occurred in domestic water use. Water for drinking and other household uses was drawn from sources physically separated from water bodies used for the carriage of domestic wastes. This was followed by the large-scale use of chlorination for bacterial disinfection, resulting in a drastic drop in the incidence of waterborne disease.

THE DELAWARE RIVER BASIN

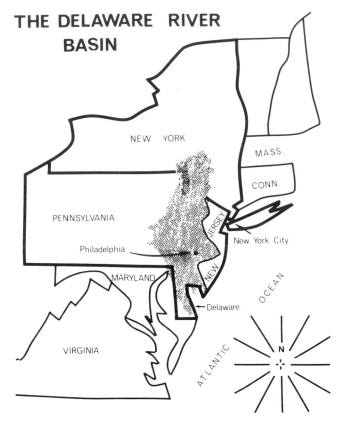

Figure 1. General Basin location

For several decades thereafter rational management and use of the waters of Delaware Basin appear to have lain dormant. Large quantities of wastes were discharged to streams and rivers, and the only cautionary note was that water for drinking purposes be drawn from relatively unpolluted sources. The country was just too busy with many more important matters than the quality of its rivers and streams.

In the post-World War II years, however, increasing prosperity, coupled with a better understanding of the role of environmental quality in man's well-being, demanded a new stage in water quality management. This was reflected in the desire for more water-based recreation and for a substantial improvement in our fisheries and aquatic wildlife resources. The first level of treatment was installed by many of the major waste discharges, but not without strong effort on the part of pollution control agencies. Again, from the late 1940's to 1960, upgrading of treatment was notably absent.

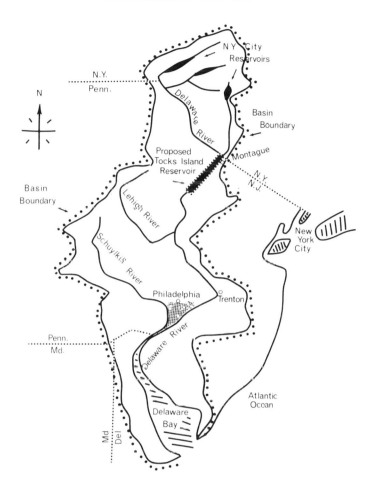

Figure 2. Basin map

The Delaware Basin is now well into the second stage in water quality management. Through the state regulatory agencies, the federal government, and the Delaware River Basin Commission, the people of the region have committed themselves to restore their waters to some higher level of quality. This is discussed more fully below.

For purposes of water quality analysis, the Delaware River can be divided at Trenton, New Jersey (see Figure 2). North of this state capital the river is generally of high quality although there are indications that some problems may be emerging in the vicinity of the confluence with the Lehigh River. The river north of Trenton flows through magnificent scenery and draws large numbers of water-based recreation enthusiasts. A large reservoir, Tocks Island, associated with a National Park area is proposed for the river north of the Delaware River Gap.

South of Trenton, New Jersey, the Delaware experiences its first reversal of current due to tidal influences. The water quality of this estuarine portion of the Delaware deteriorates rapidly as the river approaches the Philadelphia area. The poor water quality in the Delaware estuary has been a matter of concern for some years and influences activities throughout the entire basin. This is especially true for several indigenous migratory fish species, including the American shad which travels through the low dissolved oxygen level of the estuary with great difficulty. A budding sport fishery in the upper basin too often looks vainly for the arrival of significant numbers of adult shad which suffer large mortalities in the estuary. Because of the poor quality in the Delaware estuary, the concentration of population and industry, and the undoubtedly large sums of money needed even minimally to correct the situation, the water quality management of the Delaware estuary has been of particular importance.

Population and Industrial Growth. The population of the region immediately surrounding the main stem of the Delaware River has increased substantially over the past several decades (Federal Water Pollution Control Administration, 1966a). The trend for the entire river and that fraction contiguous to the estuary (Wright and Porges, 1970) is shown in Figure 3. By the year 2000 popu-

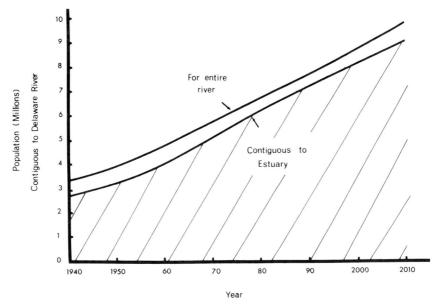

Figure 3. Population

102

lation levels of eight to nine million are expected to exert a direct impact on the main stem of the Delaware River. In the estuary region of the river the population of large cities has actually declined somewhat over the past few decades, but this has been more than offset by growth in suburban areas (Wright and Porges, 1970).

Although the major portion of the population in the basin is located adjacent to the Delaware estuary, the upper basin region is also expected to experience substantial growth if the Tocks Island reservoir and the accompanying National Recreation Area are completed. For example, it has been estimated (Weston, 1970) that the peak-season population of the drainage area surrounding the Tocks Island reservoir will increase from about 200,000 in 1970 to 790,000 in 2010, an almost fourfold increase.

Industrial growth in the area is expected to be substantial over the next forty years, as indicated in Table 1. This table shows the growth in employment expected in the Delaware Estuary Service Area. In all industrial categories between 1965 and 1980, employment is projected to increase by 1.8 times. This increase in employment with associated increases in industrial manufacturing, commodities, and services will result in increasing water withdrawals and demands by the industrial sector and increased amounts of discharged waste water.

Table 1. Projected industrial employment (thousands) for three categories of the Delaware Estuary Service Area (Federal Water Pollution Control Administration, 1966b)

Category	Year		
	1965	1980	2010
Petroleum Refinery	6,000	10,500	30,000
Chemical Products	2,500	5,000	17,000
Paper Products	1,000	1,500	3,000
Estimated Total	9,500	17,000	50,000

Hydrological Characteristics. Table 2 summarizes some of the hydrological characteristics of the Delaware River at Trenton, New Jersey. At that location the river has a drainage area of 6,780 sq miles. The mean annual flow at Trenton of 11,680 cfs is equivalent to 1.72 cfs/mi^2, a value common to the Middle Atlantic States. From a water quality standpoint, interest centers about the drought flow levels which often occur when water temperatures are still high.

Table 2. Delaware River flow at Trenton, N.J. (1913-1963 climatic year). (Anonymous, 1969)

Recurrence Internal (Years)	Probability of Occurrence (% of time flows equal to or less than)	Mean Low Flow (cfs)		
		7 days	30 days	120 days
5	20	1600	1900	2700
10	10	1500	1700	2400
25	4	1400	1600	2100
50	2	1300	1500	2000

The combination of low flows and high temperatures generally results in critical water quality conditions. The low-flow values given in Table 2 do not reflect the occurrence of a severe drought during 1964-65 when water conservation measures were practiced on a large scale throughout the basin. The estimated minimum daily regulated flow at Trenton is 2,500 cfs, assuming the New York City reservoirs release water downstream sufficient to maintain the stipulated requirement of 1,750 cfs at Montague, New Jersey. During the severe 1964-65 drought it was not possible for New York City to maintain the required flow, due partially to the fact that the third New York City reservoir—Cannonsville—was not yet fully operational. It is not yet clear whether 2,500 cfs could be maintained as a minimum flow at Trenton if another drought approximating the recent one were to occur. The maintenance of minimum flows is important to water quality in the estuary, especially with respect to the extent of sea water intrusion.

An anticipated minimum daily flow of 3,120 cfs at Trenton is expected after Beltzville reservoir (presently in final construction stages) and Tocks Island reservoir (minimal site clearing is underway) have been completed.

Below Trenton the Delaware River is tidal with a principal period of oscillation of 12.42 hr, the M_2 lunar tide. The range of tidal height increases from Liston Point—where the Delaware enters Delaware Bay—to Trenton, New Jersey, as shown in Table 3.

Tidal currents in the Delaware estuary are substantial, ranging from about one to over three knots. Tidal reversal occurs from the mouth of Delaware Bay up to the fall line at Trenton. These tidal velocities result in tidal flows on the order of 500,000 cfs at the Delaware Memorial Bridge to about 50,000 cfs in the estuary south of Trenton.

104

Table 3. Range of tidal height, Delaware estuary (FWPCA, 1966*b*)

Location	Tidal Range (ft)		
	Miles from Ocean	Mean	Spring
Trenton, New Jersey	132.0	6.8	7.1
Philadelphia, Pennsylvania	100.0	5.9	6.2
Liston Point, Delaware	48.3	5.7	6.4

Salt water intrusion from the ocean generally penetrates only to the Pennsylvania-Delaware state line—a distance of about 80 miles. Under drought conditions, however, sea salts have penetrated as far as the entrance of the Schuylkill River, a major tributary of the Delaware estuary. Figure 4 shows a plot of the extent of

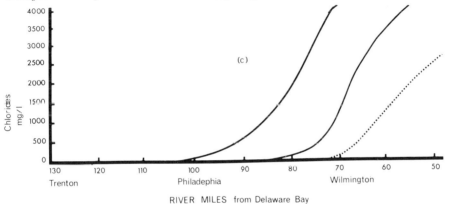

Figure 4. Salinity

sea water intrusion into the Delaware estuary during the 1964-66 drought conditions, as well as the approximate average portions of the brackish zone. The estuarine area is therefore characterized by a transition from near sea water conditions in Delaware Bay through a brackish region and finally into an entirely fresh water, tidal river situation.

Ecological Patterns and Water Uses by Man

Municipal and Industrial Water Supply. The Delaware River is presently used for municipal and industrial water supply throughout its entire length. Beginning at the headwaters the river services numerous municipalities as it travels downstream. Although the area of the basin is small (0.3% of the total land area of the

U.S.), the Delaware River system supplies part or all of the water needs of about 15 million people, or 7 1/2% of the population of the U.S., and by 2010 it is estimated that the serviced population will approach 42 million (Fish, 1968). In the upper basin area the Delaware River headwaters are impounded by the City of New York and supply a maximum diversion rate of 490 million gal per day (mgd) of flow to the New York metropolitan region. This amount of water represents a substantial out-of-basin diversion and must be compensated for by releasing flow from the up-basin reservoirs to maintain 1,525 cfs at Montague, New Jersey. The upper Delaware is also used for industrial water supply purposes, including cooling water, process water, and other uses.

The estuary provides the largest source of municipal and industrial water supply for the region. There are thirty-five municipal water systems which withdraw approximately 550 mgd directly from the estuary. The major source of withdrawal from the estuary is the City of Philadelphia at its Torresdale water treatment plant. This plant alone accounts for a withdrawal of about 200 mgd. Along the estuary industrial water use increases substantially, as indicated by a volume of withdrawal including electric utilities of almost 5,000 mgd in 1963 (FWPCA, 1966*b*).

Recreational Uses. The Delaware River supports a wide variety of recreational uses. White-water canoeing, boating, sport fishing, and swimming are common in the upper basin area. South of Trenton, in the estuarine portion of the river, extensive boating and sport fishing are practiced. There are, however, no sanctioned swimming areas in the estuary. Recreational uses are expected to increase substantially as facilities are made available or as water quality is improved through waste water treatment projects. For example, as indicated above, Tocks Island reservoir and the planned National Recreation Area in the upper Delaware (see Figure 2) are expected to accommodate more than 10 million visitor-days per year with a peak day-load of about 142,000 visitors (Weston, 1970). This vast influx of a transient population will undoubtedly produce its ecological change on the environment. The Delaware River Basin Commission is attempting to minimize any major impact of such population pressures by examining various alternatives for handling both liquid and solid wastes in regional or individual plants.

In the estuary recreational demands are substantial and in many cases are not being fulfilled because of poor water quality. Table 4 shows the estimated recreational demand in the Delaware estuary area. Almost all recreational activities are expected to double from the base year 1970 to the year 2000. Of course, some of this total recreational demand can be satisfied by the estuary itself while another portion of the demand must be satisfied by the development of recreational sources outside the Delaware area proper.

Table 4. Future recreational demand in the Delaware estuary area (Tomazinis and Gabbour, 1967)

Recreational Activity	Estimated Million Activity-Days		
	Year		
	1960	1976	2000
Swimming	26.8	45.4	79.3
Boating	5.4	9.4	16.4
Fishing	6.9	9.0	11.9
Picnicking	11.0	16.0	23.8

The water quality of the estuary plays an important role in satisfying recreational demand. As noted previously, poor water quality prevents the use of the estuary for swimming throughout its entire length. In addition, under present conditions sport fishing holds little attraction in the Delaware estuary since the probability of fishing success is so low.

This present and future recreational demand by the general public on the Delaware region is one of the major factors that has led to an increased desire to improve the quality of the waters of the Delaware.

Fisheries. Many varieties of fish were once abundant in the Delaware. In the estuary shad, sturgeon, striped bass, weakfish, and white perch were once important commercial fishes. In the upper basin area trout, bass, and pike are sought after as desirable sport fishes. From early colonial times the fishes of the Delaware, especially shad and sturgeon, were exported in great numbers. Figure 5 shows the historical trend in shad and finfish for the Delaware River and Bay and surrounding area, extending from sketchy data around 1850 to 1966. The peak period for the Delaware estuary fisheries was between 1885 and 1900. During that time the annual catch by 4,000 fishermen was on the order of 25 million lb, valued at about 4.5 million at today's prices (FWPCA, 1966b). Shortly after 1900, the annual harvest dropped rapidly, and the decline continued until today's harvest of about 80,000 lb worth about $14,000. Chittenden (1969), Morris (private communication), and Morris and Pence (1970) have reviewed the life history and behavior of the American shad with specific reference to the Delaware.

Sykes and Lehman (1957) have also studied behavioral patterns of shad in the Delaware River and the effects of poor water quality on the stock. Repeat spawners were determined to be virtually absent in the Delaware, as shown in Table 5.

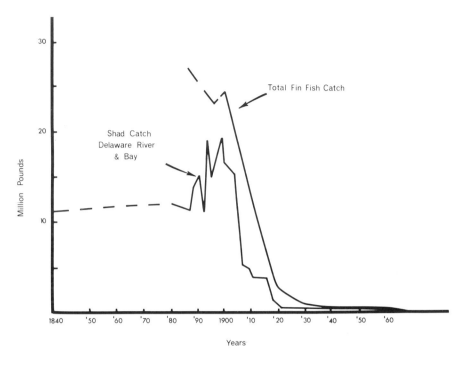

Figure 5. Shad figure

Table 5. Age distribution of shad — Delaware River (1944-47) (After Sykes and Lehman, 1957)

Age Groups	Number of First-Time Spawners	Number of Repeat Spawners
3	3	
4	68	
5	173	1
6	124	4
7	44	2
8	4	
	416	7

Thus only about 2% of the shad examined during the period were repeat spawners. This is in contrast to an estimated 35% repeat spawners for the Connecticut River and 51% for the Hudson River (Sykes and Lehman, 1957). Chittenden (1969) also found very low percentages of repeat spawners in the Delaware. During 1963 and 1964 less than 1% of the shad captured had spawned previously while for 1965 about 6% had previously spawned. Sykes and Lehman attributed the low percentages of repeat spawners in the Delaware to high mortality of adults in their fall seaward migration through the estuary. Water quality conditions in the estuary are usually still poor during the fall run. Chittenden partially attributes the low percentages to the location of upstream spawning grounds and mortality of fish resulting from apparent starvation.

Reasons for the rapid decline in the shad fishery, therefore, include overfishing resulting from improper fisheries management, industrial and municipal waste discharges into the estuary resulting in poor water quality, and siltation from runoff in farmland suburban areas.

The question of overfishing is an important one since it relates to any program for water quality management. Thus an improvement in water quality will have to be accompanied by an improvement in fisheries management techniques. For example, the 1874 report of the Commissioner of the U.S. Commission of Fish and Fisheries (1874) indicates:

> . . . fish baskets destroy millions. In 1871 I made a careful and thorough examination of the upper waters of the Delaware River, for the purpose of investigating this point. The facts elucidated were astounding. It was ascertained that a scoop-shovel, with which to shovel out the dead young shad which accumulated in the basket, was an important part of the fishermen's equipment. One proprietor acknowledged that as much as a two-horse load of the dead young shad had been shovelled from his basket during a single day . . .

There is little reason to believe that the fundamental nature of man has substantially changed in the last hundred years. If significant increases in the shad numbers were to occur, it is highly likely that man would continue to devastate the population unless adequate controls and management procedures were enacted.

Because of the importance of the American shad in the Delaware River as an indicator of man's activities, interest centered about the effects of water pollution control programs and water quality improvement schemes on the shad population. This required an initial estimation of the effects of water quality on shad

109

survival, especially in the Delaware estuary. The results of this analysis, a larger framework for analysis of shad runs and the interaction between such anadromous fish and water quality levels, are presented below.

The Need for Water Quality Management

Municipal and Industrial Wastes Loads. One of the major impacts of man's presence on water quality of the Delaware River is that of the discharge of treated and untreated waste material. The largest discharges of wastes occur along the Delaware estuary. Waste discharges into the Delaware River above Trenton are substantially less, reflecting the more sparsely populated nature of the upper basin.

About three million people along the estuary are presently served by major municipal water treatment systems. There are numerous direct discharging industries, including refineries, chemical manufacturing plants, paper mills, and steel mills. Waste reduction facilities range from secondary biological treatment in the upper portions of the estuary and north of Trenton to primary treatment or no treatment in the middle and lower portions of the estuary. During 1964 it is estimated that a level equivalent to 50% waste removal was accomplished by the combined efforts of municipalities and industries.

Table 6 shows the waste discharges to the Delaware estuary during 1964, a base year used for comparison purposes. Of the carbonaceous oxygen demand, it should be noted that the City of Philadelphia alone accounts for 450,000 lb/day or about 45% of the total 1 million lb/day. If the oxidizable nitrogen waste load (organic and ammonia nitrogen) is entirely oxidized after discharge to the estuary, requiring about 4.6 lb of oxygen per pound of nitrogen, there is an additional demand for 500,000 lb/day of oxygen. The waste discharges indicated in Table 6 have influenced the water quality of the estuary as described in the next section. The degree of influence is governed to some extent by the spatial distribution of the waste discharges. The greatest concentration of waste discharges is located along the Delaware estuary from about mile 100 to mile 75. In addition to the discharges shown in Table 6, overflows from sewer areas contribute transient discharges of raw sewage of significant amounts. Urban runoff from metropolitan streets and industrial plant sites also contribute oils, greases, and other flotsam to the estuary.

Water Quality Problems. The Delaware River north of Trenton is of a good to excellent water quality. There is a slight decrease in water quality below the entrance of the Lehigh River near Easton, Pennsylvania. Dissolved oxygen (DO) values are almost always at or above saturation, except in the reach below the entrance of the Lehigh River, where values of less than 6 mg/l during July and

110

Table 6. Waste discharges to Delaware estuary (1964 base year)

| | Pounds/day Direct Discharge | | |
	Municipal	Industrial	Total
Carbonaceous Oxygen Demand	660,000	370,000	1,030,000
Organic Nitrogen (as N)	28,500		28,500
Ammonia Nitrogen (as N)	48,500	32,500	81,000
Nitrate Nitrogen (as N)	2,000	30,500	32,500
Acidity (summer) as $CaCO_3$		1,300,000	1,300,000
Total Suspended Solids			740,000
Total Settleable Solids			260,000

August have been obtained (U.S. Department of Health, Education, and Welfare, 1960; Pollison and Craighead, 1968). The DO of the Lehigh averages less than 5.0 mg/l during the summer months. Coliform bacteria concentrations appear to be high, and counts of 110,000 MPN/100 ml have been observed below the entrance of the Lehigh (Pollison and Craighead, 1968). There is, however, evidence to indicate a more subtle effect of waste discharge on the ecology of the upper portion of the Delaware. Pollison and Craighead (1968) have observed an area of considerable weed growth below the confluence of the Lehigh and Delaware rivers. This area is evident for a distance of about 45 miles. These plants, which include *Zosteraceae* (pondweed), *Gramineae* (eel grass), *Pontederiaceae* (mud plantain), and *Haloragaceae* (water-milfoil), can adversely affect such recreational activities as boating, swimming, and water skiing and may interfere with fish populations.

The waters entering the estuary at Trenton are therefore of good quality. Oxygen values are almost always saturated, and nitrogen and phosphorus values are low. However, as one proceeds down the estuary, water quality conditions become progressively poorer. Figure 6 shows a dissolved oxygen profile during 1964 summer conditions. (The zones indicated on the figure were established for regulating purposes and are discussed below.)

The minimum daily average 1964 summer DO values dropped to less than 1 mg/l over a ten- to twenty-mile reach of the estuary. The DO is less than 3 mg/l over about a thirty- to forty-mile stretch. At the entrance to Delaware Bay, however, DO values have recovered to about the saturation level. The heavy concentration of waste discharges of oxidizable material, as indicated in Table 6, is the major cause of these low dissolved oxygen levels. Figure 7 shows the spatial

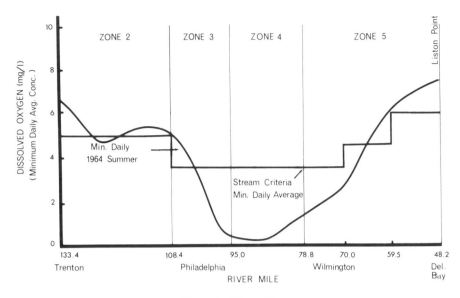

Figure 6. DO profile

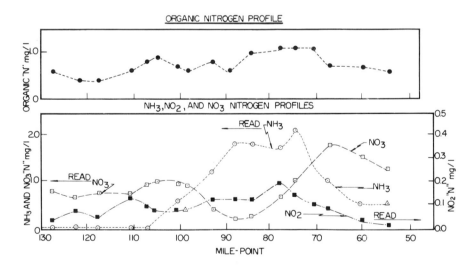

NOTE:
JULY 1967
Q = 5600 cfs
T = 24°C

Figure 7. Nitrogen

112

profile of the nitrogen components in the Delaware estuary and reflects the discharge of substantial amounts of nitrogen from municipal and industrial wastes. The oxidation of the ammonia and nitrite nitrogen components is an important sink of dissolved oxygen and must be carefully analyzed in any management scheme.

Figure 8 shows that as a result of acid discharges, some of which have subsequently been controlled, pH and alkalinity levels are depressed in the vicinity of Wilmington. In the past alkalinity has approached zero, indicating a complete exhaustion of the ability of the estuary to assimilate any further acid wastes. Figure 9 and Table 7 provide data on the 1964 chlorophyll concentrations and the time variation of phytoplankton at Torresdale, Pa., for 1961-62. The estuary is not now subject to substantial algal growth, as evidenced by the levels of chlorophyll as in Table 7. These chlorophyll levels and the counts shown in Figure 9 are well below those usually associated with bodies of water bordering on eutrophication.

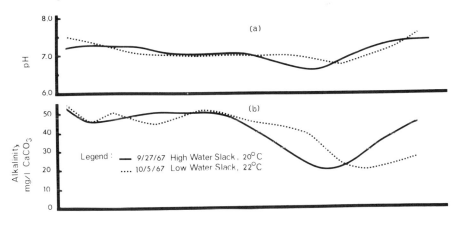

Figure 8. pH, alkalinity

Table 7. Delaware estuary chlorophyll a concentration (micrograms/liter) (After Morris, private communication)

Estuary Region	1964 Period		
	April-June	July-September	October-December
Trenton-Burlington, N.J.	24	30	17.5
Burlington-Philadelphia	14.5	22	15.5
Philadelphia-State line (Pa.-Del.)	28	63	48
State line-Wilmington	16	40	35

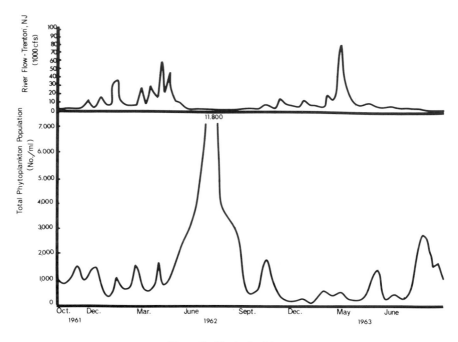

Figure 9. Phytoplankton

The water quality of the Delaware River therefore ranges from high quality in the upper headwaters to a region of extremely low quality in the Delaware estuary in the Philadelphia metropolitan region. Water quality generally recovers at the entrance to Delaware Bay where again the water is of good to excellent quality.

Socio-Economic and Technological Problems.   In addition to water quality and water use degradation, there are several serious social, economic, and technological problems that must be recognized in water quality management schemes. These problems can be summarized as follows:

1. There is a general lack of clearly defined ecological goals both in terms of desirable water quality and water use levels and in terms of desirable levels of fish populations or wildlife habitat.

2. The costs of improving our environment must be squarely recognized and such costs must compete on a priority basis with other social demands made on the economy of a region.

3. Contrary to popular belief, there are at present severe technological limitations on the ability to control waste discharges. Chief among these are:

114

i. uncontrolled urban, suburban, and agricultural drainage; ii. transient discharges from combined sewer overflows; iii. process limitations on controlling effluent levels of quality, principally in municipal waste treatment plants. For example, it should be noted that the field of water pollution control engineering does not yet have a viable proven operationally efficient means of removing nitrogen from waste waters. While many laboratory and pilot schemes have been developed and a few operated on full-scale plants, the information does not yet exist to indicate that nitrogen removal is a stable, reliable treatment method. Indeed, even conventional treatment plants with a long history of operation are still far from the reliability that may be demanded in the future.

The second point above on the costs of waste removal is all too often totally ignored or casually dismissed as a "cost we simply must bear if we are to survive." Nevertheless, the costs should be recognized and displayed to those who must ultimately pay for improvement programs. Further, it should be constantly recalled that the costs of waste removal increase nonlinearly with increases in treatment. Indeed, for some waste constituents the cost of removing 95% is equal to twice the cost of removing 90% of the waste.

The lack of specific environmental goals for a particular river system is probably one of the most difficult problems to deal with in water quality management. The key here is specificity. While general goals are useful to set a tone for the direction of environmental improvement, quantitative water quality objectives must ultimately be specified. As will be seen later, such stream quality objectives, must be coupled to waste effluent mass discharge requirements to assure the success of the program.

Institutional Management Problems. As shown in Figures 1 and 2, the Delaware River forms a boundary between several states. The federal government through its regulatory function in water pollution control also exercises an institutional role. In addition to the federal and interstate nature of the system, several large municipalities play an important role in any water quality management effort. Finally, there are numerous special interest groups representing conservationists, industrialists, and power, navigation, and recreation enthusiasts. The institutional structure in the Delaware River has therefore been quite diffuse. In the past the region has been largely ineffectual in improving water quality precisely because of the many overlapping jurisdictions.

Partly in answer to this problem, an interstate arrangement was suggested as far back as 1920 for managing the waters of the Delaware River Basin. Kneese and Bower (1968) have provided a review of the development of interstate compacts in the Delaware. The first such agency formed was the Interstate Commission on the Delaware River Basin (INCODEL) created by parallel state legislation in 1936. After a number of productive years INCODEL was replaced in 1961 by the federal-interstate compact which established the Delaware River Basin Commission (DRBC). One of the major reasons for the replacement of INCODEL was the relatively ineffective role that agency played in upgrading the quality of the waters of the Delaware, especially the estuary. Opposition to a uniform pollution control law proposed in 1941, to be administered by INCODEL, was most severe in the Pennsylvania legislature. This opposition reflected the intransigent positions of industry and the City of Philadelphia, the latter city accounting for the major portion of the untreated sewage being discharged to the Delaware at that time. In 1945, however, the proposal was passed by all states, and primary treatment was generally initiated for most municipalities. From that time on little substantive upgrading in waste treatment was accomplished by INCODEL.

As noted, the DRBC, established in 1961, operates under a state-federal compact arrangement. The compact gives broad powers to the DRBC to plan for the development of water resources in the basin and, most importantly, to implement the plans. Thus the DRBC has the authority to control pollution sources, buy and sell water, regulate low flows for water quality control, and provide for flood protection, among many other functions.

With these broad powers much hope lies with the ability of the DRBC to exercise leadership and authority in coping with the multi-faceted ecological problems of the Delaware. The DRBC has assumed the primary role in implementing a comprehensive water quality management plan for the Delaware estuary. Central to this plan and the implementation scheme for achieving the stated objectives is an analytical framework for relating waste discharge to water quality.

Mathematical Modeling of Water Quality

Need for Modeling. Recognition of the many complicating features mentioned above, especially the numerous waste sources of different magnitude and characteristics entering a body of water that ebbs and floods, makes it mandatory to construct a logical and rational framework to aid in arriving at environmental control decisions. A mathematical model which relates pollution from each source to the subsequent distribution of water quality form a major part of this analytical framework. Greater insight is thus obtained into the cause-effect relationships between man's activities and the response of the ecological system. The basic objectives of the analytical framework (Thomann, 1969) are to:

a) Determine the underlying causal relationships between waste discharges and water quality

b) Develop methods of forecasting water quality on both a long-term and a short-term basis

c) Analyze interactions between water quality and water use

d) Develop methods and guidelines for optimal water quality management

Structure of Mathematical Models. The basic component of many mathematical models of water quality systems is the balancing of the mass of a specific quality constituent. This mass balance equation forms the basis for water quality modeling and modeling of higher order interactions with quality. Because of the severity of the quality problem in the Delaware estuary, much of the modeling effort to date in the Delaware basin has been directed to that region.

A detailed exposition of mathematical modeling of streams and estuaries is given in Thomann (1971), and a review of the state of the art of estuarine water quality modeling is given by O'Connor and Thomann (1971). Only a brief review of estuary water quality models will be given here, followed by the results of the application of the modeling approach to the Delaware estuary

For a one-dimension estuary, where averages are taken over at least one tidal cycle, the pertinent mass balance equation is given by

$$\frac{\partial s}{\partial t} = \frac{1}{A}\frac{\partial}{\partial x}\left(EA\frac{\partial s}{\partial x}\right) - \frac{1}{A}\frac{\partial}{\partial x}(Qs) - Ks \pm S , \qquad (1)$$

where

$\quad$ s $\;=\;$ water quality variable
$\quad$ A $\;=\;$ cross-sectional area
$\quad$ E $\;=\;$ tidal dispersion coefficient
$\quad$ Q $\;=\;$ net fresh water flow
$\quad$ K $\;=\;$ first order decay coefficient
$\quad$ S $\;=\;$ sources and sinks of s
$\quad$ x $\;=\;$ distance downstream
$\quad$ t $\;=\;$ time in tidal cycles

If the phenomena require examination at time intervals less than one tidal cycle, the flow term in Equation (1) is considered as the total flow, and the dis-

117

persion coefficient represents only velocity gradient effects. Harleman (1971) has reviewed tidal time models versus models of the form of Equation (1). This equation forms the basic building block for more complicated models involving, for example, dissolved oxygen or different nitrogen forms.

Application to Delaware Estuary. As indicated previously, the dissolved oxygen of the Delaware estuary is an important indicator variable of water quality. The model for dissolved oxygen is given by a coupled pair of equations:

$$\frac{\partial L}{\partial t} = -\frac{1}{A}\frac{\partial}{\partial x}(QL) + \frac{1}{A}\frac{\partial}{\partial x}(EA\frac{\partial L}{\partial x}) - K_r L \tag{2}$$

$$\frac{\partial c}{\partial t} = -\frac{1}{A}\frac{\partial}{\partial x}(Qc) + \frac{1}{A}\frac{\partial}{\partial x}(EA\frac{c}{x}) + K_a(c_s - c)$$

$$- K_d L - K_{n1}N_1 - K_{n2}N_2 - B + (P-R), \tag{3}$$

where

L = carbonaceous biochemical oxygen demand

c = dissolved oxygen

c_s = dissolved oxygen saturation

K_r = BOD decay coefficient

K_d = deoxygenation coefficient

K_a = re-aeration coefficient

K_{n1} = oxygen utilization rate due to oxidation of ammonia

N_1 = ammonia nitrogen concentration

K_{n2} = oxygen utilization rate due to oxidation of nitrite

N_2 = nitrite nitrogen

B = benthal oxygen demand

P = photosynthetic oxygen production

R = oxygen demand due to respiration

Under steady-state conditions, a useful approximation for many studies, Equations (2) and (3) can be written in matrix form by using a finite difference approximation to the spatial derivatives (Thomann, 1971). Thus if the estuary is divided into n finite reaches,

$$[A]\ (L) = (W) \tag{4}$$

where

[A]   is an n x n matrix (tri-diagonal) of coefficients incorporating the geometry of the estuary (flow, cross-sectional area, volume, dispersion, reaction kinetics),

(L)   is an n x 1 vector of BOD concentrations in each segment, and

(W)   is an n x 1 vector of carbonaceous waste load inputs

Solution of Equation (4) gives

$$(L) = [A]^{-1} (W)$$

Similarly for the dissolved oxygen deficit, $D = C_s - c$ one obtains (Thomann, 1971)

$$(D) = [B]^{-1} [VK_d] [A]^{-1} (W), \tag{5}$$

where

[B]$^{-1}$ is the inverse of matrix [B], which is identical to [A] except for the re-aeration coefficient in the main diagonal, and (D) is an n x 1 vector of dissolved oxygen deficits

Equation (3) indicates, however, that the dissolved oxygen also depends on the oxidation of ammonia and nitrite due to bacterial nitrification. Therefore, consider the case of four nitrogen forms (see Figure 10) designated as

N_0 = organic nitrogen, N_1 = ammonia nitrogen, N_2 = nitrite nitrogen and N_3 = nitrate nitrogen.

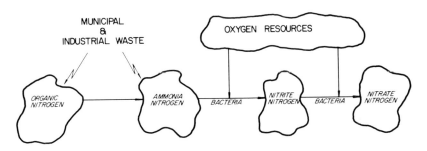

Figure 10. Representation of nitrification phenomenon

119

The equivalent matrix equations are

$$[A_o] \ (N_o) = (W_o)$$
$$[A_o] \ (N_1) = (W_1) + (VK_{o1}N_o)$$
$$[A_2] \ (N_2) = (W_2) + (VK_{12}N_1) \tag{6}$$
$$[A_3] \ (N_3) = (W_3) + (VK_{23}N_2),$$

where K_{ij} are the first order kinetic reaction coefficients, and $[A_i]$ differ only on the main diagonals.

This form of model was used together with the effect of carbonaceous oxygen-demanding loads to describe the dissolved oxygen profiles in the Delaware estuary. Details are given in Thomann, O'Connor, and Di Toro (1971) and Anon. (1969). Figure 11 is an example of a verification analysis carried out for the nitrogen species observed during July 1967. It can be seen that the mathematical model reasonably represents the spatial profile of the four nitrogen forms.

As indicated previously, the nitrification phenomenon utilizes oxygen (Figure 10) and together with other oxygen-demanding loads contributes to the total reduction of dissolved oxygen. Figure 12 shows the major component parts of the dissolved oxygen profile for July-August 1964. The upper plot shows the dissolved oxygen deficit caused by oxidation of ammonia and nitrite. The middle plot shows the total deficit, including that due to carbonaceous demand. The lower plot shows the actual observed dissolved oxygen compared with the computed values from the mathematical model. Again, the verification is sufficiently good to warrant utilization of the modeling framework for decision-making.

Estimated Effect of DO Improvement in Anadromous Fishery in Delaware Estuary. While the above water quality modeling effort provides the necessary framework for estimating the effect of waste removal programs on water quality, the impact on water use—particularly the shad fishery—is equally important. Morris and Pence (1970) have summarized the results of an analysis of the effects of several water quality improvement programs on the shad population. Three items were necessary:

a) a time variable mathematical model of dissolved oxygen in the estuary

b) a function relating DO to fish survival

c) details on arrivals and distribution of adult shad

The model described previously as Equations (2) and (3) was used. Details are given in Thomann (1971) and Pence, Jeglic, and Thomann (1968).

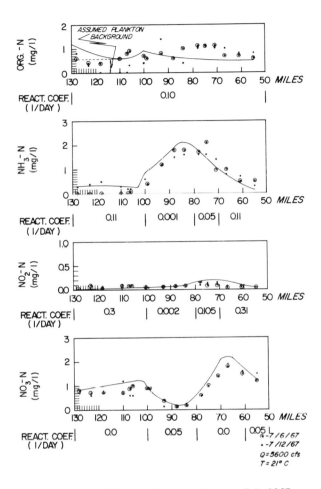

Figure 11. Verification of nitrogen forms – July 1967

A shad survival function was then constructed which related average daily DO to percent survival. Figure 13 shows the function. Essentially, survival was assumed at an instantaneous DO value of 3.0 mg/l, and it was assumed that at 2.5 mg/l instantaneous DO fish would be in distress. A random DO variation of 1 mg/l standard deviation was used, resulting in a lower bound of about 1% survival at an average daily DO of 0.5 mg/l. The logarithmic function recognizes that for higher levels of DO, survival increases by progressively lesser amounts.

The details of the shad run indicate that in general the male fish begin to arrive in April. A Gaussian distribution of arrival times was assumed, with a mean

121

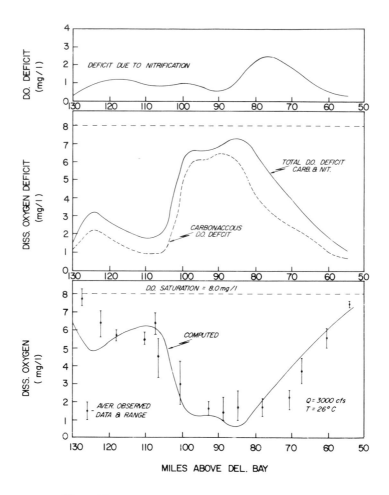

Figure 12. Verification of DO profile – July-August 1964

arrival on April 20 and a standard deviation of 8 days. Thus by May 6 about 95% of the males were assumed to have entered the lower end of the estuary. The females generally appear later than the males, and a mean arrival time of May 5 was used with a similar time spread.

These populations were then "routed" through the DO space-time surface generated by the nonsteady-state water quality model. Using the survival function and the arrival times of fish at various locations, the average percent survival of the total population could be estimated for each control program. The results are shown in Table 8.

Table 8. Estimated % survival of total upstream shad passage

Water Quality Objective	% of time — % survival Equal to or less than	
	50%	4%
II. Substantial Improvement	95	90
III. Improvement, but no special provision for migratory fish	85	80
V. Present conditions	60	20

The results indicate that under present conditions an average survival of 60% is estimated, and about once every 25 years the survival drops below 20%. Also, the difference in survival between present conditions and some improvement (level III) is substantially greater than between level III and level II. This is a reflection of the interaction between the assumed survival function and the arrival times. This greatly simplified analysis of ecological response formed at least some basis for the decision-making process. It is particularly interesting since the cost of level II was as much as 105% of the cost of level III. Depending on the implementation program, costs in 1964 dollars for level II are about $630 million, as opposed to costs for level III of about $470 million (Smith and Morris, 1969).

Of course one must recognize that the analysis did not consider a number of factors including other toxic wastes, fishing pressures, natural mortality, and mortality of fry and adults returning through the estuary to the ocean for growth at sea. It is possible to construct model frameworks that incorporate—at least to some degree—these factors. Figure 14 shows a conceptual representation of a possible model for analysis of the dynamics of the American shad population. Verified ecological models of this type would play an important role in decisions on environmental control programs.

Construction of a Comprehensive Water Quality Management Plan for the Delaware Estuary

The models discussed previously form an important input into the formulation of a comprehensive water quality management plan for the Delaware estuary. For example, the models were used to describe the achievement of water quality objectives under a series of different waste removal programs. Actually, a group of five water quality objectives were considered in the decision-making phase of constructing a comprehensive plan. The details of these objectives are given in

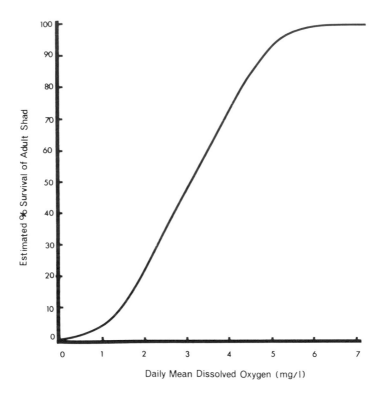

Figure 13. Percent fish survival and DO

FWPCA (1966b). For several reasons, including social and political objectives, as well as the enhancement of the anadromous fish population and technological considerations, interest quickly centered about the middle two objectives, designated objectives II and III in Table 9. The later stages of the decision-making process for comprehensive water quality management therefore utilized the output from the modeling framework discussed above in the context of determining the costs and benefits of water quality levels and the means for implementing the final program.

Costs. The implementation costs for improving the Delaware are shown in Table 9, together with the minimum daily dissolved oxygen for each water quality objective.

As seen from Table 9, a substantial increase in cost occurs between the two objectives, representing the rapid increase in cost due to increased difficulty in removing additional wastes.

124

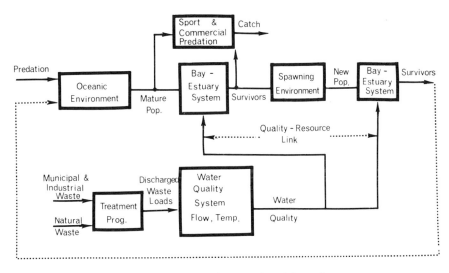

Figure 14. Block diagram of shad dynamics

Table 9. Dissolved oxygen objectives and costs of implementation (Wright and Porges, 1971)

Minimum Daily DO (mg/l) for Objective	DRBC Zones [a]				Total Costs[c]	Costs[b] (millions of dollars) Total Annual Cost
	2	3	4	5		
II	5.0	3.5	3.5	4.5	633	46
III	5.0	2.5	2.5	2.4 - 4.0	472	34

[a]See Figure 6 for location of zones and DO objectives
[b] 30 years and 6% used for annual costs
[c] Includes maintenance of present conditions at $189,000,000

After an extended series of public hearings, the staff of the Delaware River Basin Commission recommended a water quality objective essentially similar to objective II. Again, the mathematical model framework played an important input in assessing the response of the dissolved oxygen of the estuary to varying waste loads. At a formal hearing of the Commissioners of the DRBC, the staff recommendation was accepted and water quality objectives similar to level II were promulgated. As part of the implementation program for achieving these standards, a system of waste allocations was devised.

Waste-Load Allocations. The rationale of waste-load allocations is based on the fact that regions generally grow in population, industrial makeup, and consequently waste-load production. A water pollution control program based on percent waste removal will therefore not be successful since the mass of waste material to the water body will continue to increase. Thus one is led to prescribing the limits on the amounts that can be discharged. Toward this end, the model of dissolved oxygen was used to back calculate the amount of wastes that would be discharged to achieve the objective. The results of that analysis are shown in Figure 15, where for each zone the carbonaceous oxygen demand was limited to the amount shown. Nitrogenous discharges were not allocated, although the DRBC has the authority to do so at the appropriate time. Figure 16 shows the relationship between carbonaceous discharge, minimum DO and percent removal levels. Note that under projected 1975 waste loads the pollution control program being promulgated is equivalent to 90-93% removal of carbonaceous biochemical oxygen demand. As shown in Figure 15, a reserve of 10% has been set aside by the DRBC for allocation to new discharges and to act as a factor of safety in achieving the objectives.

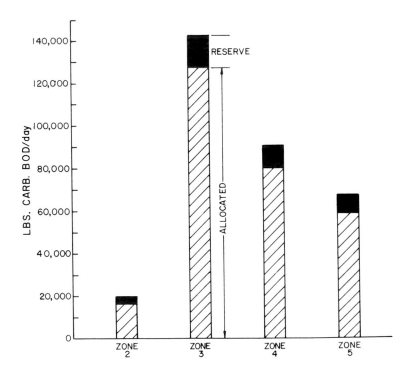

Figure 15. Waste load allocations

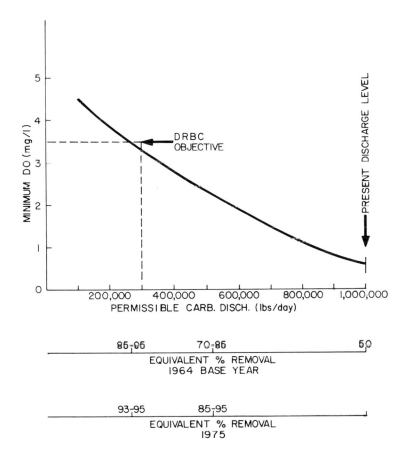

Figure 16. Carbonaceous discharge, minimum DO and percent removal levels

All waste discharges have received their stipulated allocations, and although there were a number of appeals (which have been satisfactorily resolved), progress is good in achieving the reductions in load that are required. The major dischargers have all agreed to the allocations, and time tables have been established. Because of the complexity of the problem in terms of design and construction of facilities, as well as possible local funding difficulties, the earliest date for substantial improvement in the Delaware is about 1975. At that time and for subsequent years, all look forward to a river system where a balance has been achieved between man and his activities and the ecological structure of his environment.

Literature Cited

Anonymous. 1969. Nitrification in the Delaware estuary. Hydroscience, Inc., Westwood, N.J. Prepared for Delaware River Basin Comm. 43 p. and Appendix.

Chittenden, M. E. 1969. Life history and ecology of the American shad, *Alosa sapidissima*, in the Delaware River. Ph.D. Dissertation, Rutgers Univ., New Brunswick, N.J. 485 p.

Federal Water Pollution Control Administration. 1966*a*. Water quality control study, Tocks Island Reservoir, Delaware River Basin. Dept. Interior, for Dept. Army, U.S. Army Eng. Dist., Phila, Pa.

——————. 1966*b*. Delaware estuary comprehensive study, preliminary report and findings. Dept. Interior, Phila, Pa.

Fish, R. E. 1968. Low-flow forecasting on the Delaware River. Amer. Soc. Civil Eng., J. Irr. Drainage Div. p. 223-232.

Harleman, D. J. 1971. One-Dimensional Models, 2. *In* Hydrodynamic models, state-of-the-art of estuarine modeling. TRACOR, Inc., Austin, Texas. Prepared for FWQA (in press).

Kneese, A. V., and B. T. Bower. 1968. Managing water quality: economics, technology, institutions. Res. for the Future, Inc., The Johns Hopkins Press, Baltimore, Md. 274 p.

Morris, A., and G. D. Pence. 1970. Quantitative estimation of migratory fish survival under alternative water quality control programs. Presented at Fifth Int. Water Pollution Conf., San Francisco, California.

O'Connor, J. J., and R. V. Thomann. 1971. Water quality models chemical, physical and biological constituents, ch. 3. *In* State-of-the art of estuarine modeling. TRACOR, Inc., Austin, Texas. Prepared for FWQA (in press).

Pence, G., J. Jiglic, and R. V. Thomann. 1968. Time varying dissolved oxygen model. J. San. Eng. Div., Amer. Soc. Civil Eng. 94(SAE2):381-402.

Pollison, D. P., and W. M. Craighead. 1968. Lehigh River, biological investigation. Delaware River Basin Comm., Penn. Fish Comm., and Penn. Dept. Health. 122 p.

Smith, E. T., and A. R. Morris. 1969. Systems analysis for optimal water quality management. J. Water Poll. Contr. Fed. 41:1635-1646.

Sykes, J. E., and B. A. Lehman. 1957. Past and present Delaware River shad fishery and considerations for its future. Res. Report 46, Fish and Wildlife Science, Fish Wildl. Sci. Res. Rept. 46. 25 p.

Thomann, R. V. 1969. The utility of systems analysis in estuarine water quality management. Proc. Fourth Symposium on Water Resour. Res., on Systems Analysis for Great Lakes Water Resources, Ohio State Univ. p. 119-128.

—————————. 1971. Systems Analysis and Water Quality Management. Env. Res. & Appl., Inc., New York, N.Y.

—————————, D. J. O'Connor, and D. M. Di Toro. 1971. Modeling of nitrogen and algal cycles in estuaries. Proc. Int. Water Pollution Conf., San Francisco, Cal. Pergamon Press (in press).

Tomazinis, A. R., and I. Gabbour. 1967. Water-oriented recreation benefits, phase II, projection of recreation demand and benefits. Inst. Environ. Stud. Univ. Penn. 61 p.

U.S. Commission of Fish and Fisheries. 1874. Report of the Commission for 1872 and 1873, Part II. Washington, D.C., p. 458.

U.S. Department of Health, Education, and Welfare. 1960. Report of the comprehensive survey of the Delaware River Basin. Appendix C, Municipal and industrial water use and stream quality. Prepared for U.S. Corps Eng., Phila., Pa. 241 p. and attachments.

Weston, R. F. 1970. Tocks Island region environmental study. Prepared for Delaware River Basin Comm. by Roy F. Weston, Engineers, Westchester, Pa. 117 p.

Wright, J. F., and R. Porges. 1970. Water quality planning and management experiences of the Delaware River Basin Commission. Presented at Fifth Int. Water Pollution Resources Conf., San Francisco, Cal.

—————————. 1971. Water quality planning and management experiences of the Delaware River Basin Commission. Proc. 5th Int. Water Poll. Conf., San Francisco, Cal. Pergamon Press (in press).

MAN AND THE ILLINOIS RIVER

William C. Starrett

The purpose of this paper is to discuss the important uses man has made of the Illinois River (State of Illinois, U.S.A.) and its basin and the impact these uses have had on the ecosystem of the river and its bottomland lakes.

The geographical location of the Illinois River has made it one of the important rivers in America for man and the development of some of his cultural activities (Figure 1). During the past century some of these activities have had adverse effects on the biota of the Illinois River and its adjoining bottomland lakes (Mills, Starrett, and Bellrose, 1966; Starrett, 1971). A number of biological and chemical studies were conducted on the Illinois River between 1874 and 1930 (Calkins, 1874; Forbes, 1878; Hart, 1895; Kofoid, 1903; Baker, 1906; Forbes and Richardson, 1908, 1913, and 1919; Bartow, 1913; Danglade, 1914; Malloch, 1915; Richardson, 1921a and b, 1925a and b, and 1928; Greenfield, 1925; Hoskins, Ruchhoft, and Williams, 1927; Boruff and Buswell, 1929; Purdy, 1930; and others). These and later investigations conducted on the river by the Illinois Natural History Survey and other state and federal agencies have made it possible for us to determine some of the changes which have occurred in the ecosystem of the Illinois River resulting from the impact of man.

Description of the Illinois River

The Illinois River is formed in northeastern Illinois, about 48 miles (77.2 km) southwest of Chicago at the confluence of the Des Plaines and Kankakee rivers (Figures 1 and 2) and gently flows 272.9 miles (439.2 km) across the state to Grafton, Illinois, where it empties into the Mississippi River (above St. Louis, Missouri). The former or natural drainage area of the river was about 28,200 sq miles (73,038 km^2). Man increased the drainage area to 29,010 sq miles (75,136 km^2)[1] by reversing the flow of the Chicago and Calumet rivers and

---

[1]The drainage basin of the Illinois River is divided as follows: 85.5% in Illinois, 11.0% in Indiana, and 3.5% in Wisconsin. About 44.0% of the State of Illinois is drained by the Illinois River system. Barrows (1910) states that the basin of the river is 32,081 sq miles (83,089.7 km^2).

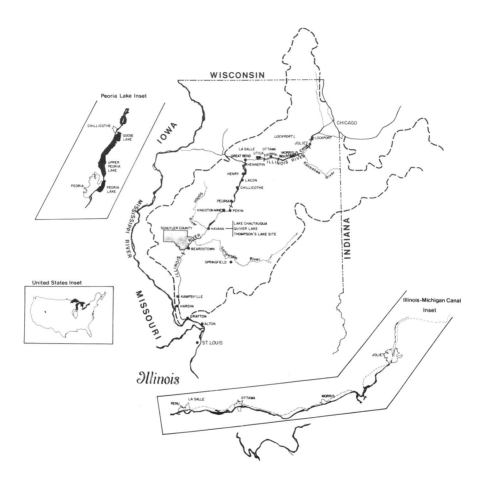

Figure 1. The Illinois River and its drainage basin

intercepting some drainage areas along the shores of Lake Michigan (U.S. Public Health Service, 1963). Today the Illinois Waterway (327 miles or 526.3 km long) has a 9-ft (2.7-m) deep navigation channel (with locks, dams, and canals) flowing from Lake Michigan at Chicago through the Chicago Sanitary and Ship Canal to the Des Plaines River and thence into the Illinois River. The Waterway forms a connecting link for water freight transportation between the St. Lawrence Waterway and the Mississippi River, which flows to the Gulf of Mexico.

Geology of the River. The Illinois River from the Great Bend (Figure 1) to its mouth occupies a preglacial channel, whereas the channel of the upper river is of more recent origin (Mulvihill and Cornish, 1929).

132

Figure 2. Aerial views of the Illinois River (1970). (A) Confluence of the Kankakee River (left) and Des Plaines River (right) forming the Illinois River (upper) about 48 miles (77.2 km) southwest of Chicago (viewing downstream). The photo shows the Dresden Nuclear Plant at the source of the Illinois River

The Illinoian glacier affected all of the region now occupied by the Illinois River and its basin, except for a small driftless area at the mouth of the river (Willman, Glass, and Fry, 1963). The Wisconsin drift extended down over more than one-half of the present basin of the river. Great volumes of water from glacial Tolleston Lake Chicago, Lake Algonquin, and later the Nipissing Great Lakes flowed through the Chicago Outlet into the Des Plaines River and thence into the Illinois (Sauer, 1916; Flint, 1957). In the Algona phase the drainage of the Great Lakes moved eastward (Niagara) and the Chicago Outlet dried up, leaving the Illinois River system a separate entity. This separation greatly reduced the flow of the Illinois. Thousands of years later another impact on the ecology of the river occurred when man rejoined the Illinois system with Lake Michigan.

Physical and Chemical Characteristics of the River. "The upper Illinois, comprising the westward-flowing section from the junction of the Des Plaines and Kankakee to the great bend near Hennepin, is independent of preglacial drainage

133

**Figure 2-B. Starved Rock lock and dam on the Illinois River below Ottawa
(viewing downstream)**

lines and is excavating a new course, its bed being usually on the rock. The lower Illinois, extending from the great bend to the mouth, occupies the preglacial channel, in which the rock bottom lies nearly 100 feet below the bed of the present stream. . . . The old river channel of the lower Illinois is much too large for the present volume of flow, so that it has been filled up by alluvial deposits forming the present bottom lands." — Hoskins, Ruchhoft, and Williams, 1927:4-5.

Originally, the wide bottomlands of the lower Illinois contained many shallow, fertile lakes and marshes which constituted an important segment of the ecology of the river. Man has drained many of these lakes and marshes for agricultural purposes.

Most of the river has a low stream gradient, as indicated in Figure 3. The rapids in the upper river and the shallow depths in parts of the lower river impeded navigation for a number of years.

Figure 2-C. The Peoria Lake (Upper and Lower) section of the Illinois River at Peoria (viewing upstream). The constricted part of the river between the two lakes is the Peoria Narrows (alluvial fan)

The Illinois River and its basin (originally prairie interspersed with oak-hickory forests) are located in the temperate zone of North America. The average rainfall at Peoria, Illinois, is 35.01 inches (88.9 cm) (U.S. Department of Agriculture, 1941). The average water temperatures of the river at Havana (1963-1969) were 82°F (27.8°C) during July and 34°F (1.1°C) during January (H. W. McFadden, personal communication). In the calendar year 1968 the discharges (flow) of the river at Kingston Mines were: total 4, 962,010 c.f.s. (140,509m^3 per second), mean 13,560 c.f.s. (384 m^3 per second), maximum 42,800 c.f.s. (1,212 m^3 per second) and minimum 4,070 c.f.s. (115 m^3 per second) (U.S. Department of the Interior, 1970).

For a number of years the Illinois River has been a polluted stream. The two main sources of domestic and industrial pollution are the Chicago-Joliet and Peoria-Pekin metropolitan areas (Figure 1). The basin of the river is in the heart of the corn-soybean (row crops) belt of America. The runoff from the fields in row crops is an important source of pollution (silt, pesticides, and fertilizers) in the Illinois. In the summer months during periods of low water levels the dis-

Figure 2-D. The Illinois River at Havana. The photo shows barges being loaded with grain

solved oxygen in some sections of the river is now less than 2.0 mg/l (Mills et al., 1966; Starrett, 1971). A summary of pertinent chemical data of the river is given in Table 1. The recent physical and chemical changes which have occurred in the ecosystem of the river have been caused by man.

Early Man and the Illinois River

An artifact of human origin was discovered imbedded in Roxana loess in Fulton County (across the river from Havana, in Figure 1) near the Illinois River valley, and its inferred age ranged from 35,000 to 40,000 years (Munson and Frye, 1965). It is believed that people of the Paleo-Indian culture first appeared in the midwestern part of the United States between the years of 15000 to 12000 B.C. (Winters, 1959).

Evidence of man's earliest use of the Illinois River (between 8000 and 6000 B.C., awaiting radiocarbon dating results) was discovered in 1969 and 1970 by Dr. Stuart Struever of Northwestern University and his associates at the Koster

Table 1. Water chemistry of the Illinois River *

Analyses	Range		
Hardness (mg/l)	204	-	444
Alkalinity (MO) (mg/l)	72	-	178
Total solids (mg/l)	330	-	550
Copper (mg/l)	0	-	0.30
Nickel (mg/l)	0.01	-	0.25
Zinc (mg/l)			0.13
Ammonia nitrogen (N) (mg/l)	0.9	-	8.2
Total phosphates (PO$_4$) (mg/l)	0.0	-	8.3
BOD (mg/l)	0.0	-	30.0
pH	7.3	-	8.7

*U.S. Public Health Service, 1963; Illinois Sanitary Water Board, 1967; Starrett, 1971 and unpublished. For detailed description of the Illinois River see Starrett, 1971.

site on the eastern margin of the Illinois River valley floodplain, about 30 miles upstream from the mouth of the river near Kampsville (Figure 1) "The Koster site experienced from 7000 to 9000 years of intermittent human occupation, and is therefore a potentially important document of changing human adaptations to the Riverine fauna and flora" (Stuart Struever, personal communication).

There are 11 definable horizons (cultural layers) at the Koster site, the oldest being horizon 11. According to Mr. Frederick C. Hill of the University of Louisville (personal communication):

I now have enough data to be able to compare the faunal remains from horizon 11 with those of horizon 6. Between the two horizons, I detect a major shift in animal utilization from that of a restricted use of mainly the river during horizon 11, to a general use of the entire river bottom, backwater lakes, and the river itself in horizon 6 times.

Between 2500 and 500 B.C. people of the Black Sand culture of the Early Woodland period lived in the central part of the Illinois River valley (Fowler, 1959). In the latter part of this period agricultural practices began to supplant the Archaic way of merely collecting plants and animals for food. The Hopewell culture in Illinois (300 B.C. to 500 A.D.) is best known from the Illinois River valley. These people required good hunting and fishing areas and practiced a

137

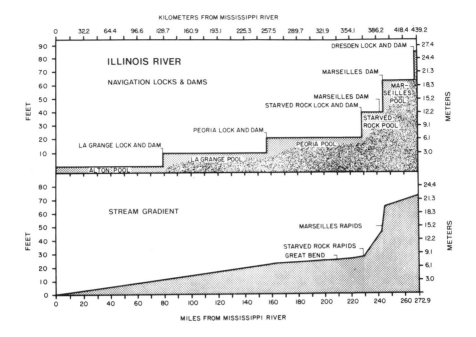

Figure 3. Impact of the present navigation dams on the natural gradient of the Illinois River.

limited amount of agriculture (McGregor, 1959). Sites of the Late Woodland period (700 or 800 A.D. to 1200 or 1300 A.D.) have been found in the vicinity of the Illinois River in Fulton and La Salle counties (Maxwell, 1959). The people inhabiting the Kingston Lake site (Middle Mississippi, 1100 to 1400 A.D.; near Kingston Mines, Figure 1) hunted and fished in the Illinois River and its bottomland lakes and bottomlands for various species of naiads (pearly mussel: Unionacea), snails, turtles, fishes, waterfowl, rails, and shorebirds (Parmalee, 1962).

In the seventeenth century the French explorers found the Illiniwek tribe (five tribes) to be the principal occupants of the Illinois country (Bauxar, 1959; Buck, 1967). The area around Lake Pimiteoui (now known as Peoria Lake, Figures 1 and 2) was a favorite wintering site for several bands of Kaskaskia Indians (Bauxar, 1959). During the eighteenth century most of the Illiniwek were annihilated by neighboring peoples. In 1818 a few remnants of the Peoria tribe still remained along the Illinois River (Buck, 1967).

The occupancy of the Illinois River valley for thousands of years by early man probably had little or no impact on the ecosystem of the Illinois River.

138

Modern Man and the Illinois River

Before 1800 a few pioneers had come into Illinois country from Kentucky, Indiana, and Tennessee and settled mainly in the forested areas south of the Illinois River and its basin (Figure 4-A). These people were hunters, rather than farmers (Barrows, 1910). By 1800 the total population of the Illinois country (later to become the State of Illinois) was only 2,458 (U.S. Department of the Interior, 1901). Illinois was admitted to statehood in 1818 with a population of about 40,000 (Buck, 1967). In 1818 only a few traders, trappers, and Indians lived on the basin of the Illinois River (Figure 4-A).

The valley of the Illinois River was, at the close of the territorial period, one of the important fur-bearing areas of the northwest. In 1816, the furs sent out from the various posts upon the Illinois River included 10,000 deer, 300 bear, 10,000 raccoon, 35,000 muskrat, 400 otter, 300 pounds of beaver, 500 cat and fox, 100 mink (Buck, 1967:26).

After Illinois became a state, settlers began to move from the southern part of the state onto the Illinois River basin (Figure 4-B).

In 1823, Springfield was a frontier village containing a dozen log cabins; the site of Peoria was occupied by a few families, and that of Chicago by a military and trading post. The rest of northern Illinois was entirely unoccupied (Barrows, 1910:66).

A new impetus for settling along the Illinois was given by the establishment of steam navigation on the Illinois River in 1828 (*ibid.*).

In 1832 the southern advance along [the] Illinois Valley was checked, and, save at Peoria, the settlers were driven south and east of the river by Black Hawk's war. Before the southern frontier had recovered from this blow, a great northern stream of immigration from New York and New England had swept into the unoccupied portions of the valley, occupying first the woodland and later the prairie (Barrows, 1910:68) (Figure 4-C).

One of the great problems which confronted the settler from the wooded hills of New England was the almost level and nearly treeless prairie, which covered much of the state (Sauer, 1916:153).

Many people believed that the prairie was a desert and the absence of timber indicated that the land of the prairie was a land of poverty (Sauer, 1916).

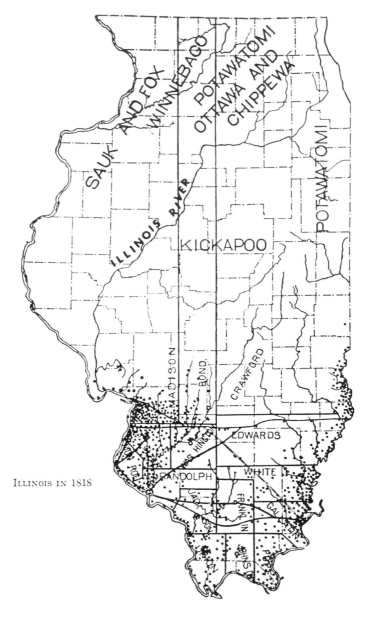

Figure 4. Maps of the State of Illinois showing the settlement of modern man on the basin of the Illinois River between 1818 and 1860. (A) Extent of settlement of modern man and general location of Indian tribes in Illinois in 1818. The dots reflect settlement areas of modern man (modified from Buck, 1967:63)

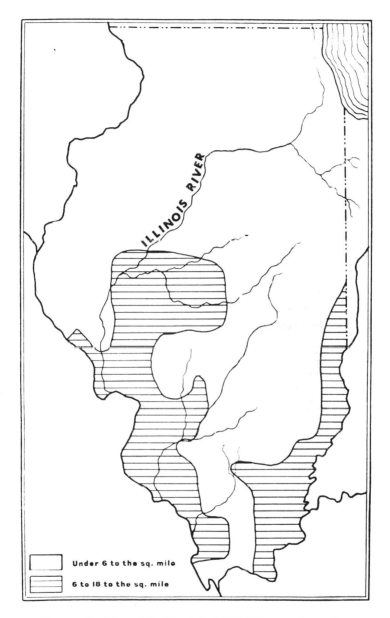

Figure 4-B. Extent of settlement in Illinois in 1830. This map shows the movement of modern man from the southern part of the state onto the lower basin of the river and his avoidance at this time of the prairie on the upper basin (modified from Barrows, 1910:67)

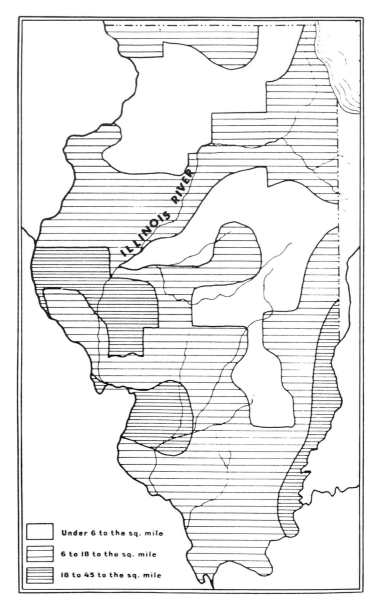

Figure 4-C. Extent of settlement in 1840 (modified from Barrows, 1910:72)

In 1840 the population of the entire basin of the Illinois River (including the portions in Indiana and Wisconsin) was about 0.19 million, excluding Chicago, which at that time was not on the watershed of the Illinois (U.S. Department of the Interior, 1901) (Figure 4-C). After 1840 more settlers began to use the prairies and further developed navigation on the river (Figure 4-D). In 1900 the population on the basin was about 3.3 million (*ibid.*). From the tentative data of the U.S. Department of Commerce (1970 *a, b, c*), the population of the Illinois River basin in 1970 was estimated at 8.6 million.[1] In 1970 the population of the city of Chicago and the remainder of Cook County amounted to about 62.9% of the entire human population on the basin of the Illinois River.

The rapid population growth which occurred on the basin of the Illinois River after 1840 was related to the geographic location of the basin; the settlers' learning to use the rich prairie soil for agriculture; the developing of coal mines and other natural resources on the basin; and the developing of various industries, rail transportation, and navigation.

Man's Uses and Impact on the Illinois River

With the rapid population growth on the basin of the river came uses by man which began to change the ecosystem of the Illinois River. One of the perplexing economic problems the early settlers faced was the lack of means for transporting commodities to the East via the Great Lakes and the Erie Canal.

Illinois and Michigan Canal. Before Illinois had obtained statehood, negotiations had been made with the Indians to acquire land to build a canal to connect Lake Michigan with the Illinois River. The Illinois and Michigan Canal[2] was completed in 1848. It extended along the Illinois River down to La Salle in order to bypass the rapids in the upper river (Figures 1 and 3).

In reference to the development of the city of Chicago and the canal, Walker (1960:5) stated:

---

[1] The human population was distributed by states on the watershed of the Illinois River as follows: (1840) Illinois 92.8%, Indiana 6.2%, and Wisconsin 1.0%; (1900) Illinois 94.7%, Indiana 3.8%, and Wisconsin 1.5%; and (1970) Illinois 93.6%, Indiana 4.3%, and Wisconsin 2.1%. The human population residing on the watershed of the Illinois River in the State of Illinois in 1970 comprised 73.6% of the entire state's population of 10,977,908 (tentative).

[2] The canal was 36 to 48 ft (10.9 to 14.6 m) in bottom width and 6 ft (1.8 m) deep (U.S. Senate, 1957). In 1933 the entire canal was closed to navigation.

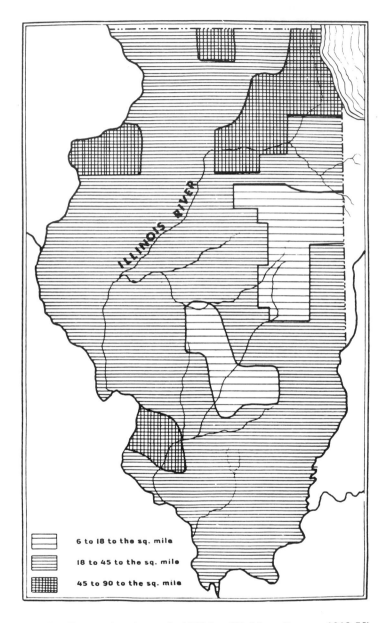

Figure 4-D. Extent of settlement in 1860 (modified from Barrows, 1910:75)

By 1840 the population had grown to 4,470 persons; in 1848, the year the canal was opened the city numbered 28,000. Within 12 years, by 1860, the population was 109,260, an increase of nearly 400% in little more than a decade.

During the early period of growth in Chicago, shallow wells supplied man's water, and outdoor privies, his sewage disposal (Walker, 1960). Soon water pollution became a serious problem, and many deaths resulted from waterborne diseases. In 1856 work was started in Chicago on an integrated sewerage system (*ibid.*). The sewers discharged untreated wastes into the Chicago River, which then emptied into Lake Michigan. This raw sewage soon created another health problem since Chicago had begun to use Lake Michigan for its water supply. A temporary solution to the problem was achieved in 1871 by deepening the canal and reversing the flow (and by pumping between 1894 and 1900) of the sewage-laden Chicago River into the Illinois and Michigan Canal and thence into the Illinois River from the Des Plaines River and the canal at La Salle.[1]

According to Kofoid (1903:202):

By the time the sewage of Chicago entered the Illinois River at La Salle it was thus already in the advanced stages of decay and available for the support of the phytoplankton or other vegetation, if, indeed, it was not already used to some extent by these agencies.

The canal apparently had only a slight impact on the ecosystem of the Illinois below La Salle.

The author believes that between 1848 and 1910 (the year the canal was abandoned between Chicago and Joliet) some of the polluted water of the Illinois and Michigan Canal flowed into the Des Plaines River at Joliet and thence into the Upper Illinois River (Figures 1 and 2-A). In the early 1870's the naiad population in the Upper Illinois at Starved Rock had not been affected by pollution (Calkins, 1874; Starrett, 1971). It is doubtful that the canal water had a serious effect on the ecosystem of the Upper Illinois River.

Flooding affected the reversal of the Chicago River, and Chicagoans were periodically endangered by waterborne diseases from Lake Michigan. A new method of diverting the Chicago River into the Illinois River system had to be developed during the 1890's to prevent human sewage from being discharged into Lake Michigan in the Chicago area.

---

[1]The diversion of Lake Michigan basin water was increased from 100 to 200 c.f.s. (2.8 to 5.7 m^3 per second) to 300 to 1,000 c.f.s. (8.5 to 28.3 m^3 per second) (U.S. Senate, 1957).

Chicago Sanitary and Ship Canal. By 1900 the Chicago Sanitary and Ship Canal[1] was completed, and water began to flow into the Des Plaines River and thence into the source of the Illinois (Figures 1 and 2). The canal extends from the West Fork of the South Branch of the Chicago River, about 6 miles (9.7 km) from Lake Michigan, to the Des Plaines River near Joliet. It was constructed primarily to carry sewage away from Lake Michigan and secondarily for navigation. Large volumes of Lake Michigan water were diverted into the system to dilute industrial wastes and the untreated sewage of a human population in 1900 of 1.6 million (this diverted water was later also used as an aid for navigation).

Diversion of Water

Between 1900 and 1938 the average amount of Lake Michigan water diverted into the Illinois River system through the Chicago Sanitary and Ship Canal was 7,222 c.f.s. (204.5 m^3 per second). The diversion during this period ranged from 2,990 c.f.s. (84.7 m^3 per second) in 1900 to 10,010 c.f.s. (283.5 m^3 per second) in 1928 (U.S. Senate, 1957).[2] A decree of the United States Supreme Court limited the amount of diversion after 1938 to 1,500 c.f.s. (42.5 m^3 per second) in additon to the domestic pumpage of Chicago (Figure 5).[3]

The diverted lake water increased the depth of the Illinois River and flooded its bottomlands. Forbes and Richardson (1919) determined that the diversion had increased the depth of the river 2.8 ft (0.9 m) at Havana (1900-1909). By 1913 about one-half of the flow of the Illinois at Peoria was Lake Michigan water (*ibid.*).

Some of the aquatic vegetation in the bottomland lakes was adversely affected by the increase in water depths (*ibid.*). Also, the inundation of the bottomlands killed large stands of trees. Years later some of the remaining

---

[1]The canal has a minimum bottom width of 160 ft (48.8 m) and a depth of 22 ft (6.7 m) (U.S. Senate, 1957). It still carries the wastes (now partly treated) away from Chicago into the Illinois and is a part of the Illinois Waterway.

[2]Between 1900 and 1910 an additional 600 to 700 c.f.s. (17.0 m^3 to 19.8 m^3) per second) of water was diverted into the Illinois and Michigan Canal and thence into the Illinois.

[3]Between 1939 and 1969 the average amount of water diverted into the Illinois River system was 3,237 c.f.s. (91.7 m^3 per second). The navigation dams completed in the 1930's enabled the Corps of Engineers to maintain pool levels in the river (Figures 3 and 5).

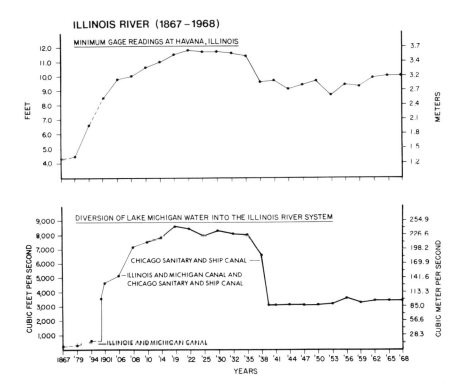

Figure 5. Effect of diversion on water depths of the Illinois River at Havana, as reflected by minimum water levels. Levees along the river, dams, and dredging also affect water depths of the river. (U.S. Senate, 1957; Corps of Engineers data furnished by H. Gordon Hanson and R. M. Mundelius, personal communications)

stumps of these trees created additional nesting sites for tree swallows (*Iridoprocne bicolor*) and prothonotary warblers (*Protonotaria citrea*).

After the opening of the canal several species of Lake Michigan diatoms were collected from the Illinois River by Forbes and Richardson (1919). In 1903 a Lake Michigan species of cisco (*Coregonus artedii*) was caught in the upper river (Illinois Natural History Survey collections).

Flooding of additional bottomlands created new habitats for fishes in the lower river. The highest recorded yield of commercial fishes from the lower river occurred in 1908 (over 20 million pounds or 9.07 million kg). Forbes and

Richardson (1919) believed that this high commercial yield was a result of the new habitats created by diversion, the recently introduced carp (*Cyprinus carpio*), and increased fertility of the water from sewage. After 1908 the fishery of the Illinois began to decline and has continued to do so (Mills et al., 1966). One of the chief factors involved in the decline of the fishery has been pollution.

Pollution in the Illinois River

The untreated sewage and industrial wastes from Chicago carried by the Chicago Sanitary and Ship Canal into the Des Plaines River and thence into the Illinois soon had a serious impact on the ecosystem of the upper river. Forbes and Richardson (1913:541-542) described the impact on the upper river (Morris to Marseilles) in 1911 as follows:

> In the seventeen-mile section of the Illinois from Morris to the upper dam the river reaches its lowest point of pollutional distress, becoming, when very hot weather coincides with a low stage of water, a thoroughly sick stream. Its oxygen is nearly all gone; its carbon dioxide rises to the maximum; its sediments become substantially like the sludge of a septic tank; its surface bubbles with the gases of decomposition escaping from sludge banks on its bottom; its odor is offensive; and its color is gray with suspended specks and larger clusters of sewage organisms carried down from the stony floor of the polluted Des Plaines, or swept from their attachments along the banks of the Illinois. On its surface are also floating masses of decaying debris borne up by the gases developing within them, and covered and fringed with the "sewage fungus" (*Sphaerotilus natans*) and the bell animalcule (*Carchesium lachmanni*) usually associated in these waters. The vegetation and drift at the edge of the stream are also everywhere slimy with these foul-water plants and minute filth-loving animals. . . .

> The normal life of the stream practically disappears in the absence of oxygen; its fishes withdraw to neighboring unpolluted waters; its mollusks, crustaceans, ordinary insect larvae and other more or less sedentary forms disappear to be replaced mainly by slime worms and *Chironomus* larvae in the sludge; and its chlorophyll-bearing plants linger only along the edges in shallow water. With the advent of cooler weather and higher river levels, most of these marked symptoms disappear, and a few fishes may even make their way into the stream, particularly along the south side in the vicinity of the mouths of creeks.

By 1912 rather severe pollution conditions extended downstream about to Spring Valley (Forbes and Richardson, 1913). Conditions worsened in the river during and after World War I. In the 1914-1920 period the canal received untreated domestic and industrial wastes having a population equivalent of about 4.2 million (Richardson, 1928). Between 1913 and 1920, Richardson (1921b) found that downstream as far as Beardstown there had been a virtual extermination by pollution of the former normal benthic organisms. Pollution from the Peoria-Pekin metropolitan area also contributed to the drastic change which occurred in the river between this area and Beardstown.

In the 1920's man began to treat some of the wastes discharged into the canal and river. The continual growth in human population and the expansion of industries on the river's basin during the past fifty years have made it difficult for engineers to maintain efficient treatment facilities. In 1960 the sewered wastes from the Chicago area before treatment had a population equivalent of 9.50 million, but the wastes discharged into the canal after treatment had a population equivalent of 1.15 million (U.S. Public Health Service, 1963). Even though the amount of untreated wastes in 1960 had doubled since 1918, the population equivalent of the discharge into the canal was approximately 77% less than in 1918 (Starrett, 1971). However, the population equivalent is based on the 5-day BOD test and is misleading for the Illinois River because of the large quantities of ammonia nitrogen (N) present in the effluents (Table 1). Butts, Schnepper, and Evans (1970) recently found that in the La Grange navigation pool of the Illinois River the nitrogenous demand imposed a greater threat to the dissolved oxygen in the river than did the carbonaceous demand. The dissolved oxygen content of the upper river is now considerably higher than it was in the 1912-1930 period, but is lower in the river from below the Peoria-Pekin area to the mouth of the river (Mills et al., 1966; Starrett, 1971). The low dissolved oxygen content in the lower river is believed by the author to be related partly to the ammonia nitrogen content. Drs. B. J. Mathis and T. C. Cummings of Bradley University (personal communication) recently found large accumulations of metals in naiads collected from the Peoria Lake section of the river.

The upper river no longer has the smell and appearance of an open sewer that it had back in 1911. In 1970 the entire river, other than for occasional oil slicks, had the appearance of a relatively clean stream. However, we have found that the plant and animal communities are still drastically affected by pollution from the Chicago Sanitary and Ship Canal and various sources along the river.

Some of the effects of pollution on the river's animal communities are perhaps best demonstrated by the numbers and kinds of benthic organisms present. The data presented in Table 2 reflect changes which have occurred in the benthic community (excluding naiads) in the Peoria Lake section of the river since 1915 (Figures 1 and 2-C). Since 1922 there has been a tremendous decline in the

Table 2. Average number of benthic organisms occurring per square meter in the Peoria Lake (Upper, Middle and Lower) section of the Illinois River in 1915, 1920, 1922, and 1964-1965 (1.2 to 2.1 m depth)

	Average number per m^2			
Kinds of Organisms	1915*	1920*	1922**	1964-1965***
Upper Peoria Lake				
Tubificidae	31	1,371	20,400	2,956
Chironomidae	16	663	-	797
Gastropoda	42	0	-	0
Sphaeriidae	244	29	51,229	0
Other	37	0	-	0
Middle Peoria Lake				
Tubificidae	0	43	431	2,354
Chironomidae	25	240	-	423
Gastropoda	48	0	-	0
Sphaeriidae	139	4	2,081	0
Other	14	0	-	22
Lower Peoria Lake				
Tubificidae	0	0	-	1,326
Chironomidae	0	397	-	534
Gastropoda	267	10	-	0
Sphaeriidae	116	0	-	0
Other	62	0	-	0

*Richardson (1921*b*)

**Richardson (1925*b*)

***Starrett and Paloumpis (unpublished)

abundance of Tubificidae worms (chiefly *Limnodrilus* spp.) in Upper Peoria Lake; however, through the years conditions have become more favorable for these worms in Lower Peoria Lake than they were in the 1920's (Table 2). Bottom fauna studies (Starrett and Paloumpis, unpublished) conducted in 1964 and 1965 indicated that the benthic community from the source of the river downstream to Beardstown was comprised almost entirely of Tubificidae worms and Chironomidae larvae. From Beardstown to the mouth of the river pollution apparently has had less effect on the benthic community. In this lower section in

1964 and 1965 *Hexagenia* nymphs, Sphaeriidae, and Gastropoda still occurred in the benthic community together with an abundance of Tubificidae worms and Chironomidae larvae. In 1915 Tubificidae worms were scarce in Richardson's (1921*a*) collections between Peoria Lake and the mouth of the river.

Fingernail clams (Sphaeriidae) virtually disappeared from the river above Beardstown in the mid-1950's (Paloumpis and Starrett, 1960 and unpublished). These organisms were an important food item in the river and its bottomland lakes for carp and diving ducks (Aythyinae), particularly the lesser scaup duck (*Aythya affinis*). Following the disappearance of the fingernail clams, a sharp decline occurred in the numbers of lesser scaups using the middle section of the river and its lakes during migration (Mills et al., 1966). In the 1960's fingernail clams formed 50.2% (volume) of the food items taken by carp collected in the river between Beardstown and its mouth (Starrett and Paloumpis, unpublished). Only one fingernail clam was found in the food contents of the carp examined from the remainder of the river. Tubificidae worms comprised only 4.3% (volume) of the food ingested by carp taken from the section of the river between Beardstown and the mouth, whereas in the upper river (source to Starved Rock dam), where there was a virtual absence of fingernail clams, carp fed heavily upon Tubificidae worms (30.5% volume). Carp collected during the early 1960's from the lower section of the river, where fingernail clams formed an important part of their diet, were deeper bodied than those taken from the remainder of the river (Mills et al., 1966).

The naiads, also, are an important group of animals in the benthic community of the river. In the early 1900's the Lower Illinois River was considered the most productive naiad stream per mile in America (Danglade, 1914). Later, pollution had a serious effect on the naiads of the entire river. A recent study indicated that at least 49 kinds of naiads were present before 1900, and by the late 1960's 25 of these had become extirpated from the river primarily by pollution (Starrett, 1971). Naiads not only are an important constituent of the benthic community, but also serve as an important source of food for the channel catfish (*Ictalurus punctatus*). Except for a short stretch of the river in the visinity of Peoria, naiads now occur abundantly only between Beardstown and the mouth of the river (*ibid.*). In this lower section of the river in the 1960's naiads comprised 25.2% (volume) of the food eaten by channel catfish (Starrett and Paloumpis, unpublished). None of the catfish examined from other parts of the river contained naiads.

The author also considers the plankton community a basic component of the ecosystem of the Illinois River. In the 1890's Kofoid (1903) computed the total annual production of plankton in the Illinois River to be 67,750 m^3. In the early 1960's Williams (1964) had extremely high phytoplankton counts from collections made from the river at the Peoria Narrows, and it appeared to him that this abundance of phytoplankton was related to enrichment and high calcium hardness. Between March and September 1964 the phytoplankton counts at the Peoria Narrows ranged from 7,000 to 233,000 plankters per milliliter (Arnold K. Cherry, personal communication). In 1964 the phytoplankton community was comprised mainly of diatoms (*Melosira, Navicula, Synedra, Cyclotella*, and *Tabellaria*), blue-greens (*Oscillatoria, Anabaena*, and *Microcystis*), and *Euglena*. The counts at the Peoria Narrows between March and September in 1969 ranged from only 6,800 to 108,000 plankters per milliliter and in 1970, from 8,500 to 117,000 (*ibid.*). This decline in phytoplankton since the mid-1960's has been associated with a sharp decrease in the abundance of blue-greens. Other than during a bloom of *Anacystis* in September 1969, the plankton community of the river at the Peoria Narrows since the mid-1960's has been comprised mainly of diatoms (*ibid.*). It has been suggested that the phytoplankton of the entire river may be limited by turbidity and the synergistic effects of toxic metals (Starrett, 1971).

The vascular aquatic plant community of the middle section of the river (principally Peoria Lake) was virtually eliminated during the severe pollution period of 1916-1922 (Thompson, 1928). After 1922 vegetation began to reappear in this section, and between the late 1930's and the mid-1950's the growth was luxuriant (Mills et al., 1966). Aquatic plants have disappeared completely from the river proper since the mid-1950's. The author believes that a pollutant has adversely affected the plant community in recent years, as well as in the 1916-1922 period. It is doubtful that turbidity and sedimentation were important factors in affecting plants in the river proper (*ibid.*). However, these factors have had important effects on aquatic plants in the bottomland lakes adjoining the river. "The combined loss of aquatic plants and bottom animals has drastically affected the numbers of diving ducks that use the Illinois during their migrations" (Mills et al., 1966:18).

Navigation on the Illinois River. The effect of the canals completed in 1848 and 1900 on the ecosystem of the Illinois River has been discussed above. The use of these canals for navigation has depended upon the development and maintenance of a navigable channel in the Illinois River for steamboats and, later, diesel-powered towboats. The Illinois Waterway (Mississippi River to Lake Michigan via the Illinois River) has evolved with the increase in human population and industry on the Illinois River basin, requiring economical and practical means of transporting various commodities (Figures 3 and 6).

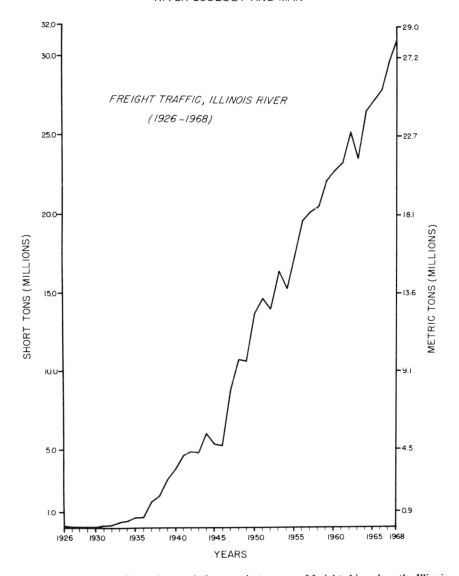

Figure 6. The graph reflects the steady increase in tonnage of freight shipped on the Illinois River and Des Plaines River between Grafton and Lockport after completion of the Illinois Waterway in the 1930's. (Corps of Engineers data furnished by H. Gordon Hanson, personal communications)

153

Since 1852 dredging (Barrows, 1910) of certain parts of the river has been required to help maintain a navigable channel. Through the years the channel benthic community (including naiads) probably has been affected by dredging. The dredge "spoils" from the Illinois are comprised largely of silt (soil eroded from the watershed) and sand and usually are pumped onto the floodplain of the river. During floods some of these "spoils" are resuspended and redeposited on the bottom of the river.

Between 1871 and 1899 four low dams and locks were erected in the river to help maintain a navigable channel (a fifth dam was built for hydroelectric power). "Because they were low, their greatest effect on the stream was during periods of low water" (Mills et al., 1966:55). During the 1930's the low dams were replaced by five higher dams and locks in the Illinois River to complete a 9-ft (2.7 m) channel connecting the Illinois with the Des Plaines River and the Chicago Sanitary and Ship Canal (Figures 1 and 2-B). In 1938 the lower 80.2 miles (129.1 km) of the Illinois were impounded by a lock and dam in the Mississippi River at Alton, Illinois. The importance for navigation of the present Illinois Waterway is reflected in Figure 6.

As mentioned earlier, most of the Illinois River was originally a sluggish stream (Figure 3). The present dams, together with the reduction in water diverted from Lake Michigan, have retarded the current so that during a period of low water levels the current in some stretches of the river is not perceptible. The current-loving Ozark minnow (*Dionda nubila*) has disappeared from the Illinois River since 1908 (Starrett and Smith, unpublished); however, factors other than the reduction of current by the dams may have been involved in the extirpation of this minnow.

The Illinois River was polluted and many drastic ecological changes already occurred before the present dams were built. The movement of polluted water is retarded upstream by dams, and the water is re-oxygenated when it passes over and through the dams (Mills et al., 1966; Starrett, 1971 and unpublished). In the absence of the dams, the low stream gradient and the nature of the pollutants in the river probably would disrupt the occurrence of the normally expected re-oxygenation of the stream and thereby increase the impact of pollution on aquatic life farther downstream.

Towboats have both chemical and mechanical effects on the ecology of the river. Sometimes oil slicks are produced on the river by a tow pumping out its bilges. In the lower river the action of a passing tow causes a large increase in the turbidity of the water and a temporary exposure of shallow bars (Starrett, 1971).

Agriculture on the Illinois River Basin. The silting problem in the navigable channel of the river is related in part to poor soil management practices on the basin and the dams.

In 1969 there were about 9.3 million acres (3.8 million ha) planted to row crops on the basin of the Illinois River[1] (Park et al., 1969; Moats, 1970; Walters et al., 1970). These row-crop plantings were: 5.7 million acres (2.3 million ha) of corn and 3.6 million acres (1.5 million ha) of soybeans (*ibid.*). The siltation problem began to worsen in the 1930's with the increase in the planting of soybeans (Figure 7) and the advent of heavy, powered farm machinery (Starrett, 1971).

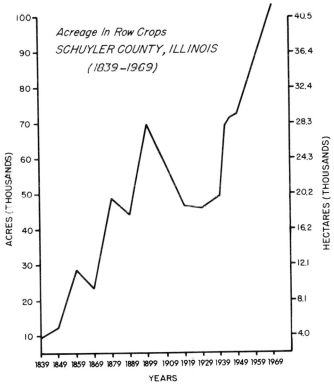

Figure 7. The row crops in Schuyler County (Figure 1) now are comprised of corn and soybeans. Before 1919 corn was the only row crop planted in the county. Soybeans were not used as a farm crop in the state of Illinois until 1914 (Ewing et al., 1949; Moats, Miller, and Zelazny, 1969; Moats, 1970)

[1] 89.5% of the row crops on the basin in 1969 were planted in the State of Illinois.

The annual sediment yield from a drainage area of 1,000 sq miles (2,590 km^2) in the upper part of the basin of the Illinois River was about 40 tons per sq mile (14 metric tons per km^2) (U.S. Department of Army, 1969). From the central portion of the basin the sediment yield was 650 tons per sq mile (228 metric tons per km^2) (*ibid.*). In one instance, a drainage area in the southern part of the basin yielded 8,000 tons per sq mile (2,802 metric tons per km^2) (*ibid.*). The distribution of particle-size sediment of the southern part of the basin averages 14% clay, 73% silt, and 13% sand (*ibid.*). As might be expected from this information, turbidities during low water level periods in the upper part of the river (15 to 47 Jackson Turbidity Units [JTU]) were lower than in the middle part of the river (15 to 140 JTU) and the lower section (36 to 320 JTU) (Starrett, 1971). In the eighteenth century Thomas Jefferson (1787) considered the Illinois a clear stream. The increase in turbidity which has occurred in the river during the past 200 years has probably favored the bottom-feeding fishes over those that feed by sight.

Sedimentation resulting from agricultural practices on the basin has had a tremendous impact on the ecology of the river's bottomland lakes during the past forty years. In Lake Chautauqua, a restored bottomland lake near Havana, sediment deposits reduced the storage capacity of this lake by 18.3% within 23.8 years (Stall and Melsted, 1951). Referring to the effects of sedimentation on Quiver Lake (Figure 1), Starrett and Fritz (1965:88) stated that:

> Today Quiver Lake is devoid of aquatic plants. The formerly deep basin of the lake has been filled in with 4- to 8-foot (1.2 to 2.4 m) deposits of silt. Turbid water at depths of over 3 feet (0.9 m) and a soft, flocculent bottom prevent the establishment of aquatic plants in the lake. Conditions in Quiver Lake are duplicated in many of the other floodplain lakes of the Illinois River; that is, in the past 35 years siltation has greatly changed the ecology of these lakes.

Since 1908 the pugnose shiner (*Notropis anogenus*), the blackchin shiner (*N. heterodon*), the blacknose shiner (*N. heterolepis*), and the weed shiner (*N. texanus*) have disappeared from the Illinois River and its bottomland lakes (Starrett and Smith, unpublished). It is our opinion that siltation (and the resulting turbidity and loss of aquatic plants) in the bottomland lakes was responsible for the extirpation of these four species of minnows.

Leveeing and Draining of Bottomlands

Modern man has not been content to limit his agricultural activities to the uplands of the Illinois River basin. Before the turn of the century he had begun to drain and levee off from the river its bottomland lakes and marshes for agricultural purposes (Mulvihill and Cornish, 1929). Originally, there were about

400,000 acres (161,878 ha) in the floodplain of the river between La Salle and Grafton (*ibid.*). Most of the draining was done between 1903 and 1920. Altogether 200,000 acres (80,939 ha) were drained, of which about 8,000 acres (3,238 ha) were abandoned and restored as lakes (Mills et al., 1966).

The draining of the lakes and marshes had an impact on the aquatic life of the Illinois River valley (Figure 8). These lakes and marshes provided spawning grounds for many fishes (including northern pike [*Esox lucius*], largemouth bass [*Micropterus salmoides*], and yellow perch [*Perca flavescens*] and feeding and resting areas for migratory waterfowl. The former marshes along the river furnished nesting sites for the least bittern (*Botaurus lentiginosus*), rails (*Rallidae*), long-billed marsh wren (*Telmatodytes palustris*), and red-winged blackbirds (*Agelaius phoeniceus*).

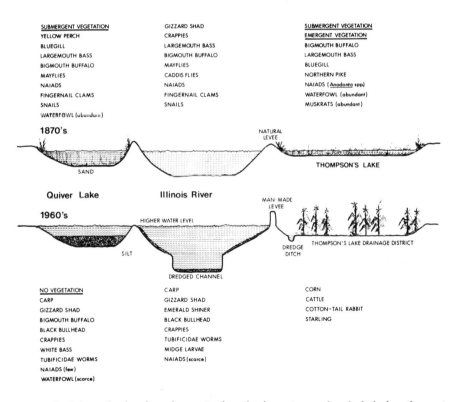

Figure 8. Schematic drawings demonstrating the impact man has had during the past century on the ecology of the Illinois River and two of its adjoining bottomland lakes near Havana (Figure 1)

157

In 1899, 546,616 lb (247,941 kg) of turtles (Testudines) were taken commercially from the Illinois River valley (Townsend, 1902). Today the annual catch amounts to only a few hundred pounds. Clark and Southall (1920) attributed the decline in the commercial take of turtles from the Illinois to the draining program. In the author's opinion sedimentation and the loss of vegetation have also contributed to the decline in the abundance of the commercial species of turtles (snapping turtle [*Chelydra serpentina*], and softshell turtle [*Trionyx* spp.) in the river's lakes. The painted turtle (*Chrysemys picta*) is now one of the most abundant turtles in the bottomland lakes, as evidenced by our wing net collections.[1]

The leveeing of almost one-half the floodplain of the Illinois River concentrated sedimentation and pollution in the river and its remaining lakes. Nevertheless, vegetation, fishes, waterfowl, and other forms of aquatic life would have been adversely affected by sedimentation and pollution even without the leveeing and draining projects.

Fertilizer and Insecticide Use

Modern agricultural practices include the application of fertilizers and insecticides on fields and crops. In 1960 Harmeson and Larson (1969) calculated the nitrate-nitrogen runoff from cultivated fields on a part of the watershed of the Sangamon River (tributary stream of the Illinois River) to be about 0.175 lb/A per day (0.031 kg/ha per day). The phosphorus input coming from upstream into the Peoria Lake section of the Illinois River in 1967 amounted to 2,050 lb (mean lb/sq mile of drainage area per year) or 2,408 kg (mean kg/km^2 of drainage area per year) (Sullivan and Hullinger, 1969). Part of the phosphorus and nitrate-nitrogen in the river originates from cultivated fields; however, the bulk of these fertilizers in the river is believed by the author to come from domestic and industrial wastes (Starrett, 1971). The effects that fertilizers from agricultural runoff have on the ecosystem of the Illinois River are not known.

The author estimates that in 1965 about 1,200 metric tons of soil insecticides (chlorinated hydrocarbons, aldrin, and heptachlor) and about 100 metric tons of foliar insecticides (almost entirely organic phosphates and carbamates) were

---

[1] Blanding's turtle (*Emydoidea blandingi*), the Illinois mud turtle (*Kinosternon flavescens*), the copper-bellied watersnake (*Natrix erythrogaster*), and the diamond-back watersnake (*Natrix rhombifera*) are now possibly extirpated from the Illinois River and its floodplain (Dr. Philip W. Smith, personal communication).

applied on the croplands of the basin of the Illinois River.[1] By 1970 organic phosphates amounted to about one-half of the insecticides applied on the upper part of the basin in Illinois; however, on the lower portion of the basin most of the soil insecticides used were chlorinated hydrocarbons (Dr. Roscoe Randell, personal communication). The average content of organochlorine insecticides in 14 naiads collected from the Illinois River proper in 1966 was 0.0331 ppm (Starrett, 1971). This relatively low amount of insecticides in naiads from the river proper indicates the possibility that a large proportion of the insecticides in the runoff from the croplands is being adsorbed in the bottom muds of the river's tributaries and bottomland lakes.

Water for Industrial and Domestic Uses. In 1959 industries (excluding power stations) on the basin of the river used 1.4 billion gallons (5.3 billion liters) of water per day from the Illinois River system (U.S. Public Health Service, 1963). The impact on the ecosystem of man's industrial use of river water is largely governed by the quality of the effluents, rather than by the volume of intake from the stream. The effects of wastes are discussed earlier under the heading of "Pollution."

The 1971 potential maximum summer usage of water from the Illinois River proper, including the mouth of the Kankakee River, by electric power stations (fossil fuel, hydroelectric and nuclear) is 2.798 million gallons (10.591 million liters) per minute (H. W. McFadden, W. T. Reid, R. R. Tanton, and R. W. Thieman, personal communications). Only 35,000 gallons (132,487 liters) per minute of this water are allocated for use by a hydroelectric plant located at Marseilles (Figure 3). Our studies have not indicated any direct effect of heated discharges on the biota of the polluted Illinois River. The upper section of the river (from the source to Ottawa) rarely is covered with ice. The average January temperature in this section of the river at Morris was 38°F (3.3°C), whereas downstream at Peoria it was 34°F (1.1°C) (Harmeson and Schnepper, 1965). The water is warmer in the upper section of the river than downstream because of the low stream flow and the heat loads now discharged into the waterway in the Chicago industrial area (*ibid.*). Possibly the increase in temperatures in the upper section of the river since 1900 has been partly responsible for the replace-

---

[1]Dr. Roscoe Randell (personal communication) furnished available data on the number of acres treated in 1965 with soil and foliar insecticides within the basin of the river in Illinois. These data included 80.4% of the counties treated with soil insecticides and 68.6% of the counties treated with foliar insecticides. He also furnished an estimate on the kinds of insecticides and rates of application. The author projected these data to include the entire basin of the river.

ment of the common shiner (*Notropis cornutus*) there by the striped shiner (*N. chrysocephalus*) (Starrett and Smith, unpublished). The common shiner is thought by Dr. Philip W. Smith (personal communication) to require cooler water than the striped shiner.

In 1969, 1.086 billion gallons (4.111 billion liters) of Illinois River water were used for domestic purposes (potable water) (Arnold K. Cherry and Frederick T. Ross, personal communications). Additional water is pumped into two pits in the Peoria area for the purpose of recharging ground water wells. Polluted water used for potable water naturally requires much more treatment than water taken from a relatively clean stream. In 1970 after a record flood upstream, the river water had a musty odor and flavor caused by actinomycetes, necessitating even further purification to make the water potable[1] (Arnold K. Cherry, personal communication). A rather similar case occurred in 1961 on the Cedar River, Iowa, after a tremendous flood (Cherry, 1962). There is a slight possibility that a relationship exists between major floods and the later occurrence in the stream of actinomycetes which affect the domestic use of the water. Intense rainfall on the basin producing a major flood might increase the runoff of fertilizers from the croplands enough to provide suitable conditions for the later development of the actinomycetes.

Recreation. Boating and water skiing are popular forms of recreation on the Illinois River. A few people swim in the river even though the water is polluted.

In spite of man's impact, the Illinois River and its remaining bottomland lakes are still the most important duck hunting and sport fishing waters in Illinois (Frank C. Bellrose, personal communication; Lopinot, 1967).

Bellrose (1947) calculated from band returns that hunters killed 165,678 migratory waterfowl in the Illinois River valley in the fall of 1943. This amounted to 48.9% of the waterfowl killed that year by hunters in the State of Illinois, including that part of the Mississippi River forming the western border of Illinois. Migratory waterfowl are a transitory community in the ecosystem of the Illinois River. They feed upon various organisms in the river's ecosystem and grains (particularly corn) in neighboring fields (Anderson, 1959). Paloumpis and Starrett (1960) estimated that waterfowl annually (1954 and 1955) deposited fertilizer materials in Lake Chautauqua (restored Illinois River bottomland lake)

---

[1]The threshold odor for potable water should be 5 or less. In 1969 the average threshold odor of the untreated river water was 52.3. In 1970 because of actinomycetes in the river water, its threshold odor rose to 500 (Arnold K. Cherry, personal communication).

on a per acre basis as follows: 12.8 lb (2.4 kg/ha) of nitrogen, 17.1 lb (3.1 kg/ha) of total phosphate, and 8.1 lb (1.5 kg/ha) of soluble phosphate. An increase in bacterial counts in Lake Chautauqua was believed to have been associated with a rise in the number of ducks using the lake (*ibid.*). In the author's opinion the transitory waterfowl community is an important part of the ecosystem of the Illinois River, and man's use of the river for killing waterfowl probably has some effect on its ecosystem.

Formerly, the Illinois River and its lakes produced excellent largemouth bass and yellow perch fishing. Today carp, freshwater drum (*Aplodinotus grunniens*), and black bullheads (*Ictalurus melas*) are the principal species of fishes taken by anglers from the river proper. In the bottomland lakes, where the bulk of the sport fishing occurs, the creel now is comprised mainly of white crappie (*Pomoxis annularis*), black crappie (*P. nigromaculatus*), bluegill (*Lepomis macrochirus*), yellow bass (*Morone mississippiensis*), white bass (*M. chrysops*), black bullhead, freshwater drum, and channel catfish. At Lake Chautauqua in 1954, 13,357 anglers caught 60,348 fish having an estimated weight of 36,442 lb (16,530 kg) (Starrett, 1957). Man's use of the river and its lakes for sport fishing probably has very little impact on its ecosystem (Starrett and Fritz, 1965).

Fishery. Carp were considered an excellent food fish in the homelands of many of the Europeans that settled in the New World. These immigrants clamored politically for the introduction of carp into the waters of this country, and their wishes were soon fulfilled; thousands of carp were stocked in the Illinois River in 1885 (Forbes and Richardson, 1908). Carp multiplied rapidly in the Illinois and in 1908 the commercial catch of this species from the river and its lakes was 15.5 million pounds (7.0 million kg), constituting 64.4% of the total catch (Forbes and Richardson, 1919). The carp has continued to be one of the most abundant fishes in the river; however, today little demand exists for it as a food fish in America (wholesale prices for Illinois River carp during the past twenty years have ranged from three to four cents a pound). Prior to the opening of the Chicago Sanitary and Ship Canal in 1900, Forbes and Richardson (1919) noted that as the carp increased in abundance, the yield of native fishes declined 22.2%. The adverse effects of carp on vegetation and the fish community have been discussed widely in American fishery and wildlife literature (Sigler, 1958; and others). In the author's opinion, man's introduction of the exotic carp into the Illinois River must be considered one of the most direct impacts he has had on the ecosystem of the river.

Forbes and Richardson (1919) believed that between 1899 and 1908 the Illinois River was being overfished. After that period the fishery declined because of pollution, the draining of many of the bottomland lakes, decrease in the number of new immigrants from Europe, and economic factors. In 1950 the commercial yield from the Illinois River was 5.6 million pounds (2.5 million kg), but in 1969 it was only 1.7 million pounds (0.8 million kg) (Starrett and Parr, 1951; Illinois Department of Conservation, unpublished). The present-day removal of commercial fishes from the river by man probably has only a minor effect on the fish community.

During the past hundred years 121 species of fishes have been collected from the Illinois River and its bottomland lakes (Starrett and Smith, unpublished). Between 1957 and 1970 we took 101 species, nine of which had not been taken in the period before 1908 (including two species not recognized in 1908 and three exotic species). Twenty species of fishes were extirpated from the river between 1908 and 1970 (Table 3). Several of the extirpated species were considered rare or accidental species in the river. On the other hand, many species of fishes common in the river before 1908, including the walleye (*Stizostedion vitreum vitreum*) and northern pike (*Esox lucius*), now are rare or limited in their distribution. The changes which have occurred in the fish fauna of the Illinois River reflect some of the drastic effects modern man has had on the ecology of the river.

Discussion

For thousands of years early man lived in the Illinois River valley and used the river for transportation and as a source of food, but he probably had little or no impact on the ecosystem of the river. However, within little over a hundred years the recent invader, modern man, with his culture, has had a tremendous impact on the ecosystem of the river and its floodplain (Figure 8). Early man was a natural part of the ecosystem of the river and its basin. Basically, modern man, too, is a part of this ecosystem, but his culture has severed his thinking, mode of living, and activities from the natural environment. This aloofness of modern man from his environment is mainly a result of such features of his culture as religion, science, technology, politics, and economics.

In the author's opinion economics and politics have played the important background role in man's cultural activities on the river and its basin. Some of these activities are even antipodal, such as man's domestic use of polluted water. Man's cultural activities are not a part of the ecosystem; nevertheless, they are the dominant influence on the ecosystem of the river. The direction this influence takes in the future largely depends upon man, rather than on the natural environment. However, man should consider the natural environment of the river and attempt to modify and adjust his cultural activities to make the Illinois River and its floodplain suitable for all of man's activities.

Table 3. Kinds of fishes extirpated from the Illinois River and its bottomland lakes between 1908 and 1970*

American brook lamprey (*Lampetra lamottei*)
Alligator gar (*Lepisosteus spatula*)
Cisco (*Coregonus artedii*)
Ozark minnow (*Dionda nubila*)
Pugnose shiner (*Notropis anogenus*)
Common shiner (*N. cornutus*)
Blackchin shiner (*N. heterodon*)
Blacknose shiner (*N. heterolepis*)
Rosyface shiner (*N. rubellus*)
Weed shiner (*N. texanus*)
Blacknose dace (*Rhinichthys atratulus*)
Creek chubsucker (*Erimyzon oblongus*)
Spotted sucker (*Minytrema melanops*)
River redhorse (*Moxostoma carinatum*)
Black redhorse (*M. duquesnei*)
Freckled madtom (*Noturus nocturnus*)
Longear sunfish (*Lepomis megalotis*)
Bantam sunfish (*L. symmetricus*)
Iowa darter (*Etheostoma exile*)
Fantail darter (*E. flabellare*)

*Starrett and Smith, unpublished.

Acknowledgments

I wish to express my appreciation to Dennis L. Dooley of the Illinois Natural History Survey for his assistance on the river and in the laboratory. I appreciate the helpful suggestions on the manuscript given to me by my colleagues with the Survey, namely: Frank C. Bellrose, George W. Bennett, R. Weldon Larimore, Philip W. Smith, George Sprugel, and Robert M. Zewadski. Richard M. Sheets, Technical Illustrator of the Survey, made the map and graphs used in this paper. The photographs were taken by Wilmer D. Zehr, Technical Photographer of the Survey.

Literature Cited

Anderson, H. G. 1959. Food habits of migratory ducks in Illinois. Ill. Nat. Hist. Surv. Bull. 27(4):289-344.

Baker, F. C. 1906. A catalogue of the mollusca of Illinois. Ill. Lab. Nat. Hist. Bull. 7(6):53-136.

Barrows, H. H. 1910. Geography of the middle Illinois valley. Ill. Geol. Surv. Bull. 15, 128 p.

Bartow, Edward. 1913. Chemical and biological survey of the waters of Illinois: Effect of Chicago sewage on the Illinois River. Univ. Ill. Bull. 10(36), Ill. Water Surv. Ser. 10:30-45.

Bauxar, J. J. 1959. The historic period. Ill. Archae. Bull. 1:40-58.

Bellrose, F. C., Jr. 1947. Analysis of methods used in determining game kill. J. Wildl. Manag. 11(2):105-119.

Boruff, C. S., and A. M. Buswell. 1929. Illinois River studies 1925-1928. Ill. Water Surv. Bull. 28, 127 p.

Buck, Solon J. 1967. Illinois in 1818. Second ed. University of Illinois Press, Urbana, Ill., 356 p.

Butts, T. A., D. H. Schnepper, and R. L. Evans. 1970. Dissolved oxygen resources and waste assimilative capacity of the La Grange Pool, Illinois River. Ill. Water Surv. Rep. Invest. 64, 28 p.

Calkins, W. W. 1874. The land & fresh water shells of La Salle County, Ill. Ottawa Acad. Nat. Sci. Proc., 48 p.

Cherry, A. K. 1962. Use of potassium permanganate in water treatment. J. Am. Water Works Assoc. 54(4):417-424.

Clark, H. W., and J. B. Southall. 1920. Fresh-water turtles: A source of meat supply. U.S. Bur. Fisheries Doc. 889:3-20.

Danglade, Ernest. 1914. The mussel resources of the Illinois River. U.S. Commer. Fisheries Rep. for 1913, App. 6, Bur. Fisheries, 48 p.

Ewing, J. A., J. H. Jacobson, B. R. Miller, W. G. Lee, and E. L. Park. 1949. Illinois agricultural statistics: State data through 1944. Ill. Coop. Crop. Rep. Serv., Ill. Dept. Agric. and U.S. Dept. Agric. Circ. 445, 164 p.

Flint, R. F. 1957. Glacial and Pleistocene geology. John Wiley & Sons, Inc., New York, 553 p.

Forbes, S. A. 1878. The food of Illinois fishes. Ill. Lab. Nat. Hist. Bull. 1(2):71-89.

——————, and R. E. Richardson. 1908. The fishes of Illinois. Illinois Nat. Hist. Surv., Urbana, Ill., cxxxvi + 357 p.

——————, and R. E. Richardson. 1913. Studies on the biology of the upper Illinois River. Ill. Lab. Nat. Hist. Bull. 9(10)·481-574

——————, and R. E. Richardson. 1919. Some recent changes in Illinois River biology. Ill. Nat. Hist. Surv. Bull. 13(6):139-156.

Fowler, M. L. 1959. The Early Woodland Period. Ill. Archae. Bull. 1:17-20

Greenfield, R. E. 1925. Comparison of chemical and bacteriological examinations made on the Illinois River during a season of low and a season of high water — 1923 — 1924. Ill. Water Surv. Bull. 20:9-33.

Harmeson, R. H., and T. E. Larson. 1969. Quality of surface water in Illinois. Ill. Water Surv. Bull. 54, 184 p.

——————, and V. M. Schnepper. 1965. Temperatures of surface waters in Illinois. Ill. Water Surv. Rep. Invest. 49, 45 p.

Hart, C. A. 1895. On the entomology of the Illinois River and adjacent waters. Ill. Lab. Nat. Hist. Bull. 4(6):149-273.

Hoskins, J. K., C. C. Ruchhoft, and L. G. Williams. 1927. A study of the pollution and natural purification of the Illinois River. I. Surveys and laboratory studies. U.S. Publ. Health Serv. Bull. 171, 208 p.

Illinois State Sanitary Water Board. 1967. Illinois water quality data 1966 with data analysis. Ill. State Sanitary Water Bd., 109 p.

Jefferson, Thomas. 1787. Notes on the state of Virginia. J. Stockdale, London, 13 p.

Kofoid, C. A. 1903. Plankton studies. IV. The plankton of the Illinois River, 1894-1899, with introductory notes upon the hydrography of the Illinois River and its basin. Part I. Quantitative investigations and general results. Ill. Lab. Nat. Hist. Bull. 6(2):95-635.

Lopinot, A. C. 1967. The Illinois angler: A survey of 1965 Illinois licensed fishermen. Ill. Dept. Conserv., Div. Fisheries, 81 p.

Malloch, J. R. 1915. The Chironomidae, or midges of Illinois, with particular reference to the species occurring in the Illinois River. Ill. Lab. Nat. Hist. Bull. 10(6):275-543.

Maxwell, M. S. 1959. The Late Woodland Period. Ill. Archae. Bull. 1:27-32.

McGregor, J. C. 1959. The Middle Woodland Period. Ill. Archae. Bull. 1:21-26.

Mills, H. B., W. C. Starrett, and F. C. Bellrose. 1966. Man's effect on the fish and wildlife of the Illinois River. Ill. Nat. Hist. Surv. Biol. Notes 57, 24 p.

Moats, R. H. 1970. Illinois agricultural statistics: Annual summary 1970. Ill. Coop. Crop Rep. Serv., Ill. Dept. Agric. and U.S. Dept. Agric. Bull. 70-1, 99 p.

—————————, B. R. Miller, and K. M. Zelazny. 1969. Illinois county agricultural statistics Schuyler county. Ill. Coop. Crop Rep. Serv., Ill. Dept. Agric. and U.S. Dept. Agric. Bull. C-32, 39 p.

Mulvihill, W. F., and L. D. Cornish. 1929. Flood control report: An engineering study of the flood situation in the state of Illinois. Illinois Division of Waterways, Springfield, Ill., 402 p.

Munson, Patrick, and J. C. Frye. 1965. Artifact from deposits of Mid-Wisconsin Age in Illinois. Science 150(3704):1722-1723.

Paloumpis, A. A., and W. C. Starrett. 1960. An ecological study of benthic organisms in three Illinois River flood plain lakes. Am. Midland Nat. 64(2): 406-435.

Park, E. L., J. R. Garrett, H. L. Castle, W. W. Adams, D. E. Wilgenbusch, P. D. Hopkins, G. F. Smith, and J. K. Sands. 1969. Indiana crops and livestock:

Annual crop summary 1969. U.S. Dept. Agric. Statist. Rep. Serv. and Purdue Univ. Agric. Exp. Sta. No. 536, 43 p.

Parmalee, P. W. 1962. Additional faunal records from the Kingston Lake Site, Illinois. Trans. Ill. Acad. Sci. 55(1):6-12.

Purdy, W. C. 1930. A study of the pollution and natural purification of the Illinois River. II. The plankton and related organisms. U.S. Publ. Health Serv. Bull. 198, 212 p.

Richardson, R. E. 1921a. The small bottom and shore fauna of the middle and lower Illinois River and its connecting lakes, Chillicothe to Grafton: its valuation; its source of food supply; and its relation to the fishery. Ill. Nat. Hist. Surv. Bull. 13(15):363-522.

_____. 1921b. Changes in the bottom and shore fauna of the middle Illinois River and its connecting lakes since 1913-1915 as a result of the increase, southward, of sewage pollution. Ill. Nat. Hist. Surv. Bull 14(4):33-75.

_____. 1925a. Changes in the small bottom fauna of Peoria Lake, 1920 to 1922. Ill. Nat. Hist. Surv. Bull. 15(5):327-388.

_____. 1925b. Illinois River bottom fauna in 1923. Ill. Nat. Hist. Surv. Bull. 15(6):391-422.

_____. 1928. The bottom fauna of the middle Illinois River, 1913-1925. Its distribution, abundance, valuation, and index value in the study of stream pollution. Ill. Nat. Hist. Surv. Bull. 17(12):387-475.

Sauer, C. O. 1916. Geography of the upper Illinois valley and history of development. Ill. Geol. Surv. Bull. 27, 208 p.

Sigler, W. F. 1958. The ecology and use of carp in Utah. Utah State Univ. Agric. Exp. Sta. Bull. 405, 63 p.

Stall, J. B., and S. W. Melsted. 1951. The silting of Lake Chautauqua, Havana, Illinois. Ill. Water Surv. Rep. Invest. 8, 15 p.

Starrett, W. C. 1957. Fishery values of a restored Illinois River bottomland lake. Trans. Ill. Acad. Sci. 50:41-48.

_____. 1971. A survey of the mussels (Unionacea) of the Illinois River:

A polluted stream. Ill. Nat. Hist. Surv. Bull. 30(5):267-403.

——————, and A. W. Fritz. 1965. A biological investigation of the fishes of Lake Chautauqua, Illinois. Ill. Nat. Hist. Surv. Bull. 29(1):1-104.

——————, and S. A. Parr. 1951. Commercial fisheries of Illinois Rivers: A statistical report for 1950. Ill. Nat. Hist. Surv. Biol. Notes 25, 35 p.

Sullivan, W. T., and D. L. Hullinger. 1969. Phosphates in Peoria Lake—a quantitative and qualitative evaluation of a nutrient in natural waters. Trans. Ill. Acad. Sci. 62(2):198-217.

Thompson, D. H. 1928. The "knothead" carp of the Illinois River. Ill. Nat. Hist. Surv. Bull. 17(8):285-320.

Townsend, C. H. 1902. Statistics of the fisheries of the Mississippi River and tributaries. U.S. Comm. Fish and Fisheries. Commissioner's Rep. for 1901, 659-740.

U.S. Department of Agriculture. 1941. Climate and man. 1941 Yearbook of Agric., 77th Congress, 1st Session, House Doc. 27, 1248 p.

U.S. Department of Army. 1969. Upper Mississippi River comprehensive basin study. Draft No. 2 of Appendix G. Fluvial Sediment in the Upper Mississippi River Basin, Corps of Engineers, Prepared by Interagency Task Force on Sedimentation, 139 p.

U.S. Department of Commerce. 1970 *a*. 1970 census of population: Illinois. Bur. Census, Preliminary Rep. PC (P1)-15, 6 p.

——————. 1970*b*. 1970 census of population: Indiana. Bur. Census, Preliminary Rep. PC (P1)-16, 4 p.

——————. 1970*c*. 1970 census of population: Wisconsin. Bur. Census, Preliminary Rep. PC (P1)-51, 4 p.

U.S. Department of Interior. 1901. United States twelfth census 1900. Census Office, Vol. 1 Population Pt. 1, 1901, Washington. ccxxix + 1006 p.

——————. 1970. 1969 water resources data for Illinois. Part 1. Surface water records. Part 2. Water Quality records. U.S. Geol. Surv., 249 p.

U.S. Public Health Service. 1963. Report on the Illinois River system, water quality conditions: Part I, Text. U.S. Dept. Health, Educ., and Welfare, Div. of Water Supply and Pollution Control, Great Lakes—Illinois River Basins Proj., 155 p.

U.S. Senate. 1957. Effects of an additional diversion of water from Lake Michigan at Chicago. 85th Congress, 1st Session, Doc. 28, 74 p.

Walker, Ward. 1960. The story of the metropolitan sanitary district of greater Chicago. The seventh wonder of America. Metropolitan Sanitary District of Greater Chicago, 39 p.

Walters, H. M., C. W. LeGrande, J. B. Goodwin, L. E. Krahn, and R. J. Ries, 1970. 1970 Wisconsin agricultural statistics. Wisc. Stat. Rep. Serv., Wisc. Dept. Agric., and U.S. Dept. Agric. Stat. Rep. Serv. 70 p.

Williams, L. G. 1964. Possible relationships between plankton-diatom species numbers and water-quality estimates. Ecology 45(4):809-823.

Willman, H. B., H. D. Glass, and J. C. Frye. 1963. Mineralogy of glacial tills and their weathering profiles in Illinois. Part I. Glacial tills. Ill. Geol. Surv. Circ. 347, 55 p.

Winters, H. D. 1959. The Paleo-Indian Period. Ill. Archae. Bull. 1:5-8.

THE NILE RIVER — A CASE HISTORY

D. Hammerton

Of the great rivers of the world, the Nile has, without question, played the most significant part in the early development of mankind. Ever since Menes, first King of Egypt, dammed the river over 5000 years ago in order to build the city of Memphis (Herodotus, ca. 440 B.C.) his successors have sought to control the river and utilise the annual flood. This year, with the completion of Nasser's High Dam at Aswan, this age-old dream has been virtually achieved.

The need to harness the waters of the Nile for the common good was perhaps one reason why Egypt devised the first system of central government in history, and, although the ancient Egyptians are more famous for their enduring pyramids and temples, perhaps their greatest achievements were in the field of irrigation engineering. Recently the Nasser government re-erected a statue of Om Hotep at Aswan as a reminder that he was the first Pharaoh to propose the use of Lake Karoun in the Fayoum Oasis for storage of flood water to irrigate the surrounding land. It is thought that Lake Karoun "regulated" the flood by reducing its peak and subsequently delaying its fall—thus being the forerunner of the present Nile reservoir system.

Not only was irrigation on a large scale practised from the times of the earliest Pharaohs, but the Nile was also the main artery of communication. Boats have been recorded from pre-Dynastic Egypt in Amratian times (4000-3500 B.C.) and were prominent from then onwards, being also used for all the main invasions southward into the land of Kush. So important was the river for this purpose that in 2330 B.C. a channel 78.5 x 10.5 x 6.35 m was cut to enable boats to pass through the first cataract. At the second cataract slipways were constructed to allow the haulage of boats around the obstacle, and there is some evidence that partial dams were used to raise the water level at cataracts to aid in navigation (Vercoutter, 1966).

Although the Nile valley is a natural highway into Africa, all expeditions to discover its source from Pharaonic times until the middle of last century met with failure. Nero, the Roman Emperor, sent a big expedition which eventually returned with the news that their way had been blocked by an impassable swamp, indicating that they may have reached the Sudd—reputedly the largest swamp in the world. This proved an impassable barrier until Baker, with a fleet of steamers and over two thousand men, forced his way through, at a great cost in lives, in 1871. His journey was commenced seven years after the location of the source of the Nile at the outlet of Lake Victoria Nyanza by Speke in 1862, who approached it after years of searching from the east coast of Africa. However, it was not until 1937 that Waldecker located the southernmost source of the Nile in the headstreams of the Kagera, largest tributary of Lake Victoria. This discovery established the Nile as the longest river in the world with a length of 6,695 km (4,160 miles).

The Blue Nile, flowing 1,600 km from Lake Tana in the Ethiopian plateau to Khartoum, provides the major part of the annual flow of the main Nile and was not fully traversed until a British military and scientific expedition conquered the notorious Blue Nile Gorge during the flood of 1968 (Blashford-Snell, 1970).

During the present century the river has been progressively brought under control and is now the only major river to be fully controlled and utilised. In the absence of any serious sources of pollution the main ecological impact of man has been seen in the effect of the series of dams on the hydrological regime and hence on the ecology of the river. Some of these effects have been considerable, as will be shown in this paper.

Physical and Geological Features

The Nile is a dominant feature of the northeast quarter of the African continent, and its basin of 2,978,000 km^2 occupies nearly one-tenth of the land surface. It is unique among rivers of the world, in that it flows due north from its source 400 km south of the equator through 35° of latitude, thus traversing several climatic zones, and it is the only tropical river to reach the Mediterranean.

The main physical features are evident from the maps (Figures 1 and 2) and only a brief description need be given here. The White Nile, comprising the Victoria Nile, Albert Nile, Bahr El Jebel, and Bahr El Abiad, has its southernmost source in Burundi at the head of the Luvironza River about 2,000 m above sea level and 6,695 km from the Mediterranean. This river joins the Kagera, principal tributary of Lake Victoria, which, with an area of approximately 69,000 km^2, is the second largest lake in the world. Occupying a down-warped depression between the great Rift Valleys, it receives drainage mainly through swamp-choked valleys on all sides but the northwest. Owen Falls Dam, on the Victoria Nile just below the lake outlet, has made Lake Victoria into the largest reservoir in the world. After a series of rapids the Victoria Nile flows through the swampy Lake Kioga before descending the western Rift in a series of cataracts and the spectac-

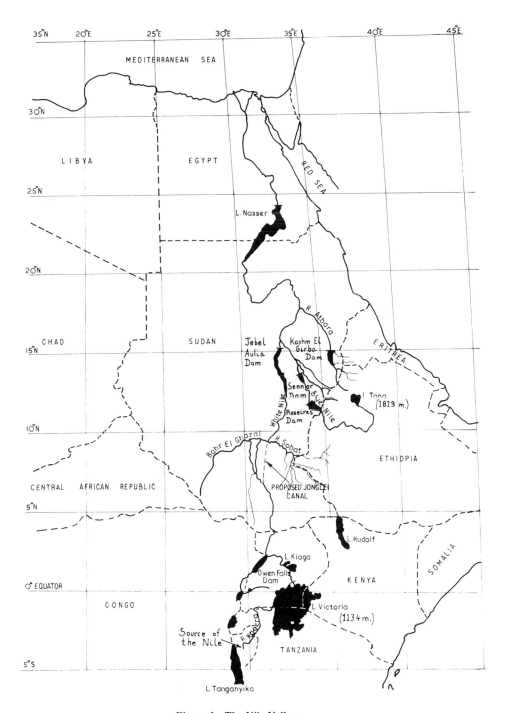

Figure 1. The Nile Valley

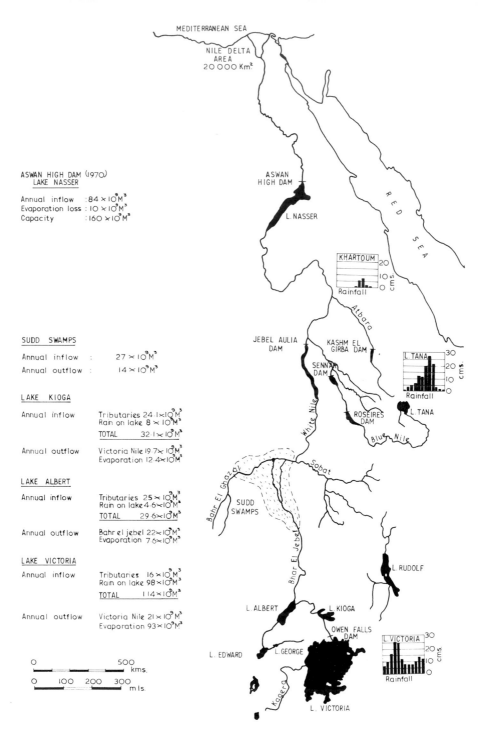

Figure 2. The River Nile — rainfall and evaporation

ular Murchison Falls to Lake Albert, which it almost immediately leaves to flow through swamps again to the Sudan border at Nimule. The river then drops again through a series of cataracts over the next 240 km to the flat plains of the Sudan. From here it is navigable by steamer for 1,600 km to Khartoum, passing first through the vast Sudd swamps.

At Khartoum the White Nile is joined by the Blue Nile which rises 1,600 km distant in Lake Tana at a height of 1,829 m. From here it flows some 800 km through the famous and, until recently, impassable Blue Nile Gorge until it emerges on the Sudan plains at Deim 740 km from Khartoum.

From Khartoum to the Mediterranean, a distance of 3,000 km, the joint river receives no perennial tributary, the longest river section in the world of which this can be said. The Atbara, which joins the Nile 325 km north of Khartoum, is dry for six months of the year but nevertheless contributes one-seventh of the annual flow.

Physical History. The physical history of the Nile, especially in the Sudan, is little known and for a long time was discussed in terms of the better known Egyptian chronology.

Berry and Whiteman (1968) have shown that this approach has led to a number of erroneous conclusions, such as the "Lake Sudd Hypothesis" (Lawson, 1927; Ball, 1939) and the idea that the Nile is a very young river (Heinzelen and Paepe, 1964). They have put forward considerable regional evidence from the middle and upper Nile systems in the Sudan to show that the Nile is a very old river. The Atbara, Blue Nile, and Sobat systems have been in existence since Early Tertiary times and the White Nile system (excluding the Albert and Uganda Niles) may be just as old.

Berry and Whiteman consider that the rock sill at Sabaloka, the "Sixth Cataract," acted as a base level of erosion for much of Tertiary and Quaternary time in central and southern Sudan. South of this sill, as far as Shambe on the White Nile and as far as Roseires on the Blue Nile, the rivers flow mainly in their alluvial deposits, whereas to the north alluvium is more restricted, and the Nile flows often directly on the rocks of the Nubian Sandstone and the Basement Complex. Although little is known about the geology and geomorphology of the great "S" bend of the Nile, they suggest that it may well have been formed by gentle updoming of the Basement and Nubian formations.

The connection with Lake Albert and Lake Victoria Nile did not take place until Late Pleistocene times, but it is of interest to note that the Nile was probably connected to a Miocene lake on the site of the present Lake Victoria. This lake,

175

named Karunga by Wayland (1931), later disappeared and the present Lake Victoria did not come into existence until Late Pliocene or Early Pleistocene times.

With regard to the Blue Nile, Berry and Whiteman have found evidence that this is an ancient river, contrary to the views of Arkell (1949a, b) and others. Formed at least as early as the Late Cretaceous-Late Eocene, its upper courses were later covered by vast quantities of lava extruded during the Oligocene in Ethiopia, and the present river system must have originated on this "vast volcanic pile." Lake Tana is supposed to have appeared in the Pliocene (Dainelli, 1943; Mohr, 1962) while the huge Blue Nile gorge was excavated in Pliocene and Pleistocene times.

By using new isotopic dates for sediments from the Wadi Halfa, Khartoum, and Gezira areas, Berry and Whiteman have provided a tentative denudation chronology for the Nile in the Sudan and new geomorphological information for the Nile as a whole.

Hydrology

Hydrologically, the Nile is the best known river in the world, and a vast amount of data has been accumulated, especially in the present century. Hurst, (1957) states that "there is no other great river in the world upon which such an accurate and extensive system of measurements is carried out" and mentions further that in Egypt records of river level extend almost continuously as far back as 660 A.D. and, with large gaps, to much earlier periods. Nileometers, for recording the height of the annual flood, existed in many parts of Ancient Egypt, the most famous—though probably not the oldest—being the Roda Nileometer near Cairo and that on Elephantine Island near Aswan. So important was the flood to the adequate production of crops that the agricultural taxes were calculated according to the height of the flood and waived if the nileometer reading was below a certain level.

The drainage area of the Nile basin was given in the preceding section as $2,978,000$ km^2, but this is undoubtedly misleading. Lebon (1965) has shown that, although maps have been drawn showing the watershed between the Nile and Congo - Lake Chad basins, the existence of such a watershed is meaningless because the Nile basin is impossible to delimit in much of the Sudan since drainage channels to the Nile do not exist. Even runoff from the Jebel Marra and Nuba mountains rarely, if ever, reaches the river, while, because of the vast swamps of the southern Sudan, virtually none of the heavy rainfall in that area (up to 1000 mm/year) reaches the river—indeed half the flow of the White Nile is lost in the Sudd.

176

The flow of the Nile is therefore derived almost entirely from two main sources, the Ethopian plateau and the East African Lakes plateau.

The well-known features of the flow of the Nile are best understood by reference to the hydrographs shown in Figure 3. From these it can be seen that the regime of the two main rivers is totally different. The White Nile provides a very steady contribution which for six months of the year is about three times that of the Blue Nile. The latter river, however, rises to a peak flow 40-70 times as great as the minimum flow and, together with the Atbara, is the source of the annual Nile Flood in Egypt.

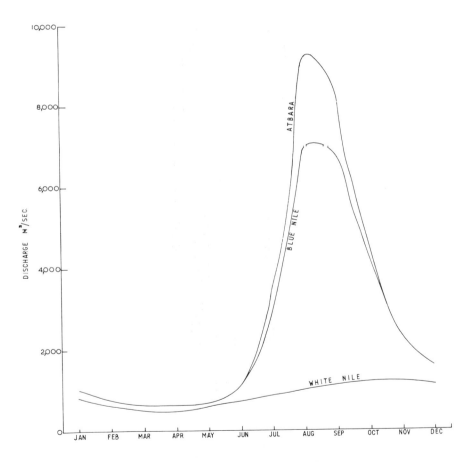

Figure 3. Hydrograph of the River Nile

177

The Blue Nile and Atbara systems derive their water from the deeply dissected Ethiopian plateau where rapid runoff and a high rate of erosion coincide with a region of high and markedly seasonal rainfall (see rainfall, Figure 2). Because of the high rate of erosion, the Blue Nile carries about 100-130 million tonnes of silt in the flood season, and the Atbara carries a proportionately smaller amount. It is this silt, transported down to Egypt and deposited in the floodplain and delta for thousands of years, that has made the lower Nile Valley one of the richest agricultural areas known. However, since the construction of the Aswan High Dam in Egypt and the Roseires Dam on the Blue Nile in the Sudan, the whole of this vast load is being deposited annually in the two reservoirs.

If we examine the catchment of the White Nile, several features are found which tend to smooth out seasonal variations in flow of the river. Firstly, the rainfall over the Lake Victoria basin is much more evenly distributed than in the Blue Nile basin, with rain in every month of the year and a double peak period. Most of this, in fact, is lost in the swampy catchment by evaporation and transpiration, and the biggest single source of inflow to the lake is rain on the lake surface. Evaporation over the year is almost equal to the precipitation, so that outflow through the Victoria Nile roughly equals the inflow from tributaries. The large surface area of the lake ensures a fairly even outflow, which is further smoothed out in Lake Kioga. Further downstream Lake Albert provides a further control of fluctuations in flow of the Nile. The discharge of the Bahr El Jebel has a pronounced autumn rise due to the contribution of a series of torrents near Nimule, while the contribution from Lake Albert is practically constant throughout the year.

Shortly after Mongalla the Bahr El Jebel enters the Sudd swamps, where half the annual flow of the river is lost. The swamps of the Sudd have a remarkable regulating effect because the river banks are above the level of the surrounding swamps. Any rise in level of the river upstream causes water to spill over the banks into the swamps, very little of this rise being felt at the downstream end of the swamps. Because the swamps are mainly below river level, very little water can drain back into the river.

Referring now to the White Nile at Malakal, seasonal variations in the flow of the river are almost entirely due to the Sobat River, which, like the Blue Nile and Atbara, derives most of its water from the Ethiopian plateau.

To summarise the foregoing, out of an annual discharge of 84 billion m^3 in the main Nile entering Egypt, 48 milliards come from the Blue Nile, 24 milliards from the White Nile and 12 milliards from the Atbara. This does not take into account normal transmission losses due to evaporation and recharge of aquifers through which the river flows. Taking into account the Sobat, it can be concluded that 84% of the discharge of the Nile comes from Ethiopia and only 16% from the African Lake Plateau.

For a full account of the Nile basin in all its aspects, the reader is referred to the monumental work of Dr. H. E. Hurst and his colleagues of the Egyptian Physical Service (now Nile Control), The Nile Basin, in eight volumes. This includes an exhaustive account of the physical and hydrological characteristics of all the tributary catchments and much information and results of research into the measurement of discharge and methods of computation.

Reference must be made here to a classic example of major research stimulated by river use policies. At the request of the Egyptian Government, Hurst and his colleagues spent twelve years assessing the storage capacity required to extend Egyptian agriculture to the limit by giving the country all the water needed every year, irrespective of variations in the annual river supply.

Two lines of approach were used (Hurst, 1957):

One was the experimental and statistical one of collecting the discharges of a number of rivers and calculating the storage capacity required to equalize the discharge in each case. The other was a mathematical approach through the theory of probability, assuming the variability of the discharge of a river is similar to chance events. The first led to an enormous amount of computation, since to deal statistically with a question one must have a mass of material. For this purpose, records extending over many years were required, and these are not numerous in the domain of rivers, so work was extended to rainfall statistics which appeared to be similar to their characteristics. Then followed the inclusion of temperatures and barometric pressures. The common characteristic of all these is that they have similar frequency distributions . . . The use of temperatures or any other statistics merely means that we go through the same arithmetical process as we should if they represented quantities of water, and we arrive at a figure which, in the case of water, would be required storage capacity . . . The longest series of river or rainfall figures, excluding those of the Roda Nileometer, do not reach 200 years, and the same is true of temperature and pressures. In order to get a longer series of figures, it was necessary to make use of the work of Dr. A. E. Douglass in measuring the rings of giant trees of America, and also of measurements made by Baron de Greer, of the thickness of varves. These are the annual layers of mud deposited on ancient lake beds. The tree rings extend to nearly 1,000 years and the varves in some cases to 4,000 years. In all, seventy-five different phenomena were analysed and 690 computations made of storage capacity. This very

179

large amount of work, however, yielded the required result for practical application.

The mathematical work was undertaken because, for some purposes, river discharges can be treated statistically as if they were chance events, which they resemble in some respects. Assuming that this is the case, it is theoretically possible to find a solution, and, with some difficulty, a solution was found which was confirmed by experiments with chance events. However, when this solution was compared with the storage capacities calculated from observations of natural events, it was clear that the storage required in the case of river discharges, rainfall, etc., was greater than that which would fit pure chance events. This is due to the tendency . . . of natural events to group themselves in runs in which high or low values preponderate.

The theoretical solution, although it was not useful in the way expected, did show clearly the difference between natural and chance events and emphasized that the old idea that thirty or forty years' records of rainfall or river discharges are enough for all purposes, is erroneous.

This work was the scientific starting point for a scheme for controlling the Nile as a whole, and as a result, Hurst, Black, and Simaika (1946) developed the idea of "century storage." They proposed that Lake Victoria and Lake Tana should be used as major reservoirs while Lakes Kioga and Albert would be used as regulating reservoirs. An essential part of this project was the Jonglei Diversion Canal (Figure 1) to by-pass the Sudd swamps which otherwise would absorb any additional water sent down from the lakes. A flood protection dam was proposed at the fourth cataract to protect Egypt from the serious dangers of a very high flood.

However, of these proposals, only the Owen Falls Dam at Lake Victoria was built, and Egypt could not afford to wait for implementation of the other proposals. In 1952 Hurst and his colleagues were asked to examine the proposal by Daninos to build a high dam at Aswan which would provide long-term storage and sufficient hydroelectric power for large-scale industrial use. They reported in favour of the dam, provided sufficient capacity could be found, and proposed a combination of the High Aswan Dam and the Equatorial Projects.

A suitable site for the dam was eventually found just upstream of the existing dam, and the dam, now completed, has a capacity of 164 billion m^3, nearly twice the annual flow of the river.

The completion of the Aswan High Dam, opened by President Podgorny of the Soviet Union and President Sadat of Egypt on January 15, 1971, has finally enabled the total utilisation of the waters of the Nile by Egypt and the Sudan. For several years, in fact, no water has entered the Mediterranean from the river, and only in the event of an exceptional flood is it likely that Nile water will again enter the sea (Halim, et al., 1967).

Any further increase in water available for irrigation now depends on construction of the proposed Jonglei Canal to by-pass the Sudd swamps in the south of the Sudan, where fifty percent of the annual flow of the White Nile is lost. This scheme is now again under active investigation.

Physico-Chemical Conditions

The major ionic composition of Nile waters at various stations is shown in Table 1:

Table 1. Chemical composition of Nile waters: results in mg/l unless otherwise stated. (Lake Victoria results from Talling and Talling, 1965)

	Lake Victoria (1952-1961)	Lake Tana (8/12/65 – 18/4/66)	White Nile (Khartoum) (1965-1967)	Blue Nile (Khartoum) (1965-1967)
Conductivity K_{20}, umho&cm	91-98	200-240	220-500	140-390
pH	7.1 - 8.5	8.1 - 8.4	8.0 - 8.9	8.2 - 9.1
Alkalinity meg/litre	0.9 - 1.1	1.52- 1.92	2.3 - 3.3	1.63- 2.66
Na^+	10 -13.5	7.2 - 9.4	25 -41	4.5 - 9.0
K^+	3.7 - 4.0	1.1 - 1.8	6.9 - 9.2	1.7 - 2.9
Ca^{++}	5 -11.8	14 -15.2	14.2 -19.0	19.6 -28.1
Mg^{++}	2.3 - 3.5	6.1 - 8.5	7.5 -11.7	5.0 - 6.5
Cl^-	3.9 - 5.0	1.0 - 2.5	6.0 - 9.5	2.0 - 7.3
SiO_2	3.0 - 9.0	18 -20	18 -28	16 -24
PO_4 (as P, µg/l)	13 -140	15 -20	5 -100	2 -120
No_3 (as N, µg/l)	5.5 -11.0	0 -80	10 -90	1 -100

Both main tributaries have fairly low conductivities, indicating a moderate amount of matter in solution, but there are considerable differences in ionic composition, first described by Beam (1906, 1908). Among the anions, bicarbonate and carbonate are predominant, as commonly found in African waters (Talling and Talling, 1965), and similar quantities are found in both rivers. Chlorides are extremely low, but sulphate is much lower in the White than the Blue Nile due to removal in the Sudd swamp (Beam, 1906; Talling, 1957).

The biggest differences between the two rivers are in the ratio of divalent to monovalent cations, high in the Blue Nile and low in the White Nile. Calcium predominates in the former and sodium in the latter.

Lake Tana has a lower calcium content but otherwise is similar to the Blue Nile in the Sudan. Lake Victoria has a somewhat low ionic content, perhaps surprising in view of the high rate of evaporation, but, on the other hand, most of the "inflow" is from rain on the lake surface. Here the low calcium and magnesium content is at or below that which is limiting to many molluscs in temperate waters, but appears to have no such effect here.

The longitudinal changes in chemical composition due to various influences in the White Nile from Lake Victoria to Khartoum have been described in detail by Talling (1957) and Bishai (1962). Seasonal changes at Khartoum in major ionic composition are not great and are mainly due to changes in the proportions of different tributaries. Slightly greater seasonal changes occur in the Blue Nile which are inversely correlated with the volume of discharge (Figure 4). Seasonal changes in nutrients in both rivers have been studied by various workers and are considered in a later section.

Fauna and Flora

The Fish Fauna in Relation to Evolution of the Nile Drainage. The fish fauna of the Nile basin is one of the richest and most diversified of any river in the world, with at least 54 genera and probably well over 300 species. This fauna is notable for its archaic elements alongside genera which are without parallel for their explosive speciation in recent geological times. The origin of this fauna is virtually unknown, though some Nilotic species are thought to have come from Asia (Hora, 1937; Menon, 1951). Fryer suggests (personal communication) that it would be interesting to make a careful analysis of the Nile fauna to see how many affinities can be detected.

There seems to be general acceptance of the view of Worthington (1954) that the major tropical rivers of Africa were interconnected in pre-rift times in the

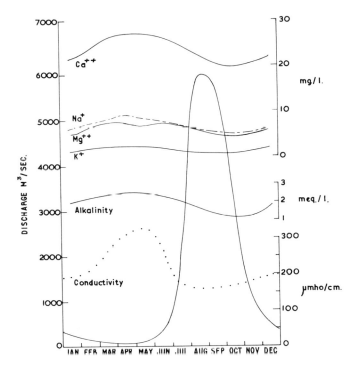

Figure 4. Relationship between water chemistry and river flow in the Blue Nile

region of their headstreams in central Africa. In Miocene times they supported a uniform fauna comprising genera whose descendants are to be found in the major rivers today. This fauna, termed "nilotic" by Beadle (1962), is characterised by the following genera: *Lates* (Nile perch), *Hydrocyon* (tiger fish), *Polypterus*, (bichir), *Distichodus*, and *Citharinus*.

After the Miocene the original drainage was disrupted by gigantic earth movements and volcanic activity which produced the Rift Valleys and their lakes. Lake Victoria appeared later (on the site of the earlier Lake Karunga of Wayland, 1931) by up-warping along its western side, bringing about a reversal of drainage in this area (Doornkamp and Temple, 1966). The lake faunas have been shown on geological grounds to have been derived from the pre-existing rivers, and it is of great interest to find that this conclusion is supported by ecological studies on present-day fish (Beadle, 1962; Corbet, 1961). A progressive adaptation was found from a fluviatile to a lacustrine life with an intermediate stage where fish feed in the lake but return to the rivers to breed.

183

A comparison of the fauna of the river with those of the Nile lakes provides some very striking contrasts, as shown in Table 2, from Fryer and Illes (1971), to which has been added data for the Nile of Egypt and Sudan taken from Sandon (1950):

Table 2. Distribution of fish species in the Nile and principal lakes. (After Fryer and Iles [1971] with data added from Sandon [1950])*

	Lake Victoria	Lakes Edward and George	Lake Albert	Lower Nile River
Non-cichlids				
Total Genera	20	10	21 + 2 ?	45
Endemic Genera	1	Nil	Nil	Nil
Total Species	38 + 1 ?	19	36 + 2 ?	98
Endemic Species	17	2	3 + 1 ?	Nil
Cichlids				
Total Genera	8	5	3	4
Endemic Genera	4	1	0	0
Total Species	170 +	28 +	9 + 1 ?	8
Endemic Species	164	19 +	4	0

*According to Dr. P. H. Greenwood (personal communication), recent work indicates the existence of over 200 species of cichlids in Lake Victoria and over 60 species in Lakes Edward and George, with over 99% endemism in all these lakes.

It should be noted that Lakes Victoria and Kioga are separated from the lower Nile by the Murchison Falls, which are impassable to fish. Lake Albert is in open communication with the river, but Lakes Edward and George are separated from Lake Albert by the Semliki Falls.

Fryer (1965) states that Lake Victoria is the second richest in species in the world, exceeded in this respect only by Lake Nyasa, while Lake Tanganyika is a close third.

The endemic species in these and other African lakes have clearly evolved within the lake basins, and this fantastic degree of speciation among the cichlid fish in recent geological times has led to fierce arguments concerning the role of predators. Because large predators, such as Lates (the Nile perch) and *Hydrocyon* (the tiger fish), were absent from lakes such as Victoria, Worthington (1937) suggested that the presence of such predators restricted speciation, a view which was later supported by Lowe-McConnell (1959) and Jackson (1965). However, this idea has been convincingly refuted by Fryer (1960*a, b,* 1965, and 1968) and Greenwood (1965), who consider that predation has accelerated, not retarded, speciation. Lowe-McConnell (1969) now appears to have changed her view, at least in part.

Lake Victoria has been quoted as a classic example of intralacustrine speciation, but Fryer (1968) has suggested that this is not the case and gives evidence to show that all the speciation which has taken place here has resulted from the physical isolation of populations in several small lakes which occurred at one or more periods in the lakes' history. Kendall (1970), however, suggested that the lake has always existed as one or, at the most, two continuous sheets of water, but nevertheless supports Fryer by pointing out that changes in level may have produced large offshore lagoons which were separated for sufficiently long periods to produce speciation and then were reconnected to the main body of water. Lake Nabugabo—a large lagoon of this type—has been in existence for only 4,000 years and has produced five endemic species of *Haplochromis* obviously derived from those in Lake Victoria (Fryer, 1968; Greenwood, 1965).

Lake Tana, at the head of the Blue Nile, does not appear to be well documented, but, according to Greenwood, contains a "Nilotic" fish fauna with one suspected endemic species *Barbus tanensis*, which has not been fully verified (personal communication from J. Rzóska).

Nile Fisheries. In common with many tropical waters, the Nile basin is not only rich in species, but also extremely productive of fish. In particular, valuable fisheries exist in the delta lakes, in the White Nile Jebel Aulia dam basin, in the southern Sudan (which exports as much as 2,000 tons of dried fish to the Congo annually), in Lake Albert (annual production up to 12,000 tons), and Lake Victoria (which can yield upwards of 70,000 tons annually) (Hickling, 1961).

Although fishing is important along most of the river and its tributaries, it is still far from fully exploited. Hickling has remarked on the astonishing powers of recovery of tropical inland fisheries due to high temperatures which accelerate all life processes and also to the abundance of herbivorous fish whose food chains are the shortest and least wasteful. Perhaps only those who have worked on tropical rivers can appreciate the density of the fish population such that in many parts of the Nile of the Sudan two or three hauls with a cast net can provide a meal for several men.

Greenwood (1966) states that our knowledge of East African fish is still in an early and exploratory stage of development, and this statement is equally, if not more, true for the fish of the lower Nile.

The importance of fish and fishery research has been recognised by the setting up of: (1) the East African Freshwater Fisheries Research Organisation at Jinja, Uganda (1947); (2) the Inland Fisheries Research Institute, Khartoum, Sudan (1956); (3) the Institute of Freshwater Biology, Cairo, U.A.R. (1955); and (4) more recently, two U.N.D.P. projects, the Lake Victoria Research Project (1966) and the Lake Nasser Development Centre, Aswan, U.A.R. (1969).

Fundamental research is also carried out at Makerere University, Zoology Department; University of Khartoum, Hydrobiological Research Unit and Zoology Department; and at Cairo University.

Fauna Other than Fish. No keys yet exist for the identification of the invertebrate fauna of the Nile, and only a few, almost certainly incomplete, faunal lists have been published.

The largest of these is that of Monakov (1969),[1] who lists 1 Decapod, 15 Copepoda, 49 Cladocera, 1 Conchostraca, 21 Chironomid larvae, 5 Trichoptera, 18 Mollusca, 9 Oligochaetae, and 35 species of Rotatoria. His paper includes an account of the distribution of zooplankton and zoobenthos, including several hundred measurements of biomasses, along 1,530 km of the White Nile south of Khartoum. Earlier faunal lists were published by Rzóska for Hemiptera-Heteroptera and parasitic Copepoda of the Upper Nile, Crustacean plankton and Rotatoria of the Blue and White Nile. An account of the Crustacean plankton of Lake Victoria is given by Rzóska (1957). Proszynska (1967) lists 237 papers on African Cladocera and Copepoda, many of which refer to Nile waters. Investigations into the seasonal cycle and succession of zooplankton in relation to Nile dams is considered later in this paper.

The river-crabs of the Nile in the Sudan have been described by Williams (1969). *Potamonautes nilotica* is the most widely distributed—from the delta to the affluent streams of Lake Victoria. A key to the Potamidae is in course of preparation.

The distribution of snail transmitters of schistosomiasis—*Bulinus* a n d *Biomphalaria*—has been described by Williams and Hunter (1968) for the White Nile near Khartoum, and a key to the aquatic molluscs is in preparation by Williams.

Helminth parasites of Nile fish have been studied in detail by Khalil (1969) who also compared the Sudanese helminth fauna with that of other African countries.

[1] Unfortunately some of his identifications are incorrect.

Nile Flora

Algae. Brook (1954) listed over 160 species of plankton algae from the Blue and White Niles, several species being new records for Africa. However, ecological studies by Rzoska, Brook, and Prowse (1955); Brook and Rzoska (1954); Prowse and Talling (1958); and Talling and Rzoska (1967) have shown that very few of these are important in the plankton. The dominant species in the Blue Nile and White Nile are *Melosira granulata* and its variety *angustissima*, *Anabaena flos-aquae* f. *spiroides*, *Lyngbya limnetica*, and, as noted in the Blue Nile after the construction of the Roseires Dam, *Microcystis flos-aquae* and its epiplankter *Phormidium mucicola*.

Talling (1965, 1966) and Talling and Rzoska (1967) have shown that the dominant algae in the various lakes in the headwaters of the Nile differ considerably from the above and from each other. Thus, although the Lake Tana plankton was dominated by the genus *Melosira*, *M. granulata* was not present, and the four species found did not occur in the Blue Nile.

Lake Albert was dominated by a diatom, probably *Nitzschia bacata*, with *Stephanodiscus astrea* (now *S. rotula*) and *Anabaena flos-aquae* also important. Lake Edward was dominated by another diatom, *Surirella engleri*, with a species of *Glenodinium* abundant.

The Kazinga channel (between Lake George and Lake Edward) came nearest in composition to the lower Nile, with a plankton dominated by *Melosira granulata* var. *angustissima*, *Microcystis aeruginosa*, and *Anabaena flos-aquae*, while Lake George itself was dominated by *Microcystis aeruginosa*.

By far the most varied phytoplankton occurred in Lake Victoria. Talling (1966) found that at least 24 species were important with none really dominant, apart from small coccoid Myxophyceae and diatoms, notably *Stephanodiscus astraea*, *Melosira nyassensis* var. *victoriae* and *Nitzschia acicularis*.

The most distinctive algal flora has been found in two lakes in the Sudd swamps, Lake No and Lake Ambadi. Gronblad, Prowse, and Scott (1958) and Gronblad (1962) described a total of about 211 species of desmids, of which no less than 21 were entirely new species and 48 new varieties and formae. Many remarkable species were described, including two new species of *Micrasterias*, totally unlike anything previously known with unlike hemicells on all the specimens seen. No further collections have been made in this area, and it seems likely that further new species await discovery in these swamp waters.

One other alga deserves comment here—the terrestrial Chaetophoraceous alga, *Fritschiella tuberosa*, discovered by Brook (1952, 1956). This alga makes an annual appearance on the drying silt of the Blue Nile as the river level drops after each flood. Bower (1935) attached considerable importance to *Fritschiella* in retracing the origin of the land flora, and Brook comments:

> The exacting environmental conditions under which the alga grows in Khartoum would seem to substantiate the claim that this alga may well have been one of the first successful colonisers in the landward migration of the aquatic flora.

Riverain Flora. Gay (1956, 1957) has described the riverain vegetation of the White Nile up to above Lake Albert, has related the flora to the main phases of the river, and has also made observations on the ecology of *Cyperus papyrus, Vossia cuspidata,* and "swamp grassland." Halwagy (1963) describes the vegetational succession on islands and shifting sand bars of the main Nile in the vicinity of Khartoum, starting with first colonisation with a thin cover of *Tamarix nilotica* through a succession of communities to the climax represented by dense *Acacietum.* The successive stages were shown to be associated with a progressive increase in silt content of the soil, improved water relations, and a greated degree of soil stabilisation.

Detailed studies of the ecology of the Sudd swamps were made by the Jonglei Investigation Team (Anon., 1954) and will be referred to later. Thornton (1957) studied the arthropod fauna of the papyrus *Cyperus papyrus* and found a well-defined succession of species within the umbel as it grew, matured, and died.

Rzóska, who has contributed so much to the hydrobiology of the Nile in the Sudan, in a forthcoming major paper on the Sudd swamps describes their historical origin, limnological characteristics, and detailed hydrobiology.

This account has, of necessity, only touched on a few aspects of the studies which have been carried out on the flora and fauna of the Nile. In addition to the references cited above, some of the more important sources of information have been included in the bibliography at the end of this paper.

The Impact of Man

Human impact on the ecology of the River Nile stands in marked contrast to the situation in the highly industrialised river basins of Europe and North America, where pollution by sewage and industrial wastes has frequently altered the aquatic environment beyond recognition. In the Nile valley urbanisation and, more particularly, industrialisation have scarcely begun, while population density in relation to the size of the river is very low (with the exception of the delta region, which has the highest density in the world). Water, abstracted primarily for irrigation, is rarely returned to the river, and cities with waterborne sewage disposal systems, such as Khartoum and Cairo, use the partially treated effluent for irrigation purposes. Thus pollution, with the possible exception of the delta, is either totally absent or negligible, by virtue of the vast dilution afforded by the river. An exception to this is at Jinja, where sewage from a population of 100,000 and some trade effluents affect the Victoria Nile for a distance of 6 km downstream.

During the present century the flow of the Nile has been increasingly brought under control by means of dams and barrages, and it is the former which, by altering the hydrological regime, have brought about profound changes in the biology of the river, and thus undoubtedly contributed the major ecological impact. Most of this section will therefore be devoted to the studies which have been made on dams in the Nile valley. The location of these dams is shown on the maps (Figures 1 and 2), while some background data is set out in Table 3:

Table 3. Hydrological features of Nile dams

Dam	Year of Opening	Area of Reservoir at T.W.L. (km^2)	Altitude at T.W.L.	Maximum depth (m)	Capacity (million) (m^3)	Maximum drawdown (m)	Ratio of outflow to volume
Aswan III	1933	-	121	5.3	5.3	-	16.1
Sadd El Ali*	1971	5,000	185	90	164	20	1.2
Sennar	1925	160	422	17	0.9	8	50.1
Jebel Aulia	1937	600	377	15	3.5	6	8.1
Kashm El Girba	1964	150?	473	20	2.0	13	-
Roseires	1966	290	480	50	3.0	13	16.1

*Aswan High Dam

Dams and River Ecology. Although the first three dams, at Aswan, Sennar, and Jebel Aulia, were built in 1902, 1925, and 1937, respectively, unfortunately no studies of a biological nature were carried out until long afterwards, and therefore the original physico-chemical and biological conditions can only be deduced from our present knowledge of the river. The earliest studies appear to have been made on the algae of the Aswan Reservoir and lower Nile at Cairo University around 1942-1945 (Abdin, 1948*a, b, c*), but most of the subsequent work which has contributed greatly to our knowledge of the ecology of the river has been carried out by members of the Hydrobiological Research Unit, which was set up in the University of Khartoum in 1953 to promote fundamental limnological research on the Nile.

Abdin (1948*a, b, c*), in studies of the seasonally filled Aswan Reservoir and the river at Cairo, showed that the former was a moderately eutrophic reservoir with a species succession closely related to the annual flood. Shortly after filling and settlement of silt, the diatom *Melosira granulata* appeared with a maximum growth in December to February when it disappeared due to nutrient depletion. This was followed by a variety of chlorophyceae, which were dominant from March to April or May, when *Melosira granulata* reappeared as the approaching flood brought an increase in nutrients, *M. varians* becoming dominant just before the rapid rise in silt eliminated all plankton in July. In view of the later studies mentioned below, it is of interest to note that Abdin commented on the noticeable absence of blue-green algae. No quantitative measurements were made, and no comparison was made with the plankton population upstream of the dam basin.

Brook and Rzóska (1954), in a longitudinal and seasonal survey of the influence of the Jebel Aulia Reservoir (a narrow river-lake over 500 km long showed that both phyto- and zooplankton increased more than 100-fold, and that this increase was clearly associated with the change from river to lake-like conditions. The sparse plankton upstream comprised mainly diatoms and Copepoda, with some adventitious forms, while the pure, dense plankton at the dam was dominated by blue-green algae and Cladocera. Despite settlement of silt, turbidity actually increased near the dam because of the dense phytoplankton, while pH and dissolved oxygen increased as a result of photosynthetic activity.

Rzóska, Brook, and Prowse (1955) then studied seasonal plankton development in the Blue and White Nile near their junction and showed that dense seasonal growths of a pure plankton formation were present in both rivers and that the annual fluctuations were closely correlated with the hydrological regime of each river. They noted that the few observations from previous workers, e.g., Beam (1906) indicated that before the Sennar and Jebel Aulia dams were built, the plankton was very scanty and mixed with detritus and adventitious forms.

Prowse and Talling (1958) continued and enlarged the scope of these studies in a five-year study of the phytoplankton of the White Nile. They found that the sequence of species in the Jebel Aulia Reservoir was closely related to the relative abundance of algae in the river upstream and that active growth of algae resulted in depletion of nitrate, phosphate, and silicate. They also suggested that deficiency of inorganic nitrogen was the main limiting factor for the growth of the principal diatom, *Melosira granulata*. Over the five-year period a remarkably regular seasonal succession was noted, *M. granulata* being invariably followed by the blue-green alga, *Anabaena flos-aquae* var. *intermedia* f. *spiroides,* sometimes accompanied by *Lyngbya limnetica* and usually followed by a second maximum of *Melosira*. Primary production, as measured by photosynthesis rates, was found to be high, both near the dam and in the river 40 km downstream, with values in the region of 2.2 g carbon/m^2/day.

Talling and Rzóska (1967), in a valuable study of the Blue Nile, again emphasised the impact of the Sennar Dam in enhancing plankton development. In the first longitudinal survey made of this river, they found only a sparse and rudimentary plankton in the upper course of the river, but demonstrated that the dense plankton produced in the reservoir not only maintained itself through several generations for 350 km (30-40 days in time) downstream to the White Nile junction, but at times an actual increase took place. It was thus evident that most of the plankton biomass at Khartoum was produced in free-flowing river conditions below the reservoir. The seasonal succession of algae was remarkably similar to that earlier found in the White Nile, and again it was suggested that deficiency of nitrogen limited the growth of *Melosira*.

Monakov (1968), summarising some aspects of the work of the Soviet fisheries scientific expedition to the White Nile in 1963-64, made the first assessments (based on 200 samples) of the zooplankton and zoobenthos biomass over 1,500 km of the river. His results again emphasise the importance of the dam, but it is of interest to note that whereas the maximum biomass of zooplankton is close to the dam, the maximum development of zoobenthos is near the southern limit of the dam basin.

Impact of a New Dam on the Blue Nile. Until 1964, as already mentioned, all research had been conducted on reservoirs long after they were built and after their biological regimes had become thoroughly stabilised (see Talling and Rzoska's comments in Lowe-McConnell [1966], p. 47).

With the construction, between 1961 and 1966, of the Roseires Dam nearly 300 km upstream of the earlier Sennar Dam, the opportunity arose to study the impact of a new dam on the river with a survey of conditions, both before and after it was put into use.

This work—in fact, a long-term study of the whole 740-km length of the Blue Nile within the Sudan—has formed the main part of the writer's research activities since 1963. At the request of the Board for Hydrobiological Research the H.R.U. (Hydrobiological Research Unit) was asked to assess the impact of the Roseires Dam on physico-chemical and biological conditions in the river down to Khartoum, to study the stratification and changes in the reservoir itself, and to try to assess the effect on insects of medical importance which breed in the river, e.g., Chironomidae, Simulidae (vectors of Onchocerciasis), and mosquitoes. Unfortunately, because of lack of staff, it has not been possible to carry out all of this programme, and no studies could be made on fish or on insects and bottom fauna.

Since 1964 no less than 35 longitudinal surveys by means of a truck and portable dinghy have been made as far as the frontier with Ethiopia, i.e., for three full seasons before and three seasons after first filling of the dam. This has been supplemented by more frequent sampling of the river downstream of Sennar. The work has included observations on thermal stratification, primary production, in relation to solar radiation and light penetration, and fully quantitative studies of the plankton. Since 1966 the writer has been assisted by A. M. Moghraby, who has made a special study of the zooplankton of the Blue Nile, and also by O. Bedri, who has studied silt transportation and sedimentation.

It is hoped that the full results of this very detailed work will be published in due course, but some of the more important conclusions, so far mentioned only in the Annual Reports of the Hydrobiological Research Unit (Hammerton, 1968-70) will now be discussed.

Surveys in the three years before completion of the dam confirmed the influence of the Sennar Reservoir which brought about a seasonal 100- to 200-fold increase in both phyto- and zooplankton, which then persisted, with little reduction in numbers, for 350 km to the junction with the White Nile. Considerable losses took place in the siltier waters of the main Nile, but the principal components of the Blue Nile plankton were still recognisable 100-200 km north of Khartoum. Samples in the region extending 300 km upstream of the influence of the Sennar Reservoir invariably revealed only traces of a rudimentary plankton with usually fewer than 50-100 algae per ml and less than 10 zooplankton individuals in 10 litres (Figure 5).

First filling of the dam in October 1966 brought about noticeable changes in the whole 650 km of river to Khartoum. Secchi disc transparency increased rapidly to 2.8 m at a buoy near the dam, compared with 0.2 m at Abu Shendi, 60 km upstream, and a previous maximum of around 0.5 m at Roseires. In Sennar Reservoir and at Khartoum values of 1.8 and 0.6 m were found—both

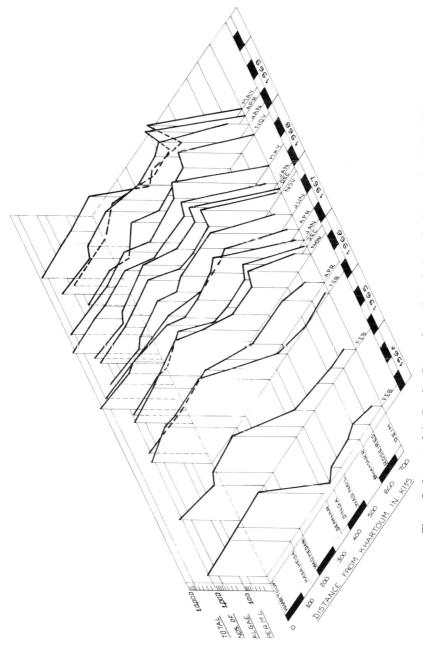

Figure 5. Impact of the Roseires Reservoir on phytoplankton development in the Blue Nile. (This 3-dimensional graph shows total phytoplankton numbers per ml in a series of longitudinal surveys before and after completion of the dam in June 1966)

values greater than in the whole 13-year period the river had been under study. Later increased turbidity resulted from algal growth, but at times even greater transparencies than these have been measured.

The main features of plankton development since 1966, both seasonally and longitudinally, are presented in graphical form in Figures 5, 6, and 7. From these it can be seen that:

(1) The main development of plankton now takes place in Roseires Reservoir, where increases ranging from 10- to over 200-fold have been measured. However, the average density of plankton has not been as great as previously recorded in Sennar. It is thought that this is due to the much greater depth of the new reservoir while the euphotic zone, in which algal growth takes place, is only slightly deeper than at Sennar. If plankton density is related to surface area, then Roseires is undoubtedly the more productive.

(2) Contrary to experience at Sennar, the river below the dam frequently only carries a fraction of the plankton density of the upstream side. This is because of the great depth and the fact that intermittent stratification tends to restrict the plankton to the upper layers, while the water is drawn off from the bottom at a depth of 30-40 m.

(3) Although peak growths at Sennar have been higher than in the period before the advent of Roseires Reservoir, the overall increase in the dam basin has been much less, especially when high growths were carried down from Roseires. This clearly indicates some form of nutrient limitation—a situation which was foreseen by Talling and Rzóska (1967), who said:

> It seems likely that the introduction of additional water storage, and hence time available for plankton increase, will increase the significance of factors other than development time—e.g., nutrition—in limiting the downstream populations at Khartoum.

(4) The new reservoir has brought about a marked change in the seasonal cycle of species succession in the river (Figure 7). The remarkable predictability of this succession in the Sennar and Jebel Aulia dams over a period of more than fifteen years has been noted already, and it was of great interest to find that within a month of first filling, the blue-green alga *Microcystis flos-aquae* with the epiplankter *Phormidium mucicola* became dominant at Roseires. During the next two seasons it became increasingly prominent in the new reservoir and also spread downstream, first to Sennar and finally, at the end of the third season, it became dominant in the whole river down to Khartoum (Figure 7). During this

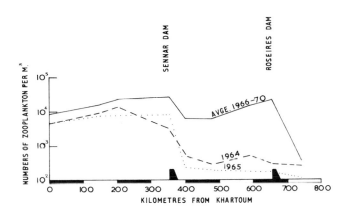

Figure 6. Impact of the Roseires Reservoir on zooplankton development in the Blue Nile

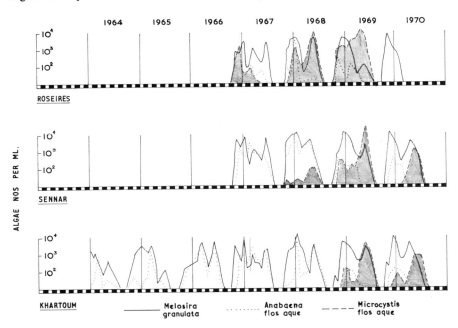

Figure 7. Changes in the seasonal succession of the principal species of algae at three stations in the Blue Nile, 1964-1970

time *Anabaena*, prominent for so many years, became less important but, more surprising still, in the fourth season (1969-70) blue-green algae practically disappeared from the river, and chlorophyceaceae, notably *Volvox aureus* and *Pediastrum* spp., became prominent, especially at Roseires.

(5) The ubiquitous diatom *Melosira granulata* has retained its position as the dominant alga and produced larger growths, almost invariably dominant at the beginning and end of each season. The importance of this species, as well as others of this genus, in the Nile has already been noted, and it is suggested here that the well-known ability of species of *Melosira* to live in the bottom mud of water bodies may be highly significant. Lund (1966) has surveyed the important role played by water movements in the seasonal cycle of the planktonic species of *Melosira* and has shown that in many lakes the appearance of the diatom coincides with periods of water turbulence. Abdin (1948*a, b, c*), Prowse and Talling (1958), and Talling and Rzóska (1967) have attributed the decline of *Melosira* after its first appearance to nutrient deficiency and its reappearance later to a renewal of nutrients with the approaching flood. However, during the writer's studies *Melosira* has on a number of occasions produced large growths at Sennar when nitrate nitrogen was almost undetectable—below 5 µg/l, usually as the river was beginning to rise. An increase in turbulence, due either to a rise in level or to wind-induced currents, could be the initial cause of such growths by bringing *Melosira* filaments into circulation.

(6) The large increase in plankton in the river between Sennar and Khartoum has not been observed during the present studies. All the evidence points to a decrease in both phyto- and zooplankton, at least as far as Hasa Heisa, with some evidence of recovery near Khartoum, possibly due to reduced current where the Blue Nile is dammed up by the White Nile during the productive season. There is one exception to this—some species of Rotifera have at times increased considerably between Sennar and Khartoum.

From the foregoing it can be seen that a large increase in the biological productivity of the river has resulted from the construction of the Roseires Dam, primarily because it has extended the productive length of the river from about 400 km to over 700 km, and also because the average productivity has increased. This is shown not only by the measurements of plankton density, but also by measurements of primary production rates before and after the advent of the new dam, showing an increase and including some of the highest rates yet measured in a free-flowing river.

So far we have discussed the effect of dams on plankton production. Consideration will now be given to some other effects which have been either studied, or at least noted, particularly the effects on bottom fauna, incidence of pest insects, and fish productivity.

Effect on Bottom Fauna. The only serious study of bottom fauna in the Nile was that reported by Monakov (1968), who found a maximum biomass of zoobenthos near the upper end of the Jebel Aulia dam and a big decrease near the dam. Mollusca and Chironomidae were better represented in the reservoir, both in species and in actual numbers. However, in a longitudinal survey of the White Nile in April 1967 the writer found that the bivalve mollusc *Corbicula fluminalis* was abundant from Malakal to Renk, while from Jebelein to Jebel Aulia no live specimens were found in numerous grab samples, though old eroded shells were present before 1937 when the dam was completed.

Furthermore, in a survey of the Roseires Dam basin in May 1968 it was found that, as a result of massive silt deposition on the rocky bed of the river near Fazughli, the whole population of the giant freshwater mussel *Etheria elliptica* had been wiped out over a distance of several kilometres. This species was extremely abundant in 1966, but none have survived within the reservoir, though the species is still present in rocky areas below the dam.

Effects on Pest Insects. As already mentioned, Lewis carried out much valuable work on the emergent insect pests of the Nile, admirably summarised by Rzóska (1963). He showed that chironomid midges of the genus *Tanytarsus*, a serious nuisance in Khartoum and the Gezira irrigation scheme, increased in abundance after the construction of the Sennar dam. *Tanytarsus* became such an important pest at Wadi Halfa after construction of the Aswan reservoir that at times an asthma camp was built in the desert, and removal of the town was considered (Lewis, 1958; Lowe-McConnell, 1966).

There is no doubt that chironomidae have increased in the Roseires area, while *Simulium damnosum*, the transmitter of human onchocerciasis, breeds in the turbulent waters near the sluice gates. The reservoir has also flooded considerable areas of shallows with dense vegetation in which enormous populations of Chironomidae, Simulidae, and mosquitoes developed, causing considerable suffering to the nearby villagers (see also below).

Fish Productivity. Unfortunately, no attempt was made to measure the fish population of the Blue Nile before the construction of the Roseires Dam or, indeed, of any of the other Nile dams. However, according to the local fishermen, there has been a big increase in the size and abundance of fish in the Roseires area, and it is also well known that the best fishing in the White Nile is in the Jebel Aulia reservoir. Conditions for the development of fisheries in seasonally filled reservoirs such as these are far from ideal because of the big annual rise and fall in level. Jackson (in Lowe-McConnell, 1966) points out that a big annual draw-down does not allow for the growth of marginal vegetation which provides food and shelter for fish fry and also that deposition of silt smothers the vegetation and chokes and smothers eggs on spawning areas. It has been estimated that an average of 20 million tonnes of silt, out of a total of 130 million tonnes carried by the Blue Nile, settles out each year in Roseires Reservoir.

One unexpected result of the construction of Roseires Dam is that the greatly increased transparency of the water in the reservoir and in the river for 20-50 km downstream has made it impossible for fishermen to use the cast-net, one of the traditional methods of fishing in the Nile. The method is highly successful in the normally silty river, but with increased transparency the fish see the net being thrown and take evasive action.

Ecological Effects of Lake Nasser. The lake now rising behind the Aswan High Dam, which was officially opened on January 15, 1971, differs in a number of respects from the preceding Nile reservoirs and therefore merits special comment.

The dam, the largest earth-fill structure in the world, has been slowly filling since 1964, and the reservoir will, when full, cover an area of 5,000 km^2. Its total capacity of 164 billion m^3 is equal to twice the average annual flow of the Nile entering Egypt, but because most of the annual flow is now abstracted for irrigation in the Sudan and Egypt, the level is only slowly rising, and big annual fluctuations in level are recorded. Since 1964 virtually no water has flowed into the Mediterranean. Moreover, the Nile Waters Agreement of 1959 provides for the eventual complete utilisation of the river by means of irrigation schemes now projected in both countries, by which time the Sudan will receive 18.5 and Egypt 65.5 milliards out of a total of 84 billion m^3. However, these amounts will be proportionately reduced by the annual evaporation in Lake Nasser, which has been estimated at 10 billion m^3—twice the storage capacity of the old Aswan Reservoir.

The dam has increased Egypt's potential arable land from 7 million to over 9 million acres and already has made the country self-sufficient in wheat and an exporter of rice to the value of $200 million in 1970. When the hydroelectric capacity of 10 billion kilowatt hours per annum is fully realised, it has been estimated that the dam will increase the national income by $500 million per annum.

However, ecological repercussions have already been reported and will have to be offset against the obvious benefits.

At a meeting in December 1968, entitled "Conference on Ecological Aspects of International Development," which was held at Warrenton, Virginia, several references were made to the effects of the Aswan High Dam (UNESCO, 1969).

Dr. M. Kassas of Cairo University pointed out that since the deposition of Nile sediment had ceased along the shore of the Nile Delta, the shoreline had started to retreat by several millimetres a year, and that sand bars separating two lakes from the sea were likely to collapse, thus transforming them into bays and endangering land reclamation projects further inland.

The rich sardine fishery off the Nile delta was reported by Dr. C. J. George of Union College, Schenectady, N.Y., to have been destroyed by the halt of nutrient flow into the Mediterranean. Prior to 1964, 18,000 tons of sardines were fished annually, while in 1969 the catch was 500 tons. Five delta lakes had also become less productive. However, fisheries in Lake Nasser are yielding 5,000 tons annually and could eventually compensate for this loss if there is not a subsequent decline, as occurred in Lake Kariba.

Dr. H. van der Schalie of Michigan University, on the basis of his part in a schistosomiasis control project sponsored by W.H.O., expressed the opinion that the spread of schistosomiasis in the areas irrigated from the dam could outweigh the benefits. Other speakers at this conference reported increased incidence of schistosomiasis following new irrigation projects in many parts of Africa, including Bambia, Ghana, Nigeria, South Africa, Rhodesia, Tanzania, Uganda, and Upper Volta.

It has also been recently widely reported that salt water is penetrating the delta margin with consequent loss of irrigable areas.

Finally, the settlement of all Nile silt—around 50 to 100 million tonnes per annum—in Lake Nasser has led to the need for a large fertiliser industry to compensate for the loss of nutrients in the irrigation water, while the age-old industry of brick-making from Nile mud will have to find an alternative raw material.

One other possible consequence of the Aswan High Dam deserves brief mention, though it is outside the field of river ecology. Previously the discharge of the Nile into the Mediterranean had a general eastward drift, which provided a freshwater barrier at the north end of the Suez Canal. The absence of this barrier means that Red Sea fauna can henceforth migrate into the Mediterranean and vice versa, with possibly serious ecological repercussions.

Impact of Dams on Man. Although not strictly within the field of river ecology, a brief reference should be made here to the human aspect of dam construction. This subject has received serious attention elsewhere, and Scudder (1966) mentions that in the case of the Kariba, Volta, and Aswan High Dam little attention was paid to the problems of resettling the peoples displaced by the flooding of the reservoir basins until after dam site preparations were initiated.

At Roseires Reservoir it was not until a few months before first filling that several thousand villagers were told by a government survey team to move to new areas which were above the top water level of the new lake. Unfortunately, the officials underestimated this level by one or two metres at the southern end of the reservoir. As a result, hundreds of people were flooded out of their hastily re-erected homes, and their newly planted fields and crops were destroyed. The people were thus deprived of both shelter and food just as the heaviest rains started, while, to add to their miseries, the flooded vegetation provided a breeding ground for one of the densest swarms of chironomid midges, mosquitoes, and biting insects that the writer has experienced, with a consequent outbreak of malaria. Little was known about this outside of the area affected at the time, because the only road to these villages had been submerged in the new lake. Road transport had to await the fall in level the following spring, and a new road was not built until more than a year after the dam was completed. Compensation was promised to the people, but, as they bitterly informed the writer, no amount of money would make up for the loss of their former lands and the miseries they had suffered.

In the Sudanese section of Lake Nasser some 50,000 Nubians from the Wadi Halfa area were resettled in a new agricultural scheme irrigated from the Khashm El Girba Dam (opened in 1964) on the Atbara River close to the Ethiopian frontier. However, about 5,000 refused to move, and later several hundreds who found that they could not adjust to the new way of life at Khashm El Girba also joined them to live in a shantytown at the edges of the rising waters of Lake Nasser. Here they await government plans to redevelop agriculture, fishing, forestry, and tourism, but in the meantime eke out a bare subsistence with smuggling, which is the only source of income available to most.

Plant and Animal Invasions. The almost explosive spread of the notorious water hyacinth, *Eichhornia crassipes*, in the Nile has been documented by Gay (1958), Gay and Berry (1959), Davies (1959), and Heinen and Ahmed (1964) and is cited here as an example of man's unintentional impact on the environment.

E. crassipes is a native floating weed of South America which was unknown in the Sudan before 1957, when it is believed to have entered from the already infested waters of the Congo by unknown means via the Bahr El Ghazal system. By the end of 1962 it had spread some 1,800 km up and down the White Nile and also up many of the tributaries, such as the Sobat and Pibor system, where it is still slowly advancing up the headstreams. Recently it was estimated that the annual growth of this pest in the White Nile was of the order of one to 1 1/2 million tonnes wet weight per annum.

Apart from seriously interfering with shipping, it increases evaporation rates, probably by a significant amount (Little, 1966), and has harmed the fisheries by (a) making it difficult for fishermen to get their boats in and out of the river, (b) blocking fish nets, and (c) probably reducing the crop of fish in various ways (Bishnai, 1962). A further harmful effect is that it supports the snail transmitters of schistosomiasis and has extended the northern limit of two snail species (Williams and Hunter, 1968).

The cost of control measures to the Sudan Government has increased from about £S.1/2 to £S.1 million per annum between 1962 and 1970.

Animal Invasions.   While the spread of water weeds in Africa has been accidental in origin, there is another type of invasion which deserves mention, i.e., the deliberate introduction of fish species from one body of water to another, a policy which has been widely practised in Africa.

Lake Victoria probably contains over two hundred species of fish, but is notable for the absence of two important African predators—*Hydrocyon* (tiger fish) and *Lates niloticus* (Nile perch), though the fossil record shows that *Lates* existed 25 million years ago in the much earlier Miocene Lake Karunga which occupied part of the area of the present lake.

About fifteen years ago it was suggested that Nile perch should be introduced to Lake Victoria with the intention that they should exploit the large shoals of the relatively uneconomic *Haplochromis*. In 1955 and 1956 *Lates* from Lake Albert were introduced into the Nile below Owen Falls Dam, and the fish spread into Lake Kioga, being common throughout that lake by December 1962.

In May 1960 a perch was caught in Lake Victoria for the first time and a second one in November of that year. It has been suggested that *Lates* may have gained access to Lake Victoria through the turbines of the Owen Falls Dam. Since then it has spread through the north end of the lake, its progress having been studied by Gee (1965), who showed that the introduced fish were fatter and heavier per unit weight than in their native Lake Albert. During 1964 the population increased by 500-700% in one part of the lake. So far it does not appear to have harmed the established *Tilapia* fisheries, but the long-term effects of this introduction, as well as that of the herbivorous *Tilapia zillii*, are yet unknown.

Fryer (1960*a, b*; 1965; 1968) argued eloquently, but unfortunately too late, against the introduction of *Lates* and showed that the idea that the fisheries would benefit was based on ignorance of several fundamental biological concepts. He stressed the disastrous effects of animal introductions elsewhere—notably that of the sea lamprey into the Great Lakes of North America, which destroyed the trout fishery of Lake Michigan. It remains to be seen whether Fryer's fears are realised, or whether—as hoped by the government departments concerned—the fisheries will benefit through the utilisation by *Lates* of the vast numbers of *Haplochromis*.

Future Developments — Jonglei Diversion Canal. Reference has already been to the proposal of Hurst et al. (1946) that a canal to by-pass the Sudd swamps should be built as part of the comprehensive Equatorial Nile Project. The purpose of this canal, which is now being re-examined by the Sudan and Egyptian governments, is to increase the flow of the White Nile, which is at present lost in the swamps between Mongalla and Lake No, by an estimated 7.8 billion m^3.

At present there is an estimated area of 13-14,000 km^2 of permanent swamp in this section, while the total area of swamps in the rainy season, including the Bahr El Ghazal, is of the order of 40,000 km^2.

Construction of the canal, which is intended to take about half the flow in the river, would require a regulating barrage at Jonglei, and would eventually result in the drainage of most of the permanent swamp. The vast ecological changes which would result from this drainage and the impact on the 700,000 Nilotic people of the area were investigated between 1946 and 1954 by a special Commission, known as the Jonglei Investigation Team, set up by the Sudan government. The team included irrigation engineers, surveyors, political officers, and a veterinary expert and was assisted by biologists and fishery experts. Their study (Anon., 1953, 1954) has added greatly to our knowledge of the whole area and analyses in detail the anticipated ecological changes likely to take place.

The livelihood of the Nilotic tribes depends on cattle rearing, agriculture, and fishing, and their whole existence is adapted to the regime of the river with regular seasonal movements towards and away from the river. Fishing methods are primitive, but adequate, in an area where the annual flooding provides numerous swamps, lagoons, and lakes in which fish are so plentiful that spears, harpoons, and traps yield large catches. The principal fish are *Lates* (Nile perch),

Citharinus, Tilappia spp., Bagridae, Claridae, and the lungfish *Protopterus aethiopicus.* Fish markets are found at Juba, Wau, and Malakal, and there is a considerable export of dried fish to the Congo.

The Jonglei Team estimated that the loss of the flood fisheries would mean a loss in yield of 3,000 tonnes per annum, with a potential loss (i.e., if efficient methods were used) of 20,000 tonnes.

A particular feature is that although the natural grasses of the area are value-less as fodder when dry, large herds of cattle can be maintained because owners can find fresh grazing in the dry season in areas liable to prolonged floods. The canal would eliminate floods and destroy this livelihood, profoundly altering the way of life of the Nilotics, which is intimately bound up with cattle.

The Jonglei Investigation Team proposed various remedies with the aim of preserving the way of life of the Nilotics as far as possible. To counter the loss in fishing, they proposed training in efficient methods and the development of fish farming, using depressions, lagoons, etc., totalling 3,000 acres to produce an estimated 3,000 tonnes of fish per annum. Hickling (1961) thought that such a yield was optimistic and suggested that double this area would be required. Various remedies were also proposed to enable cattle rearing to be continued.

Barbour (1961) has pointed out that this region could one day become a rice granary comparable to the great deltas of Southeast Asia, but that the way of life of the Nilotics stands as an obstacle, not merely to the extension of agriculture, but even to the adoption of a commercial attitude to the rearing of sheep and cattle.

The Retreat of Wildlife in the Nile Valley. There is considerable evidence that a marked retreat of both vegetation and the larger wildlife has taken place in the lower Nile valley during historic times. Many scenes depicted in the tombs and temples of Ancient Egypt suggest that lions and large game animals flourished in lower Egypt and the Nile Delta.

Cloudsley-Thompson (1967) has summarised much evidence that a rich fauna and flora existed as far as the delta until the last 150-200 years, since when the process of retreat has been greatly speeded up. The elephant disappeared from Egypt in early dynastic times. The hippopotamus was common even in the delta up to the end of the seventeenth century, but today it is rarely seen north of Kosti, some 2,800 km from the delta. The Nile crocodile was worshiped in Pharaonic times and was abundant in Upper Egypt, according to travellers, up to the end of the nineteenth century; today it has nearly disappeared from Egypt, is rare in many parts of the Sudan, and only common south of Roseires on the Blue Nile and south of Renk on the White Nile. Various travellers in the early nineteenth century commented on the abundance of woodland in the northern Sudan and the presence of lions. Even 25 years ago lions were occasionally heard near Sennar, but today it is rare to find them outside the Dinder Game Reserve, or the south and extreme west.

The retreat of terrestrial wildlife is as much due to deforestation as to hunting. Trees are very slow-growing in the arid climate, unless watered from the river, and increased population has led to tree felling for building and firewood at a rate far above the regeneration capacity of the trees.

Elimination of some species has, of course, its own ecological repercussions. The effect of crocodile destruction has not been studied in the Sudan, but in Rhodesia elimination of the reptile led to a big increase in otters, with consequent damage to fisheries. In the Congo disappearance of crocodiles led to an increase in predatory fish. In the Sudan crocodiles are officially regarded as vermin, and their eventual disappearance from most of the river seems certain. In the Dinder Game Reserve, the writer was surprised to be informed by a game warden that all the crocodiles in the Dinder River had been shot so that bathing was safe for visitors!

The Future Outlook

Since the earliest recorded times the Nile River has been used increasingly, and almost exclusively, for irrigation, to an extent unmatched by any other major river. Other important uses such as fisheries, transport, and the recent development of hydroelectric power must always remain secondary to the need for irrigation water, on which depends the food supply of all Egypt and much of the Sudan.

We have seen that the impact of man became of major significance only during the present century, with the construction of large dams which have brought about profound changes in biological productivity. The magnitude of such changes can be gauged from the fact that as a result of the Sennar and Roseires dams, the production of phytoplankton has at times increased by 100- to 200-fold in the whole length of the Blue Nile in the Sudan—a distance of over 700 km. Increased plankton production and bottom fauna have been noted in both the Blue and White Niles, and there is good evidence that fish productivity has also increased. The rich plankton growth reaching the main Nile at Khartoum has furthermore been traced for considerable distances north of Khartoum. It can be concluded that, apart from increased nuisance from some pest insects, the main effect of dams in the Sudan has been beneficial and has resulted in considerable increases in the biological productivity of the river.

The Aswan High Dam has also increased biological productivity, both within the lake and probably in the whole river down to the delta, and a commercial fishery is being established in the lake. However, because the new dam traps all the Nile silt, repercussions have been felt in the delta, which is suffering from marine erosion, penetration of sea water, and the loss of the valuable offshore sardine fishery. Other effects have been seen in the erosion of the Nile banks and loss of fertility in the irrigation water, necessitating increased use of fertilisers.

Turning now to the future, it can be seen that, with increasing human populations in all the countries of the Nile basin, two developments affecting the ecology of the rivers are inevitable in the foreseeable future. These are: (1) further Nile control works, and (2) urbanisation and industrialisation.

New Nile Control Works. The eventual need for yet more irrigation water in the right place at the right time will require new control works, notably the Jonglei Canal, and further dams, possible sites being Lake Tana, Lake Kioga, and Lake Albert. These dams would be primarily for either storage or flood control and river regulation, but further dams solely designed for hydroelectric power have been suggested, for example, at the sixth cataract in the Sudan.

We have already seen that the Jonglei Canal—to increase the flow of the White Nile—will, by draining the enormous Sudd swamps, have far-reaching effects, not only on the aquatic habitat, but especially on 700,000 Nilotics whose way of life is so intimately bound up with the regime of the river.

The impact of further control works in the White Nile basin is aimed at both increasing the discharge and equalising the flow. Clearly, it is desirable, from the irrigation point of view, to even out the flow of the river, but experience elsewhere would suggest that the elimination of high flows could lead to silting up of parts of the river bed, with harmful effects on navigation. Elimination of the swamps could slightly affect the water chemistry—in particular there might be an increase in sulphate.

With regard to the Blue Nile, it is unlikely that any control works would greatly affect the huge annual flood. The construction of the Aswan High Dam has, in any case, removed the need for this, and a dam at Lake Tana would merely be used to augment the flow in a dry year.

205

Urbanisation and Industrialisation. As stated earlier, the Nile is remarkably free of pollution. However, as urbanisation and industrialisation take place, it seems inevitable that sooner or later sewage and industrial effluents will be discharged to the river and perhaps the East African lakes. In Egypt and the Sudan such effluents are unlikely to be discharged to the river because they can be used for irrigation. However, this will not be the case in countries such as Ethiopia, Kenya, and Uganda. It is hoped that these countries will profit by the experience in Europe and America, where neglect and inadequate legislation led to some appalling examples of river and lake pollution. Industrialisation on such a scale is, of course, only a remote possibility, and all that one can foresee is the possibility of a mild degree of eutrophication (with possibly gross local pollution, if adequate preventive measures are not taken).

It must be pointed out here, however, that even a mild degree of eutrophication, much lower than experienced in most European rivers, could have serious effects in the Blue and White Nile reservoirs, and particularly in Lake Nasser. Already, as we have seen, considerable plankton growths occur in both the river and lake sections, and low levels of nutrients, such as phosphate and nitrate, are probably an important limiting factor. It would, therefore, appear that embarrassingly large growths of algae, particularly when one bears in mind the strong sunlight, almost complete absence of cloud cover, and high water temperatures (25-30°C) prevailing at these latitudes. The present level of algal growths can at times be troublesome in the Khartoum water supply treatment works, so that any big increase could lead to the need for expensive modifications. Heavier growths in the reservoirs could lead to severe deoxygenation of the lower layers, which might at times result in fish mortalities. A further point to bear in mind is that blue-green algae can cause the death of cattle drinking water rich in certain species.

Gradual eutrophication of a reservoir of the size and depth of Lake Nasser could have more serious results because the possibilities of semi-permanent stratification are much greater. Prolonged deoxygenation can lead to chemically or biologically corrosive water, which may damage pumps, as happened at Chew Valley Lake (Hammerton, 1959), or seriously damage turbines, as happened at Kariba.

In conclusion, the foregoing argument stresses the need for further and more detailed studies on every aspect of the impact of man on the Nile system. It is perhaps worth pointing out that very large sums are often spent investigating the engineering aspects of river control works, but little, if any, attention is given to the biological and social implications of such works. This has been particularly true of some dam schemes, but it is to the credit of the Sudan government that the ecological and sociological implications of the proposed Jonglei Canal were thoroughly investigated, though the scheme has yet to be carried out.

Clearly, there is a need to start thinking of the Nile and its tributaries as a complete ecosystem; hydrologically this is almost the case with the Nile Waters Agreement and the generally good collaboration between all the governments of the Nile basin. However, the most carefully integrated system of Nile control could run into trouble if eutrophication were allowed to take place on any scale. Until now most problems of eutrophication have occurred in temperate regions, but in countries such as the Sudan and Egypt such problems could be magnified several times. Fortunately, the problem is not imminent, but it is nevertheless one which should be increasingly borne in mind by those whose responsibility it is to control and develop the utilisation of the Nile waters to the maximum advantage of all concerned.

Acknowledgements

I wish to express my thanks to members of the Board for Hydrobiological Research, University of Khartoum, and especially to its former Chairman, Professor Mustafa Hassan (now Vice-Chancellor, University of Khartoum) for his valuable advice and encouragement and for permission to publish this paper.

My grateful thanks are due to the following, who provided information and advice: Dr. G. Fryer, Dr. P. H. Greenwood, Professor M. Kassas, Dr. J. Okedi, Dr.T. Petr, Dr. J. Rzoska, and Dr. K. Thompson.

Thanks are also due to the following for their comments on the manuscript: Professor J. L. Cloudsley-Thompson, Professor J. R. Vail, Dr. F. Stansfield, Dr. J. W. G. Lund, and Dr. J. F. Talling.

Finally, I would like to thank my former colleagues in the Hydrobiological Research Unit, University of Khartoum, and especially Ustaz Asim Ibrahim El Moghraby, who shared in so much of the work.

Literature Cited

Abdin, G. 1948a. Physical and chemical investigations relating to algal growth in the River Nile, Cairo. Bull. Inst. Egypt 29:19-44.

—————————. 1948b. Seasonal distribution of phytoplankton and sessile algae in the River Nile, Cairo. Bull. Inst. Egypt 29:369-382.

—————————. 1948c. The conditions of growth and periodicity of the algal flora of the Aswan Reservoir (Upper Egypt). Bull. Fac. Sci. Cairo Univ. 27:157-75.

Anonymous. 1953. The Equatorial Nile Project and its effects in the Sudan. Jonglei Investigation Team. Geog. J. 119:33-48.

—————————. 1954. The Equatorial Nile Project. (Report of Jonglei Investigation Team) Sudan Gov. V. I-IV, Index, Introduction.

Arkell, A. J. 1949a. The Old Stone Age in the Anglo-Egyptian Sudan. Sudan Antiquities Serv. 1:1-52.

—————————. 1949b. Early Khartoum. Oxford Univ. Press.

Ball, J. 1939. Contribution to the geology of Egypt. Cairo.

Barbour, K. M. 1961. The Republic of the Sudan. Univ. London Press. 292 p.

Beadle, L. C. 1962. The evolution of species in the lakes of East Africa. Uganda J. 26:44-54.

Beam, W. 1906. Chemical composition of the Nile waters. Second Rep. Wellcome Res. Lab., Khartoum, p. 206-214.

—————————. 1908. Nile waters. Third Rep. Wellcome Res. Lab., Khartoum, p. 386-395.

Berry, L., and A. J. Whiteman. 1968. The Nile in the Sudan. Geogr. J. 134(I):1-37.

Bishai, H. M. 1962. The water characteristics of the Nile in the Sudan with a note on the effect of *Eichhornia crassipes* on the hydrobiology of the Nile. Hydrobiologia 19:357-382.

Blashford-Snell, J. N. 1970. The conquest of the Blue Nile. Geogr. J. 136:42-51.

Bower, F. 1935. Primitive land plants. London.

Brook, A. J. 1952. Occurrence of the terrestrial alga *Fritschiella tuberosa* Iyengar in Africa Nature 164:754.

——————. 1954. A systematic account of the phytoplankton of the Blue and White Nile at Khartoum. Ann. Mag. Natur. Hist. 12(7):648-656.

——————, and J. Rzoska. 1954. The influence of the Gebel Aulia Dam on the development of Nile plankton. J. Anim. Ecol. 23:101-114.

Brook, A. J. 1956. A note on the ecology of the terrestrial alga *Fritschiella tuberosa* in the Sudan. New Phyt. 55:130-132.

——————, and H. Kufferath. 1957. A bibliography of African freshwater algae. Rev. Algologique 4:207-238.

Cloudsley-Thompson, J. L. 1967. Animal twilight. Foulis, London. 204 p.

Corbet, P. S. 1961. The food of non-cichlid fishes in the Lake Victoria basin, with remarks on their evolution and adaptation to lacustrine conditions. Proc. Zool. Soc. London 136:1-101.

Danielli, G. 1943. Geologia dell' Africa Orientale. R. Acc. Italia Roma.

Davies, H. R. J. 1959. Effects of the water hyacinth (*Eichhornia crassipes*) in the Nile Valley, Nature, London 184:1085.

Doornkamp, J. C., and P. H. Temple. 1966. Surface drainage and tectonic instability in part of southern Uganda. Geogr. J. 132:238-252.

Elster, H. J., and S. Gorgy. 1959. Der Nilschlamm als Nahrstoffregulator im Nildelta. Naturwissenschaften 46:147.

Elster, H. J., and R. A. Vollenweider. 1961. Beitrage zur Limnologie Agyptens. Arch Hydrobiol. 57:241-343.

Fryer, G. 1960a. Some controversial aspects of speciation of African cichlid fishes. Proc. Zool. Soc. Lond. 135:569-578.

——————. 1960b. Concerning the proposed introduction of Nile perch into Lake Victoria. East African Agr. J. 25(4):267-270.

Fryer, G. 1965. Predation and its effects on migration and speciation in African fishes: a comment. Proc. Zool. Soc. London 144:301-310.

——————. 1968. Speciation and adaptive radiation in African lakes. Verb. Internat. Verein. Limnol. 17:303-322.

——————, and T. D. Iles. 1969. Alternative route to evolutionary success as exhibited by African cichlid fishes of the genus *Tilapia* and the species flocks of the great lakes. Evolution 23:359-369.

——————. 1971. The cichlid fishes of the great lakes of Africa: their biology and evolution. Oliver & Boyd, London (in press).

Gay, P. A. 1956. Riverain flora of the White Nile. *In* Third Annu. Rept. Hydrobiol. Res. Unit. Khartoum Univ.

——————. 1957. Riverain flora of the White Nile. *In* Fourth Annu. Rept. Hydrobiol. Res. Unit. Khartoum Univ.

——————. 1958. *Eichhornia crassipes* in the Nile of the Sudan. Nature 182: 538-539.

——————, and L. Berry. 1959. The water hyacinth: a new problem on the Nile. Geogr. J. 125:89-91.

Gay, P. A. 1960. Ecological studies of *Eichhornia crassipes* in the Sudan. I. Analysis of spread in the Nile. J. Ecol. 48:183-191.

Gee, J. H. 1965. The spread of Nile perch (*Lates niloticus*) in East Africa with comparative biological notes. J. Appl. Ecol. 2(2):407-408.

Greenwood, P. H. 1965a. The cichlid fishes of Lake Nabugabo, Uganda. Bull. Brit. Mus. Nat. Hist. Zool., 12(9):313-357.

——————, 1965b. *In* G. Fryer. 1965. A further comment (on predation and its effects). Proc. Zool. Soc. London 144:301-310.

——————, 1966. The fishes of Uganda Rev. ed. Kampala-Uganda Society.

Gronblad, R. G., G. A. Prowse, and A. M. Scott. 1958. Sudanese Desmids. Acta Bot. Fenmica 58:1-58.

Gronblad, R. G. 1962. Sudanese Desmids II. Acta Bot Fenmica 63:1-19.

Halim, Y., S. K. Guergues, and H. H. Saleh. 1967. Hydrographic conditions and plankton in the South East Mediterranean during the last normal Nile flood. Int. Revue Ges. Hydrobiol. 52:301-320.

Halwagy, R. 1963. Studies on the succession of vegetation on some islands and sandbanks in the Nile near Khartoum, Sudan. Vegetation 11:217-234.

Hammerton, D. 1959. A biological and chemical study of Chew Valley Lake. Proc. Soc. Water Treatment Exam. 8:87-132.

——————, [ed.]. 1962-1971. Annu. Rep. Hydrobiol. Res. Unit. Univ. Khartoum.

Hammerton, D. 1964. Hydrobiological research in the Sudan. Sudan Phil. Soc. Annu. Symp. 12:181-204.

Heinen, E. T., and S. Ahmed. 1964. Water hyacinth control on the Nile River, Sudan. Dept. Agr., Khartoum. 56 p.

Herodotus, ca. 440 B.C. The Histories [Transl. by Aubrey de Selincourt]. Penguin Classics, London, 1954.

Hickling, C. F. 1961. Tropical inland fisheries. Longmans, London.

Hora, S. L. 1937. Geographical distribution of Indian freshwater fishes and its bearing on the probable land connections between India and the adjacent countries. Curr. Sci. 5:351-356.

Hurst, H. E., R. P. Black, and Y. M. Simaika. 1946. The Nile Basin. Vol. VII. The future conservation of the Nile. Phys. Dep. Cairo.

Hurst, H. E. 1957. The Nile. London Constable. 331 p.

Jackson, P. B. N. 1965. Reply to G. Fryer and P. H. Greenwood. In G. Fryer, 1965: Predation and its effects on migration and speciation in African fishes: a comment. Proc. Zool. Soc. Lond. 144:301-322.

Kendall, R. L. 1970. An ecological history of the Lake Victoria basin. Ecol. Monogr. 39:121-176.

Khalil, L. F. 1969. Studies on the helminth parasites of freshwater fishes of the Sudan. J. Zool. Lond. 158:143-170.

Lawson, A. C. 1927. The Valley of the Nile. Univ. Calif. Chron. 29.

Lebon, J. H. G. 1965. Land use in Sudan. Bude (Cornwall) Geographical Publications (World Land Use Survey. Regional Monographs, Sir L. Dudley Stamp [ed.]). 191 p.

Lewis, D. J. 1958. Trans. R. Entomol. Soc. London 110:81-98.

——————. 1966. Nile control and its effects on insects of medical importance. *In* R. H. Lowe-McConnell [ed.], Man-made lakes. Inst. Biol. Symp. No. 15. Acad. Press, London.

Little, E. C. S. 1966. The invasion of man-made lakes by plants. *In* R. H. Lowe-McConnell [ed.], Man-made lakes. Inst. Biol. Symp. No. 15. Acad. Press, London.

Lowe-McConnell, R. H. 1959. Breeding behaviour patterns and ecological differences between *Tilapia*. Proc. Zool. Soc. London 132(1):1-30.

——————, [ed.]. 1966. Man-made lakes. Inst. of Biol. Symp. No. 15. Acad. Press, London.

Lowe-McConnell, R. H. 1969. Speciation in tropical freshwater fishes. Biol. J. Linn. Soc. 1:51-75.

Lund, J. W. G. 1966. Znachenie turbulentnosti vody v periodichnosty razvitiya nekotorykh presnovodnykh vidor roda *Melosira* (algae). Bot. Zhurn 51: 176-178.

Menon, A. G. K. 1951. Distribution of clariid fishes, and its significance in zoogeographical studies. Proc. Nat. Inst. Sci. India 17:292-299.

Mohr, P. 1962. The geology of Ethiopia. Univ. Coll. Addis Ababa Press.

Monakov, A. V. 1968. The zooplankton and zoobenthos of the White Nile and adjoining waters in the Republic of the Sudan. Hydrobiologia 33:161-185.

Morrice, H. A. W. 1956. A plan for the Nile Valley. Sudan Eng. Soc. J. 22-30.

Proszynska, M. 1967. Bibliography of Cladocera and Copepoda of African Inland Waters. Ghana J. Sci. 7:37-49.

Prowse, G. A., and J. F. Talling. 1958. The seasonal growth and succession of plankton algae in the White Nile. Limnol. Oceanogr. 3:222-237.

Rzoska, J., A. J. Brook, and G. A. Prowse. 1955. Seasonal plankton development in the White and Blue Nile at Khartoum. Verh. int. Ver. Limnol. 12:327-334.

Rzoska, J. 1957. Notes on the Crustacean plankton of Lake Victoria. Proc. Linn. Soc. London 168:116-125.

—————. 1961. Some aspects of the hydrobiology of the River Nile. Verh. int. Ver. Limnol. 13:505-507.

—————. 1963. Mass outbreaks of insects in the Sudanese Nile basin. Verh. int. Ver. Limnol. 15:194-200.

Sandon, H. 1950. Illustrated guide to the freshwater fishes of the Sudan. Sudan Notes and Records. Monograph. 61 p.

Scudder, T. 1966. Man-made lakes and population resettlement in Africa. *In* R. H. Lowe-McConnell [ed.], Man-made lakes. Inst. Biol. Symp. No. 15. Acad. Press, London.

Talling, J. F. 1957. The longitudinal succession of water characteristics in the White Nile. Hydrobiol. 11:73-89.

—————. 1965. The photosynthetic activity of phytoplankton in East African lakes. Int. Rev. ges. Hydrobiol. 50:1-32.

—————, and I. B. Talling. 1965. The chemical composition of African lake waters. Int. Rev. ges. Hydrobiol. 50:421-463.

Talling, J. F. 1966. The annual cycle of stratification and phytoplankton growth in Lake Victoria. Int. Rev. ges. Hydrobiol. 51:545-621.

—————, and J. Rzoska. 1967. The development of plankton in relation to hydrological regime in the Blue Nile. J. Ecol. 55:637-662.

Temple, P. H. 1969. Some biological implications of a revised geological history for Lake Victoria. Biol. J. Linn. Soc. 1:363-371.

Thornton, I. W. B. 1957. Faunal succession in numbers of *Cyperus papyrus* L. on the Upper White Nile. Proc. Roy. Entomol. Soc. London 32:119-131.

UNESCO. 1969. Conference on the ecological aspects of international development. Bull. Int. Hydrol. Decade. V. 5, No. 2.

Vercoutter, J. 1966. Semna South Fort and the records of the Nile levels at Kumma. Kush 14:125-164.

Wayland, E. J. 1931. Summary of progress of the Geographical Survey of Uganda (1919-29). Gov. Printer Entebbe.

Welcomme, R. L. 1965. Recent changes in the *Tilapia* stocks of Lake Victoria. J. Appl. Ecol. 2(2):410.

Williams, S. N., and P. J. Hunter. 1968. The distribution of *Bulinus* and *Biomphaleria* in Khartoum and Blue Nile provinces, Sudan. Bull. World Health Org. 39:949-954.

Williams, T. R. 1969. Freshwater crabs of the River Nile. 16th Annu. Report Hydrobiol. Res. Unit, Univ. Khartoum (in press).

Worral, G. 1958. Deposition of silt by the irrigation waters of the Nile at Khartoum, Geog. J. 124:219-222.

Worthington, E. B. 1937. On the evolution of fish in the great lakes of Africa. Int. Rev. Hydrobiol. 35:304-317.

——————. 1954. Speciation of fishes in African lakes. Nature, London 173: 1064-1067.

CASE HISTORY: THE RIVER THAMES

K. H. Mann

The River Thames meanders through the farmland and towns of southern England for a distance of 147 miles (237 km) before meeting the tidal waters of the Thames estuary at London (Figure 1). The total fall is only 370 ft (113 m), so that the average slope is 2.5 ft per mile or 0.47 per thousand. The rainfall in the catchment area averages 29 inches (74 cm), and the flow at Teddington Weir, the limit of tidal influence, averages about 800 mgd (million gal per day, or 3 million m^3day^{-1}). The catchment area is 3,845 sq miles (9,951 km^2), of which almost half is chalk and limestone. Thus the water is hard, with a calcium content of the order of 100 mg/l.

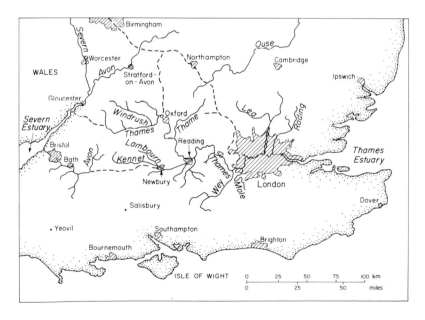

Figure 1. Map of southeast England showing the River Thames, its tributaries, and the major canals (broken lines) connecting with other watersheds

215

The most distinctive character of the river is that around the end of the 18th century it was engineered for navigation, i.e., provided with weirs and locks every few miles to maintain a constant depth of water (minimum 6 ft or 2 m). This has the effect of slowing the flow, controlling floods, and holding a large volume of water as reserve for London's drinking water. More important than the water in the riverbed itself is the water held in the porous rocks on either side of it, for the groundwater level in the whole valley is raised by the hydrostatic pressure of the water in the riverbed.

Brief Historical Review

Long before the river was engineered in any way, while there were still deep pools and shallow rapids, primitive man developed trackways which crossed the river at suitable fords. Settlements developed there, reflected in place names like Oxford and Wallingford. Some early human artefacts were the water mills, through which water was channeled by wooden weirs. Associated with these weirs were various kinds of fish traps. As early as 1075, William the Conqueror ordered that "milles and fisheries be destroyed," apparently because they interfered with navigation and salmon migration. In Magna Carta (1215) appeared the phrase "let all kiddles be abolished"; this also referred to weirs and fish traps.

The first double lock, or pound lock, was built in 1620, but all others were single gates in a weir and remained so until about 1750. The engineering of the river for horse-drawn barge traffic was carried out around 1800. About the same time a complex pattern of canals was developed between watersheds, providing transport for bulky materials between manufacturing centres, before the advent of railways.

Naturally, sewage pollution occurred wherever man settled. Oxford was a seat of learning and was a sizable community all through the Middle Ages. In 1644 John Taylor, a minor poet, found himself penniless in Oxford and took on the job of River Inspector. I quote:

> *Then by the Lords Commissioners, and also*
> *By my good King (whom all true subjects call so)*
> *I was commanded, with the Water Baylie*
> *To see the rivers cleaned, both night and dayly.*
> *Dead Hogges, Dogges, Cates and well flayed Carryon Horses,*
> *Their noysom Corpses soyld the Water Courses;*
> *Both Swines' and Stable dunge, Beasts guts and garbage,*
> *Street durt, with Gardners' Weeds and Rotten Herbage.*
> *And from those Waters' filthy putrification*
> *Our meat and Drinke were made, which bred Infection*

Myself and partner, with cost paines and Travell,
Saw all made clean, from Carryon, Mud and Gravell,
And now and then was punisht a Delinquent,
By which good meanes away the filth and stink went.

—(Quoted in Rolt, 1951)

In a report of the "Select Committee on the River Thames" in 1865 occurred the statement:

It was also proved to your committee that the discharge of sewage has of late years became a serious impediment to navigation, as at times the accumulation of solid matter is so great in the locks that it has to be removed at considerable expense, and delay occurs before the lock gates can be opened.

—(Barclay, 1963)

Soon after this report appeared, the Conservators of the River Thames, a body armed with powers to control the river in all its aspects, including pollution, was given jurisdiction over the main river along most of its length. By 1893 they had power to control pollution in the whole watershed. There are records from 1866 onwards of prosecutions under the Thames Conservancy Act against large numbers of factories, and of Statutory Notices being served on municipalities. Within forty years some form of sewage treatment was provided for almost every town and village. Most of the sewers laid and the treatment sites selected are still functioning.

A recent refinement, acquired by the Conservancy in 1965, is the power to prevent discharge of polluting substances into any underground strata within the catchment area. However, it is stated that consent to a discharge must not be unreasonably withheld.

The river receives at present about 180 mgd (0.68 million m^3day^{-1}) of sewage effluent and 42 mgd (0.16 million m^3day^{-1}) of industrial effluent. The minimum standard applied to sewage effluent is 30 mg/l suspended solids and 20 mg/l BOD but where the volume of effluent is large in relation to river flow, higher standards are required, in some cases 10 mg/l suspended solids and 10 mg/l BOD. The 30/20 criterion is also applied to a pulp mill effluent, and oil is required to be absent from any effluent.

The general picture is of a river basin with a population of 3.5 million people, receiving all the effluent which this implies, yet supporting more angling and pleasure boating than any other river in the United Kingdom. Many people swim in the river, although there are those that hesitate to do so, and it is more than 100 years since a salmon was caught in the Thames, although they were once plentiful.

Water Abstraction

The annual flow of the river at the junction with the tidal estuary is of the order of 1100 million m^3 per annum. From the catchment area is taken two-thirds of London's water supply and water for 24 other communities. The demand is increasing continually. The total abstraction from surface waters in the catchment area is about 710 million gal; in addition, there is considerable abstraction from groundwaters, currently about 471 million gal. Hence the total abstraction, much of which is returned to the river somewhere downstream of the point of abstraction, is of about the same magnitude as the flow of the river. The breakdown between the various users is given in Table 1. Water supply undertakings are responsible for about three-quarters of all abstraction.

Table 1. Analysis of water abstraction in the Thames Catchment Area, from data supplied by the Thames Conservancy

Purpose of Abstraction	Annual volume from surface waters (million m^3)	Percentage of total from surface waters	Annual volume from ground- waters (million m^3)	Percentage of total from ground- waters
Public and private water supply	521.349	73	367.426	78
Power station cooling	134.548	19 ⎫		
Industrial process and cooling waters	52.715	7 ⎬	76.028	16
Agricultural (spray irrigation)	1.073 ⎫		1.427	1
Agricultural (other uses)	0.027 ⎬	1	13.670	3
Miscellaneous	0.227 ⎭		12.192	2
	709.939	100	470.743	100

In the London area, water is held in large reservoirs for use in dry weather. However, it is predicted that in years of exceptionally low rainfall (such as have occurred three times in the past fifty years) the reserves will be inadequate. There are no more acceptable sites for reservoirs in the Thames Valley, so a new scheme for meeting drought conditions has been devised and tested (Nugent, 1971). The principle is that boreholes are sunk into chalk or limestone within the catchment area, and the flow of the river and its tributaries is augmented by pumping water from the underground aquifer storage. The best results were obtained from boreholes situated one to three miles from a stream. Pumping caused the surface of the water table to drop about 20 ft (6.5 m) in the boreholes, and test holes showed that the water table surface sloped upwards and outwards from the boreholes in a funnel-like manner, so that a one- to two-foot drop was found at a distance of two miles. Four boreholes in the valley of the River Lambourn (northwest of Newbury, Figure 1) enabled 280 million gal to be pumped from the aquifer storage in six months. It has been estimated that a ten-year development of this scheme will enable the flow of the Thames to be augmented by 100 million gal a day (375×10^3 $m^3 day^{-1}$) at the end of a severe drought, and after pumping for six months. The cost will be only one-quarter to one-third the cost of conventional surface reservoirs.

The University of Reading Study

One might have expected that a river heavily loaded with the effluents of sewage treatment plants and used intensively by water supply undertaking would be biologically unproductive, or would produce aesthetically undesirable materials. However, there was evidence that fish were abundant in all parts of the river. Staff and students of the Department of Zoology of the University of Reading undertook to study the productivity, at all trophic levels, of a section of the river located about midway between Oxford and London.

The area selected for intensive study was at the confluence of the Thames and its largest tributary, the Kennet (Figure 2a). In "dry weather" conditions, the Kennet accounts for 23% of the total flow of the Thames basin. The main sewage effluents upstream of the study area are Swindon (6.6 mgd) and Oxford (8 mgd), both situated many miles away along the main river, and Newbury (2.1 mgd) and Reading (9.45 mgd) on the Kennet. The Reading effluent is only a few miles upstream and markedly affected the study area. The study is reported in detail in Mann, et al., 1971.

Fish Production. The fish populations were sampled by seine nets (Williams, 1963, 1965, 1967; Mathews, 1970), and showed every indication of stability, with high population densities, relatively slow growth rates, and no particular year class having an exceptionally high or low abundance. Mann (1964, 1965) solved the balanced energy equation for the two most abundant species, roach (*Rutilus rutilus* [L.]) and bleak (*Alburnus alburnus* [L.]), using the method of Winberg (1956). This was later modified (Mann, et al., 1971) to take account of

219

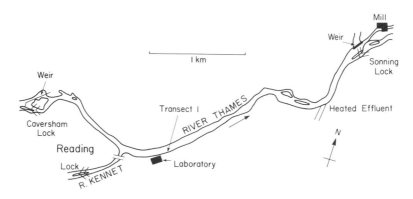

Figure 2a. Sketch map showing the section of the River Thames, near the junction with the River Kennet, which was the subject of intensive ecological study by staff and students of the University of Reading

the seasonal cycle of fat storage and metabolism. A separate study was made of young fish in their first two years of life (Mathews, 1970), and the fecundity of roach and bleak was plotted in relation to age (MacKay and Mann, 1969). Production of the roach and bleak was calculated by the Allen curve method (Allen, 1951), using estimates of (1) population fecundity, 600 eggs/m² for roach and 3000 eggs/m² for bleak; (2) age class strengths from the population studies mentioned above; and (3) estimates of numbers and sizes present between hatching and becoming fully vulnerable to the nets, assuming constant relative rates of growth and mortality. It was concluded that production by fish in their first year of life amounted to 70% of the total population production, and that all species together were producing nearly 200 kcal/m². In terms of fish, the Thames is a very productive river indeed.

The next phase of the study was to attempt to unravel the food web leading from primary production to fish.

Primary Production. (a) Phytoplankton. Since the river is modified from its original state to produce a series of relatively slow-flowing basins, phytoplankton is the major source of primary production. In summer the water becomes a greenish soup with light intensity at 1 m less than 40% of surface illumination, and at the bottom (3-4m) less than 0.01% of the surface value. In the spring and summer of 1967 diatoms predominated (80-90% of total cells). Centric (planktonic) forms (*Stephanodiscus, Cyclotella*) predominated over pennate (benthic) forms (*Navicula, Nitschia*) in the proportions 8 or 9 to 1. In winter Chlorophyceae were more important (70% of the population in December). The total cell number ranged from 72,000/ml on April 16 to 195/ml in January. In the Kennet, plankton was less rich, with a maximum of 9,700 cells/ml in May. Benthic diatoms were more important, rising to a maximum in strong currents.

The horizontal and vertical distribution of chlorophyll was checked and found to be uniform ±2%, indicating that the water is well mixed and one station in each river would be representative of that river. Taking chlorophyll *a* as 2% of dry weight, phytoplankton biomass was estimated and compared with total suspended organic matter. In the Thames above its confluence with the Kennet phytoplankton varied from 75% of total organic matter in suspension in June to 2% in November. The annual mean was 7.05 g/m^3 dry weight of organic matter in suspension, of which 2.03 g/m^3 (28.8%) was phytoplankton. In the Kennet, by contrast, the annual mean was 7.84 g/m^3 of organic matter in suspension, of which only 5.1% was phytoplankton. The difference was attributable to the presence of the main Reading sewage outfall about 4 km up the Kennet.

Photosynthesis and respiration in the Thames and Kennet above their confluence were investigated by monitoring oxygen changes in light and dark bottles suspended at various depths. Calculations were made for 5 depth strata: 0-30, 30-90, 90-150, 150-210, and below 210 cm. From profiles drawn at 100-m intervals through Kennet and Thames the average volume of each stratum was determined. The production and respiration values in each stratum on each occasion were then weighted to give an integrated figure representative of that section of the river, according to the shape of the riverbed. These data were plotted to give annual curves, and the year's total production and respiration estimated from the area under each curve. The production of the Thames below the confluence was calculated by considering it as a mixture of waters of two different levels of productivity, in proportion to their relative rates of discharge on each occasion.

On 18 occasions bottles were exposed at two depths on a continuously rotating frame (Ministry of Technology, 1964). It was found that net and gross production was 1.38 ± 0.23 times higher than in stationary bottles, presumably because rotation prevented algae from settling out and broke down gradients of nutrients, carbon dioxide, and extra-cellular secretions. The conditions in the rotating bottles were thought to approximate more closely those in the river, so all values from stationary bottles were increased by 38%.

Photosynthesis in the plankton community of the Thames above Kennet mouth greatly exceeded respiration, leaving a balance of net community production of 4,388 kcal/m^2/year. This figure represents the gross photosynthesis less the respiration of the plants and the respiration of the heterotrophs, particularly bacteria. No figure is available for true net primary production. In the Kennet there was an excess of respiration over photosynthesis, giving a net community production of -275 kcal/m²/year. The discharge of the Thames was normally considerably greater than that of the Kennet, and the water below the confluence had a positive net community production of 1,907 kcal/m²/year.

(b) Macrophytes. As it turned out, macrophyte production was trivial compared with plankton production. This was not obvious at the beginning, because the north bank of the river has an almost continuous line of willow trees overhanging the water, and the rushes (*Acorus*) and water lilies (*Nuphar*) are abundant along the southern bank. Leaf fall from the trees was monitored in quadrats on the bank, and total leaf fall into the river determined from the area of trees over river, as seen in aerial photographs.

Production of aquatic macrophytes was estimated by following seasonal changes in biomass. Averaged over the river surface, production was 79 kcal/m²/year for trees, and 44 kcal/m²/year for aquatic macrophytes.

(c) Epibenthic algae. This study is still in progress. In shallow water production is very intense in summer, but beyond one meter light is severely limiting. Since only 15% of the riverbed lies at less than one meter depth, it is likely that periphyton production is considerably less than phytoplankton production.

Production by Benthic Animals. Two sites were chosen for study, one in the Thames above Kennet mouth and one below it. At each, two transects were laid out at right angles to the bank and sampled 12 times in 13 months. Four zones were recognized (Figure 2b), and random stratified sampling carried out in each. Samples were obtained by diving, except in the marginal vegetation, where a large "corer" was used. As far as possible, the samples were used to identify age classes of animals, and for each cohort estimates were made of the population density and mean from the products of numbers present and growth rates, either algebraically or graphically using the method of Allen (1951). The results are summarized in Table 2.

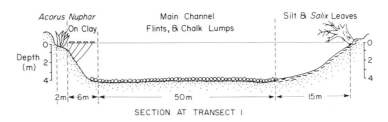

Figure 2b. Schematic section through the river showing the four main zones of benthos

Table 2. Production of benthic animals ($kcal/m^2/year$)

	Above Kennet	Below Kennet
Browsers and Grazers		
Gastropods (*Viviparus, Lymnaea, Bithynia, Acroloxus*)	13.8	8.6
Crustacea (*Asellus*)	2.1	1.6
Tubificids (5 times mean annual biomass)	3.2	5.1
Insect larvae (*Caenis, Cyrnus*)	0.8	0.6
	19.9	15.9
Filter Feeders		
Bivalves (*Anodonta, Unio, Sphaerium*)	61.3	25.6
Sponge (*Spongilla*)	37.8	72.3
Bryozoan (*Plumatella*)	8.1	18.4
	107.2	116.3
Predators		
Leeches (*Erpobdella, Glossiphonia, Helobdella*)	2.0	1.7
Chironomid (= Tendepedid) larvae	15.4	3.1
	17.4	4.8
Grand Total	144.5	137.0

Tubificid worms could not be sorted to age groups, and *Spongilla* and *Plumatella* could not be counted as individuals, so the production of all these groups was estimated from biomass changes (details in Mann, et al., 1971).

Sources From Which Fish Obtained Their Energy. By solving the balanced energy equation for roach and bleak (see above) it was estimated that the populations require 183 $kcal/m^2/year$ for the first year of life and 877 $kcal/m^2/year$ for all other age classes. The next question was: From what sources do the fish obtain this energy? Stomach contents were analyzed on a volumetric basis for a range of size classes and seasons. The observed gut contents were corrected for relative rates of digestion and for assimilability, to give the proportion of energy derived from each food source. The results, shown in Figure 3, show that the young fish rely heavily on zooplankton (rotifers, cladocerans) and chironomid larvae, while the older fish derive most energy from organic detritus, insects taken at the water surface, and algae.

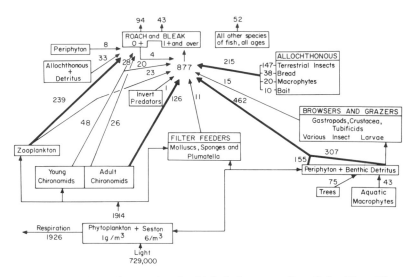

Figure 3. Energy flow diagram for the biological community of the River Thames at Reading, England. All fluxes are in kcal/m²/year. The organisms are arranged approximately by trophic levels. At the base of the diagram is shown production by phytoplankton periphyton, trees, and macrophytes. There is also an input of suspended organic matter from upstream. All these sources contribute to the tripton and benthic detritus. Invertebrates are shown across the centre of the page; they are mainly herbivores and detritus feeders; the invertebrate carnivores (predators) are leeches and carnivorous insects. The dominant fish species are roach (<u>Rutilus</u>) and bleak (<u>Alburnus</u>), with other species (perch, <u>Perca</u>; gudgeon, <u>Gobio</u>; etc.) grouped in the top right-hand corner. To reduce the number of crossed lines, compartments for periphyton, allochthonous material, and detritus are duplicated in the top left corner

Pathways of Energy Flow. The data reviewed in the previous sections are summarized in Figure 3. The figures obtained by direct estimate of the production of the browsing and grazing benthic invertebrates below Kennet mouth are almost exactly those obtained by back-calculation from the food requirements of fish. The same is true for invertebrate predators, suggesting that the two methods give reliable estimates. However, the productivity of the filter feeders, particularly the bivalves, sponges, and *Plumatella* is much greater than the calculated fish large for the fish to eat. The fate of their production, of the order of 50 kcal/m²/ year, is in doubt. At death they may be consumed by scavenging fish, in kcal/m²/year, is in doubt. At death they may be consumed by scavenging fish, in which case the autolyzed remains of their bodies would be unrecognizable in fish guts, or they may be consumed by invertebrates or by microorganisms. The production estimates for sponges and *Plumatella* are likely to err on the low side, rather than the reverse, and it is probable that fish consumption has been underestimated because of the difficulty in identifying the remains of these soft-bodied animals.

The discrepancy between chironomid production and fish consumption is in the other direction: Far more chironomids were taken by the fish than could be accounted for by benthic sampling. The frequency of sampling and the mesh of the sampler were inadequate for chironomids, so the direct estimate is undoubtedly too low. Moreover, the estimate of adult chironomids taken at the water surface greatly exceeds the estimate of larval production. In part this may be due to immigration of adults from other areas. A conservative estimate of benthic production is:

	$kcal/m^2/year$
Direct estimate of production of benthic invertebrates other than chironomids	134
Calculated consumption of chironomid larvae and adults	197
	331

With these reservations, the energy flow chart in Figure 3 may be interpreted. Heavy lines indicate the major pathways of energy flow. The largest single source of food for fish is benthic detritus and periphyton. This is consumed in large quantities in winter. The next most important source is allochthonous material, produced in great quantity and mostly taken as insects at the surface in summer, but cladocera and rotifers are the major component of the diet of fish in their first year of life. Contrary to the more usual situation, benthic invertebrates are a minor source of energy for the fish.

Management of Fish Stocks

The fish populations (Table 3) are dominated by the cyprinids roach and bleak, with lesser numbers of gobiids and perciformes. Salmonids are not uncommon in the upper reaches and the tributaries, but Atlantic salmon, once common, have been barred from migration by the pollution of the Thames estuary for about a hundred years. The eels that occur in the river are thought to migrate by way of other watersheds and the interconnecting canal system (Figure 1). The major fish predator is the pike, *Esox lucius*, which is locally abundant.

These "coarse fish" stocks are much prized by British anglers. The Thames Conservancy protects the stocks by enforcing closed seasons, during which no fishing is legal, as follows:

Trout:	14 October to 31 March
Pike:	15 March to 30 September
All other coarse fish, including eels:	15 March to 15 June

The Conservancy also imposes a size limit, varying from species to species, by which it is required that fish caught of less length than that specified must be returned to the river with as little injury as possible. Fishing competitions, involving large numbers of people, in which the aim is to land the maximum weight of legal-sized fish, are held frequently on many parts of the river.

Table 3. Species of fish found in the Thames between Caversham and Sonning locks and the number caught in 14 seine nettings from October 1958 to October 1959, opposite the University of Reading River Studies Centre

Species	Number Caught
Rutilus rutilus (L.) (Roach)	4,838
Alburnus alburnus (L.) (Bleak	4,022
Gobio gobio (L.) (Gudgeon)	914
Perca fluviatilis L. (Perch)	546
Leuciscus leuciscus (L.) (Dace)	522
Abramis brama (L.) (Bream)	130
Acerina cernua (L.) (Pope/Ruffe)	52
Esox lucius L. (Pike)	16
Gasterosteus aculeatus L. (Stickleback)	2
Cottus gobio L. (Miller's thumb)	2
Phoxinus phoxinus (L.) (Minnow)	1
Squalius cephalus (L.) (Chub)	1
Anguilla anguilla L. (Eel)	-
Barbus barbus (L.) (Barbel)	-
Salmo trutta L. (Trout)	-
Tinca tinca (L.) (Tench)	-
Roach x Bream hybrid	-
Total Catch	11,046

Williams (1967) reported on the growth and mortality of the more abundant species in the Reading area. He showed that the average roach was ten years of age by the time it reached the legal minimum size, and less than 1% of the fish vulnerable to a net with a mesh of 0.5 inch (12 mm) knot to knot reached this size. Bleak reached the size limit by about three years of age, but because they are such small fish (the size limit is 4 inches), they are not highly regarded by competitive fishermen. For dace and perch the situation was similar to that for roach.

The density of the fish populations was extremely high (Williams, 1965). The composite results of studies of fecundity, survival of young fish, and age composition of the older stocks suggest that a "typical" situation for roach would be as follows: In an area of 10 m x 10 m of riverbed, 60,000 eggs would be liberated in May, 2,000 young fish would be present in July, about 100 in fall, and about 50 at one year of age. From that time on there would be a mortality of about 30% per annum. The density of roach vulnerable to a net with 0.5-inch (12 mm) mesh was estimated by mark recapture experiments as 100 to 150 per 100 m^2. Bleak were much more abundant, and the total stock vulnerable to the net was estimated at about 1,000 fish per 100 m^2, or roughly a million fish for every 2 km of river.

Undoubtedly, the slow growth rates, leading to fish taking almost the whole of their maximum life span to reach "legal" size, are a reflection of the high population density in relation to food resources. The ecosystem study showed that particularly during the winter the fish were deriving most of their energy from organic detritus, a food which they assimilated with extremely low efficiency. The fecundity of roach was low compared with other situations (MacKay and Mann, 1969), and the growth rate was much slower than in all other British rivers for which data are available.

There are two possible explanations: Either the fish are genetically adapted to slow growth on a low-grade diet, or the slow growth is imposed phenotypically by the high population densith in relation to a limited food supply. The experiment to follow changes in growth rate when population densities are reduced has not been performed, but this type of experiment was carried out for perch in Windermere (LeCren, 1958) and indicated that the stock had the potential to respond rapidly to a drastic reduction in numbers.

The question then arises: Does this high population density and slow growth result from man's interference, or might this be expected to occur in a natural lowland river? A factor worth considering is the relatively low numbers of piscivorous fish. Small perch feed mainly on invertebrates, and few survived to reach the size at which they were predominantly piscivorous. Pike, which become piscivorous at an early age, were relatively scarce (Table 3). Pike normally spawn in shallow, marshy situations. The engineering of the river for navigation, with dredging, construction of a towpath, and consolidation of the banks to limit flooding, has resulted in the virtual elimination of these shallow waters and probably inhibits the breeding of pike. There is strong anecdotal evidence to suggest that anglers kill pike regardless of the legal limit, while they conscientiously return other species to the river, in the belief that they are wisely conserving the stocks and removing predators. Williams (1967) recommended drastic thinning of cyprinid stocks to improve growth rates, but I am not aware that any action has been taken. Over most of the river the fisheries are privately owned.

227

The Thames Estuary

The aspect of pollution control in the Thames which has been reported in the press of many parts of the world is concerned with the estuary, rather than the purely freshwater reaches of the river. Marked deterioration of the estuary began towards the end of the 18th century when the use of water closets became widespread in London and sewage began to reach the river by way of surface water drains. Before this time estuarine and migratory fish were abundant: In 1766, 130 Thames salmon were sold in Billingsgate market in one day. By 1850 all commercial fishing in the London area had ceased, and in 1852, after bad cholera outbreaks, an act was passed forbidding abstraction of drinking water from the river below Teddington Weir (the limit of tidal influence). After considering many schemes, the municipal authorities installed a system to collect sewage effluents and discharge them on the north and south sides of the river, on the falling tide, well downstream of central London. Chemical precipitation was installed at the northern treatment plant, with the solids being barged about 60 miles out to sea for dumping. In the 1890's the fish began to return to the estuary, but population growth led to serious overloading and the development of anaerobic conditions in the river. By 1953 anaerobic or near anaerobic conditions persisted for much of the year in a section about 30 miles long downstream of London Bridge. It was reported˝ (Department of Scientific and Industrial Research, 1964) that the river was completely anaerobic, from surface to bottom, for at least twenty miles, during at least one month in summer. The water was black with ferrous sulphide, and hydrogen sulphide was discharged into the atmosphere.

In 1960 the London County Council began the virtual rebuilding of the treatment plant on the south bank, and that on the north bank is under reconstruction. A biological survey in 1957 revealed no fish in the tidal reaches of the London area, but in 1965 fish were caught in the screens at the cooling water intake of an electricity-generating station. A regular program of recording such catches was begun in 1967 (Wheeler, 1970), and since that time over 50 species of fish have been taken, mainly marine species in the lower half of the estuary, and predominantly freshwater species upstream of London. Each year the smelt have migrated further upstream until they now occur west of central London.

Discussion

Reviewing the history of the river, we can trace it back to the stage before man's gross interference, when the water was in places fast-running, shallow, and clear and probably supported a predominantly salmonid population. As human settlement increased, pollution with sewage and garbage increased to the point where, in the 17th century, the waters at Oxford were revolting. But even at this stage there were penalties for polluting the watercourse that were enforced. Engineering of the river for navigation preceded a strict general control of sewage pollution, so that lock gates became blocked by sewage solids. In the second half of the 19th century the last salmon was caught, but since that time there has been a steady increase in pollution prevention. However, it has not been practicable to restore the water to its pristine condition. A load of suspended solids of 30 mg/l at source, which was found at Reading to average 7 mg/l after dilution, is considered acceptable.

Although many effluents have a BOD of 20 mg/l, photosynthesis, surface aeration, and the presence of weirs every few miles seem to supply the oxygen requirements, for there is no evidence of serious oxygen deficits. A typical 24-hour cycle in the Kennet, carrying the Reading sewage effluent, showed a peak of 9.6 mg/l by day and a minimum of 8.8 mg/l at night. The mineralization of the suspended solids by bacteria results in the liberation of abundant nutrients, and these lead to the high planktonic production observed in the main river.

The volume of the river at "normal" levels is about 4,500 million gal (17 million m^3), and the average flow at the junction with the estuary is 800 mgd (3 million m^{3}day^{-1}). Hence the water in the river basin is changed, on average, once in six days. Under these conditions it is surprising to find a rich phytoplankton development in the middle reaches of the river. Moreover, the species of phytoplankton concerned are the truly planktonic forms, rather than benthic forms carried in suspension. On the other hand, truly planktonic cladocera and rotifers are not common, and the link between phytoplankton and young fish tends to be through forms whose habitats are the dense clumps of vegetation, or the interstices of the coarse gravel of the bottom deposits. Presumably, cladocerans could not complete their life histories in plankton which reaches the sea within a week. It is not entirely clear what is the source of cells which continually replenish the phytoplankton under these conditions. One assumes there are pockets of stagnant water providing the seed for this vast flowing chemostat.

The productivity of this system, as measured about midway along its course from source to mouth, is impressive. The density of the fish stocks is one of the highest on record, except for fertilized fish ponds, and although the production/biomass ratio is not exceptionally high (1.12 for roach; 1.92 for bleak), there is a total production of fish flesh of about 200 kcal/m^2 or 2,000 kg/ha. About 70% of this is attributable to young fish in their first year of life and is not realizable as yield. However, with skillful management it is likely that a yield of 500 kg/ha could be obtained.

The benthos is under very heavy grazing pressure, and in conditions of dense phytoplankton and suspended organic matter the important producers are the filter feeders. Unionid mussels had a population density of 28 per m^2, their biomass amounting to about one metric ton per 100 m of river; the total production of filter feeders was at least 100 kcal/m^2.

The benthic production is by no means large enough to support the fish populations. The larger fish make extensive use of terrestrial insects at the water surface and organic detritus, to which they turn when other food is in short supply. The young fish rely on rotifers and cladocerans.

In summary, the river is an example of eutrophication, which in a lake is undesirable because the products of photosynthesis accumulate and decay. The Thames, because of its modification for navigation, consists of a chain of shallow, almost lake-like basins in which there is turbulent mixing, adequate aeration, and in which the water has a residence time measured in hours. The water is, on the whole, pleasant for boating but less so for swimming. It is more or less odourless and has few accumulations of unsightly algal blooms. Above all, it is biologically highly productive and has a considerable potential for the harvest of animal protein, if this could be used economically. It purifies the sewage effluent from a population of 3.5 million, provides most of London's drinking water, and provides extensive boating and fishing in an area where natural lakes do not exist.

Heavy expenditure on sewage treatment plants to improve the quality of the effluent could undoubtedly increase the recreation potential, but under present circumstances I doubt if there are any rivers in areas of comparable population density which perform their function better. If rivers are the renal systems of landscapes, this kidney is working well, under stress.

Acknowledgments

The Thames Conservancy, through a number of its officers, assisted in several ways. Much of the benthic sampling was carried out by their professional divers based on Reading. Mr. H. Fish, Chief Purification Officer, kindly supplied valuable background information. In addition to the colleagues named in the list of references, the author is indebted to Professor G. Williams and Dr. A. D. Berrie of the University of Reading for assistance in supervision and in interpretation of results. Professor A. Graham supported the work by advice, encouragement, and making available the facilities of the Department of Zoology of the University of Reading. Financial support was provided by the Natural Environment Research Council.

References Cited

Allen, K. R. 1951. The Horokiwi Stream. Bull. Mar. Dep. N.Z. Fish 10.

Barclay, W. G. 1963. Pollution of the Thames Valley during the last century. Thunderbird Enterprises Ltd., London. Mimeo.

Department of Scientific and Industrial Research. 1964. Effects of polluting discharge on the Thames Estuary. Water Pollution Res. Tech. Paper 11. Her Majesty's Stationery Office, London. xxviii + 609 p.

LeCren, E. D. 1958. Observations on the growth of perch (*Perca fluviatilis* L.) over twenty-two years with special reference to the effects of temperature and changes in population density. J. Anim. Ecol. 27:287-334.

MacKay, I., and K. H. Mann. 1969. Fecundity of two cyprinid fishes in the River Thames, Reading, England. J. Fish. Res. Board. Can. 26:2795-2805.

Mann, K. H. 1964. The pattern of energy flow in the fish and invertebrate fauna of the River Thames. Verh. Intern. Verein. Limnol. 15:485-495.

—————. 1965. Energy transformations by a population of fish in the River Thames. J. Anim. Ecol. 34:253-275.

—————, R. H. Britton, A. Kowalczewski. T. J. Lack, C. P. Mathews, and I. McDonald. 1971. Productivity and energy flow at all trophic levels in the River Thames, England. Proceedings UNESCO/IBP Symposium on Productivity Problems of Freshwaters. Polish Academy of Sciences, Warsaw (in press).

Mathews, C. P. 1970. Estimates of production with reference to general surveys. Oikos 21:129-133.

Ministry of Technology (U.K.). 1964. Water Pollution Research 1964:129-130.

Nugent, The Rt. Hon. The Lord. 1971. The Lambourn Valley Pilot Scheme. Salmon Trout Mag. 191:34-38.

Rolt, L. T. C. 1951. The Thames from mouth to source. Batsford, London.

Wheeler, A. 1970. Fish return to the Thames. Sci. J. November, p. 28-32.

Williams, W. P. 1963. A study of freshwater fish in the Thames. Ph.D. Thesis, University of Reading, England.

Williams, W. P. 1963. The population density of four species of freshwater fish, roach (*Rutilus rutilus* L.), bleak (*Alburnus alburnus* L.), dace (*Leuciscus leuciscus* L.) and perch (*Perca fluviatilis* L.) in the River Thames at Reading. J. Anim. Ecol. 34:173-85.

—————————. 1967. The growth and mortality of four species of fish in the River Thames at Reading. J. Anim. Ecol. 36:695-720.

Winberg, G. G. 1956. Rate of metabolism and food requirements of fish. Fish Res. Board. Can. Transl. Ser. 194 (from Intensivnost obmena i pischevye petrebrosti ryd. Nauchnye Trudy Belorusskovo Gosudarst vennovo Universiteta imeni V. I. Lenina, Minsk.).

Additional References

Cornish, C. J. 1902. The naturalist on the Thames. Seeley & Co. Ltd., London.

Fritsch, F. E. 1902. Preliminary report on the phytoplankton of the Thames. Ann. Bot. London 16:1-9.

—————————. 1903. Further observations on the phytoplankton of the River Thames. Ann. Bot. London 17:631-647.

—————————. 1905. The plankton of some English rivers. Ann. Bot. London 19:163-167.

Gameson, A. L. H., and M. J. Barrett. 1958. Oxidation, reaeration, and mixing in the Thames Estuary, p. 63. *In* Oxygen relationships in streams. U.S. Public Health Serv. Robt. A. Taft Sanitary Engineering Center Tech. Rep. W58-2.

Mathews, C. P., and A. Kowalczewski. 1969. The decomposition of leaf litter in the River Thames. J. Ecol. 57:543:552.

Metropolitan Water Board (London). Annual Reports. Yearly.

Ministry of Housing and Local Government and Scottish Office. The surface water year book of Great Britain's. H. M. Stationery Office, London: 1952, 1955, and each year thereafter.

Negus, C. L. 1966. A quantitative study of growth and production of unionid mussels in the River Thames at Reading. J. Anim. Ecol. 35:513-532.

Taylor, J. R., G. W. Humphreys, and T. P. Frank. 1935. Report on Greater London drainage. Ministry of Health, H. M. Stationery Office, London.

USES OF THE DANUBE RIVER

Reinhard Liepolt

The Danube River is one of the greatest rivers of the world. With a length of 1,888 km, it is the eleventh longest in the world and the second longest in Europe, exceeded only by the Volga.

The drainage area of the Danube comprises 805,000 km^2. Its source (Figure 1) lies 1,000 m high in the Black Forest in West Germany, and its mouth in the Black Sea, with two arms of the delta lying in Rumania and one in the Soviet Union. The annual mean discharge to the delta is 6,500 m^3/sec or 200 billion m^3. The annual sediment discharge into the Black Sea can be estimated at 67.5 million tonnes (Figure 2).

The Danube River is a link joining eight states of Europe, and its tributaries lie in twelve countries. In the drainage area live nearly 70 million people, and in all twelve countries, about 447 million. These figures show the extraordinary international importance of this river.

The uses of the Danube have been numerous and varied in the past. Now and in the future the versatile uses become more and more important. In the following, detailed information is given about the complex uses of this great European stream and its tributaries.

Navigation

The historic beginning of navigation is to be found in towage navigation, where the ships were drawn upstream by men or horses on small paths along the banks, while the transport downstream was accomplished by the river current. Even in the years between World War I and II the ships were drawn upstream by steam engines to pass the rapids of the cataract stretch between Yugoslavia and Rumania, well known as the "Iron Gate." This practice was given up some years ago after completion of the power plant "Djerdap."

The use of steam power in navigation made it possible to use iron vessels for transport instead of wooden ships—thus higher transport speed and greater transport volume. New economic areas and more goods were made accessible for Danube navigation.

Figure 1. Source of the Danube at the Black Forest, 1087 m, Fed. Germ. Rep.

Figure 3 shows the development of cargo traffic in the last fifty years (1914-1969). The two World Wars brought severe setbacks but since World War II the advance has been enormous, and now there are long chains of barges and other pushed, towed, and self-propelled cargo craft. Steamboats with coal for fuel are gradually being removed from service. In 1969 the total fleet of all Danube countries had a tonnage capacity of 2,698,974, with 589,196 hp.

These figures indicate the great economic importance of the Danube for all adjacent countries, as well as the role of the river cultural exchange.

Figure 2. Mouth of the Sulina-channel into the Black Sea, Rumania

Table 1. Amount of cargo handled in Danube ports in 1969

Country	Service in 1000 tonnes
Hungary	16,838
Soviet Union	14,114
Bulgaria	12,691
Yugoslavia	12,115
Rumania	9,406
Austria	7,086
Czechoslovakia	6,477
West Germany	4,659
Total	83,386

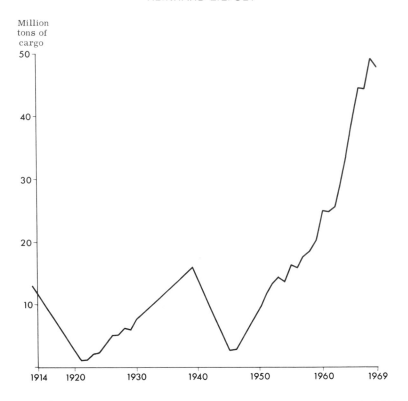

Figure 3. The development of cargo traffic on the Danube in the years 1914-1969

Pollution of the Danube by navigation is brought about by handling of oil products, use of detergents and other materials for cleaning the vessel, and discharge of water from cooling systems. In some ports tanks for waste oils have been ordered to help solve the problem of oil pollution. Water from the river is used on board for drinking after it has been purified, for washing and bathing after rude purification over gravel filter, for engine cooling, and to clean the ship. As many Danube ships also navigate in the Black Sea, corrosion of the ships' hull is inevitable. Periphyton and settlement of suspended matters are hardly noticeable.

An agreement was signed in 1948 to ensure free navigation and the maintenance of this international waterway; by this agreement the international Danube Commission was founded. This Commission submits recommendations to the governments of the member states, which pass them into law, thus ensuring general guidelines and specifications for all Danube ships. The agreement encompasses the total navigable part of the Danube, from Ulm in the Federal Republic of Germany to the mouth of the Danube in the Black Sea.

This important European waterway to the East will get a strong impetus from the completion of the Rhine- Main-Danube Canal and by the Danube-Oder Canal.

Flood Control and Land Drainage

The management of the Danube undertaken in the past century has not always considered the river's natural, geological evolution. Interference by man has caused great changes which, together with the cutting down of large forest areas, has led to a decline in the ground-water level and thus the desiccation of large areas. On the other hand, fertile land has been reclaimed from swamps and marshes.

The straightening of the meandering tributaries and the regulation of the riverbed by strangulations have resulted in increased floods, as have the many stream trenches which gave the river a new bed.

Dams along the river have been built to consider the different water levels of many years in the various districts and the economic value of the areas which should be protected. For these reasons and because there are many areas not protected by dams, areas of more than 100,000 ha are sometimes flooded. Dams which are built up too near at the riverbed increase the danger of floods in the downstream countries. During severe flooding the Danube may be as wide as 10 to 20 km.

Management of the river has favorably influenced ice conditions. Only in the severe winter of 1928-29 was the river frozen for nearly 2,000 km. In other years only short distances are covered with ice and only for 20 to 45 days.

More ice is found upstream of power plants at river dams and in recent years waste-water discharges have brought about an increase in water temperature.

Hydrotechnic measurements have concerned not only flood protection, but drainage, dewatering, and irrigation as well. Through these improvements large areas have been reclaimed for agricultural use, especially in the eastern Danube countries. It is intended that all floodplains, including the delta, will be drained and protected from floods. These floodplains, covering an area of 2,645,000 ha, together with the land now used for agriculture, would bring the total area needing irrigation to about 5,000,000 ha in the eastern Danube countries. If the necessary water demand is calculated at 0.5 liter/sec/ha, 2500 m^3/sec would be needed. As the annual mean flow of the Danube at its mouth is about 6500 m^3/sec, an irrigation of this extent at minimum flow could endanger the navigation. These problems can be solved only through the cooperation of all Danube countries.

Power Generation

The time of water mills (Figure 4) on the Danube River has passed, although some of these small, floating generators can be found in the eastern Danube countries.

The Kachlet power plant near Passau in West Germany, which began operation in 1927, was the first hydroelectric plant on the Danube. The next plants followed after 1950, when the energy demand began to grow rapidly. It was soon recognized that the Danube River, with a mean flow of 6500 m^3/sec at its mouth and a total fall of 700 m, is a mighty energy source. The development of the waterway for navigation was combined with the construction of power plants. Because of navigation, the number of dams must be small, resulting in a high usable fall for each plant.

Along the upper part of the Danube the construction of hydroelectric plants will be economical only when navigation can pay for the dam and the locks and the owner of the power plant has to pay only for the erection of the energy plant itself. The Danube is not navigable by larger vessels (1350 tonnes) at low flow; a participation in the costs would therefore be justified for navigation.

Figure 4. Old ship-mill on the Rumanian Danube

Approximations of the usable waterpower energy after completion of the energy plants are (current calculations):

Section of the Danube

German	2,700 GWh
Austrian	14,200 GWh
Remaining	30,000 GWh
Total	46,900 GWh

Besides the many hydroelectric plants which have been built on the Danube, in West Germany and Austria only the large energy plant at the "Iron Gates," built by Yugoslavia and Rumania, is finished. With its annual production of 10,700 GWh, it is among the ten largest hydroelectric plants in the world. Its generation is five times greater than the production of all 28 German plants together (not including Jochenstein) and one and a half times greater than the projected generation of all Austrian Danube power plants (completed and planned.

Two more plants at borders are planned, to be built cooperatively by the adjoining nations—one by Czechoslovakia and Hungary (Gabcikovo/Nagymaros) and the other by Rumania and Bulgaria (Izlas/Somovit).

The continually growing demand for electricity intensifies the interest in further development of the Danube. The electrical energy of the completed and planned waterpower plants on the Danube and its tributaries will be essential for the electric current net in east and west.

Uses of Water Removed from the River

Domestic (Municipal) Use. The Danube River is not so polluted by interference of man that it cannot be used as drinking water after purification. The water quality corresponds to the biological class 2 (B-mesosaprob) for long stretches. Even after heavy discharges of sewage, quality class 3 or 4 is registered for only short stretches because of the ability of the river to purify itself. However, it must be said that downstream of large cities, because there exist only a few purification plants, the fecal load with *Escherichia coli* and *Salmonella* is noticeable.

In surface water which is to be used as drinking water not more than ten colonies of ml/enterobacteriaceae are allowed. In all investigated stretches of the Danube the bacterial pollution is above this amount.

Since 1960 only at Budapest has it been possible to take drinking water directly from the Danube. The water plant on Margaret Island pumps more than 10 million m^3/year into the municipal water supply. It is also intended that the water supply of Württemberg in West Germany be taken directly from the river downstream of the Leipheim reservoir.

Ice cover and a reduction of the ability to self-purify may lead to an enrichment of such chemical compositions as phenols and naphthenes, which give a disagreeable taste and smell to the water.

Industrial Use. Industries on the Danube need especially large quantities of cool water, crudely purified to eliminate the suspended matter. These industries include steam power plants, refineries, iron and steel industry, chemical industry, chemical industry, dockyards, sugar production, mines, and paper and cellulose industry. These works may use to 15 m^3/sec, and the waste water is usually discharged into the Danube again.

Industries which need water of a better quality use ground water from near the river bank. The growing pollution increases the cost for water improvement and thus the costs for industrial production.

Nuclear Power Generation. At the moment two nuclear power plants on the Danube are in operation: one in Gundremmingen, West Germany, with a cool water demand of about 14 m^3/sec, and the other in Vinca near Belgrade in Yugoslavia. The temperature increase of the Danube water caused by these two nuclear power plants is insignificant and to date no damages have been reported. More nuclear plants are planned in other Danube countries.

Agricultural Irrigation. The demand for Danube water for the irrigation of agricultural areas depends on the economic conditions in each country. In general, in the dry areas of the eastern Danube countries irrigation may result in an increase in production of up to 300%. Precipitation in summer and late autumn in these regions is not sufficient to ensure a good harvest. Irrigation plants now exist for a substantial area in the Danube countries. Land which could benefit from irrigation covers an area of 2,645,000 ha.

Table 2. Areas in demand of irrigation in the Danube countries

Country	In demand of irrigation (ha)	Already irrigated (ha)
West Germany	unknown	100
Austria	49,000	49,000
Czechoslovakia	240,000	240,000
Hungary	321,550	245,800
Yugoslavia	300,000	100
Bulgaria	731,200	143,800
Rumania	1,003,900	600,000
Soviet Union	unknown	unknown
Total	2,645,650	1,278,800

For these areas a demand of 0.5 liter/sec/ha is necessary; thus a removal of 1300 m^3/sec Danube water. Estimates for future development indicate the same rate of removal. At this time there are no complications concerning water quality; the abundance of nutrients has a favorable effect.

Waste Disposal

Domestic and Agricultural Sewage.   There is significant pollution of the Danube and its tributaries with domestic sewage. As mentioned above, 70 million people live in the drainage area, and only a small part of the sewage is purified. In most cases, the larger towns and cities on the Danube have no sewage plants; however, because of its good stream velocity, its low temperature, and its dissolved oxygen content, the river has an extraordinary self-purification power. This is especially important for the upper part of the Danube, which is rather loaded with fecal sewage. Oxygen values are between 80 and 100% but range as low as 40% and as high as 130% (Hungary).

Ammonium (NH_4^+) normally varies between 0.1 and 0.4 mg/l, nitrates (NO_3^-) between 4 and 15 mg/l, and phosphate (PO_4^{3-}) between 0.02 and 0.200 mg/l).

Below the cities extreme values are found, but only for short distances. The Danube water is rich in nutrients, and when the stream velocity is reduced, enrichment of plankton and a good development of the bottom fauna takes place. In many reservoirs algal blooms and decaying sludge are to be found. These parameters of heavy eutrophication and hygienic considerations make mechanical and biological purification necessary. Such projects are beginning in many places.

241

Industrial Effluents. Waste waters of industry influence the water quality of the Danube, especially in the upper course because of the small dilution.

In the Brigach, one of the sources of the Danube, repeated fish kills indicate uncontrolled discharge of waste waters of iron industry. Forty-five km downstream of these factories the two tributaries Breg and Brigach join with the still-small Danube. Brewery wastes and wastes of a sulfite-process wood pulp factory (500 liter/sec) bring about an intensive growth of *Sphaerotilus*, which settles in the downstream reservoirs and causes a remarkable sludge decay. A part of the acid liquor from the wood pulp factory is evaporated. Therefore, the dissolved oxygen content does not decrease below 50%. Oil refineries on the German Danube (40,000 tonnes/day), because of their efficient production, use less water, and their refineries have modern biological purification plants which provide the best possible protection for the Danube water. A downstream-situated wood pulp industry has no evaporation plant for its acid liquor and therefore causes a significant pollution of the Kelheim reservoir. Waste waters of a sugar factory increase the organic load.

In Austria there is relatively little industrial pollution. A factory for nitrogen increases the ammonium content (NH_4^-) to 5 mg/l but only in the waste-water flume. The wastes of a large iron and steel factory containing cyanogen, phenols, ammonium, and oil are discharged in the same section of the Danube. These wastes particularly influence the litoral biocoenosis.

In the Danube port of Linz an oil combustion plant was erected. Below this town a significant tributary, which carries the partly decomposed wastes of two large wood pulp factories and of a chemical fibre factory, along with troublesome sewage fungus, joins with the Danube.

The effluents of starch and sugar industries are easily decomposed, and their pollution is only local.

In the area of Vienna domestic sewage (1.7 million inhabitants) and many sorts of industrial wastes are discharged into the Danube. A large sewage disposal plant is under construction. Together with the wastes of a great oil refinery, the sewages of Vienna pollute the Danube significantly up to the border (water quality class 3 to 4).

In Bratislava an oil refinery and a chemical factory add to this pollution.

The next heaviest pollution is found in Hungary, where waste waters containing oil from several oil refineries and waste waters containing phenol from a brown coal mine are a heavy load for the Danube. But the rapid-flowing Danube, rich in oxygen, is able to digest all these industrial pollutions through its great self-purification ability so that its water can be used for drinking water after clarification in Budapest. Downstream of this municipal water plant the sewage of the city, with 1.8 million inhabitants, causes the water quality to be of class 3. The cold season and minimum flow stop the biological decomposing processes. Except for these pollution centres, the Danube water in Hungary is only moderately polluted.

In Yugoslavia 130 factories discharge their waste water into the Danube River and its tributaries. The oxygen content of the water corresponds to a saturation of 80-100% in most cases, and toxic matters have not yet been found. The great dilution given by the mighty tributaries Tisza, Drava, and Sava improves the water quality of the Danube. Chemical parameters move in a normal range. The biological classification shows quality class 2.

While in Bulgaria the pollution is due to food manufacture, the pollution in Rumania comes from iron and steel industries (Galati) and a reed-manufacturing paper and cellulose plant (Braila). Other industries of significance are wharves, oil ports, distilleries, canneries, sugar production, cellulose factories, and fertilizer production. Further industrial plants are under construction, but a purification of the wastes is intended. The high dilution prevents a deterioration of the Danube water quality. The same can be said of the Danube in the Soviet Union.

Radionuclides. In most of the Danube states regulatory investigations of and -activity have been carried out since 1958. The results show that values never rise above 10 pCi/liter. From the Danube source to its mouth there is an increasing tendency from 3 pCi/liter to 9 pCi/liter, as the radioactivity is bound not to the water but to the dissolved and undissolved matter. The greatest enrichment of radioactivity was found in seston and water flora (*Cladophora*, *Fontinalis*). Fish, the last link in the food chain, showed the lowest radioactivity. Sludge activity was traceable to 179 pCi/g dry weight.

The nuclear tests of the United States and the Soviet Union have caused in nearly all states along the Danube a noticeable, but not dangerous, increase in the radioactivity level in water, water organisms, and in sediments.

At the present time there exist three reactor plants on the Danube: the nuclear power plant in Gundremmingen (27 MWel), the reactor plant in Vienna, Austria (0, 1 MWel), and the nuclear power plant at Vinca, Yugoslavia (10 MWel).

The internationally recommended maximum permissible concentration of 100 pCi/liter for any mixture of -, -, and -emitters in drinking water has never been reached in the Danube.

Commercial Fisheries and Other Biological Exploitations. In the past, fishery on the Danube was a productive business. Centuries ago *Huso huso, Cyprinus carpio,* and *Silurus glanis* were objects of a blossoming trade in the eastern Danube countries; and the upper region of the Danube also had a rich fish population.

Interference by man and a changed limnological situation brought a regress of fishery in the past century. Of a hundred fish species known today in the Danube, 75% are used commercially or for sport. The average commercial catch is 4,400 tonnes annually, shared among the riparian states as follows:

West Germany	100 tonnes
Austria	200 tonnes
Czechoslovakia	350 tonnes
Hungary	450 tonnes
Yugoslavia	700 tonnes
Bulgaria	600 tonnes
Rumania	1,100 tonnes
Soviet Union	900 tonnes
	4,400 tonnes

In addition, 45,000 tonnes—mainly cyprinids—are caught in ponds and lakes in the floodplain and in the delta.

In the upper region of the Danube, where fishery is nearly exclusively used for sport, the main fish are *Chondrostoma nasus* (50%), *Barbus barbus* (10-25%), *Vimba vimba* (10%), *Esox lucius* (5%), and Salmonidae (5%).

The modern river dams without fish passes prevent fish migration. Since the completion of the 24-m-high dam of the water power plant "Iron Gate" (Yugoslavia/Rumania), Acipenseridae and *Alosa pontica* (Danube herring) can migrate only up to this point (Figure 5).

In the reservoirs of the upper Danube an increase of fish produce was achieved by introducing an artificially bred fish stock.

Acipenser ruthenus is the fish of highest quality in the middle region. The great mass of fish stock consists of Cyprinidae. By damming up the flood areas, fishery in Yugoslavia has decreased from 20,000 tonnes/year to 1,200 tonnes/year. At the lower Danube, too, where floodplains have been dammed for agricultural use, the food and spawning places in an area of 80,000 ha have been damaged.

244

Figure 5. Danube power plant Ybbs-Persenbeug, Lower Austria

Cyprinids (*Cyprinus carpio* L. and *Abramis brama* L.), *Lucioperca lucioperca, E. lucius, Perca fluviatilis, S. glanis,* especially *Alosa pontica,* find their best living conditions in the floodplains along the Danube. Their annual production surpasses 100 kg/ha.

In the mouth area freshwater fish, as well as migrating fish from the Black Sea, find good conditions. The prevailing species are *H. huso, Acipenser guldenstadti, A. stellatus, A. ruthenus,* and *Alosa pontica.*

Before the erection of the power plant "Iron Gate" the first three species migrated upstream 1,000-1,500 km. *Acipenser sturio* is caught in a quantity of up to 800 tonnes annually at the Danube mouth.

In the delta the harvest of reed (*Phragmites communis*) is commercially important. In Rumania reed is an important raw material used in the manufacture of cellulose and its by-products in a factory at Braila.

The total area of the delta, with an area of 5,640 km^2, is the second largest delta in Europe, the first being the Volga delta (18,000 km^2). Eighty percent of the delta area belongs to Rumania and 20% to the Soviet Union. Fifty-one percent of the delta lies below sea level.

An area of 4,900 km^2 (87% of the delta) is used for the following industries:

	Fishery	Reed	Agriculture	Forestry
Rumania	1,260	2,140	620	200
Soviet Union	160	360	160	-
Total (km^2)	1,420	2,500	780	200

Hygiene

Microbiology is of great importance in assessing the degree of contamination. The most considerable contamination of the Danube by sewage is found below the cities, with the coli-titer (bacteria of the Escherichieae) and the pathogenic species of the Enterobacteriaceae particularly noticeable.

In Austrian investigations of wastes below sewage effluents, *Salmonella* species can always be detected. This makes the spread of disease (Salmonellosis) through water possible. In such contaminated Danube areas a latent epidemic danger must be considered.

Recreation and Culture

The Danube basin is one of the most beautiful areas for recreation in all Europe (Figures 6, 7, and 8). Despite civilizational interferences, the river has remained a harmonious picture in its upper course.

After joining with the glacial waters of the Inn River, the Danube becomes an alpine river, passing pretty wood-covered valleys with great velocity. After entrance into the Pannonian plain an open landscape offers serenity.

In the district of the "Iron Gate" the Danube breaks through the Banat region and East Serbian mountains, with many riffles and rapids which will disappear after completion of the power plant. The stream becomes mightier and mightier until it divides into three branches flowing through a unique paradise of nature accompanied by lakes, reedwoods, and a versatile bird population until it reaches the Black Sea.

The total drainage area of the Danube, and especially the river itself with its different tributaries, offers many possibilities for recreation to its inhabitants and visitors. People like fishing, boating, swimming, water-skiing, and hunting. The best recreation is offered by a passenger journey on the river from Passau to the Black Sea.

Figure 6. Upper Danube, Fed. Germ. Rep.

The Danube is the most important stream in Europe with regard to history and culture. In prehistoric times the space along the Danube was a favoured area for settlement. The oldest traces go back to 180,000 years before Christ (glacial epoch). From the later Paleolithic (50,000-20,000 B.C.) we have two statues of a goddess of fertility, made of limestone and ivory ("Venus von Willendorf"), which belong to the oldest plastic art of Europe. About 4000-5000 B.C. (Neolithic) the bearers of the "Donauländische" Culture were wandering up the Danube from western Asia, pursuing commerce along the whole length of the Danube. During the Bronze Age Illyrian and Tracic peoples were supplanting their culture.

About 400 B.C. the Celts from the West settled here, founding towns and managing merchantmen's navigation. The Romans ruled over the whole of the Danube from the beginning of the Christian Era until the fourth century.

Thrice the Turks threatened Vienna. In the eighteenth century the Turks' influence was driven back, and during the nineteenth and twentieth centuries the eastern peoples obtained their independence and self-determination.

Figure 7. Breakthrough of the Danube, Schwäbischer Jura at Weltenburg, Fed. Germ. Rep.

European culture is in debt to the territory of the Danube for important artistic creations. In the Middle Ages innumerable monasteries were founded where painting and applied art flourished. During the eighteenth century there was a flowering of painting, architecture, and sculpture—the "Donaubarock."

In the first half of the nineteenth century the most important works of music were composed in the capital of the monarchy at the Danube, and these works obtained world-wide renown (Mozart, Haydn, Beethoven, Schubert, and Bruckner). During the second half of the same century the waltzes and operettas of Josef Lanner and Johann Strauss were spreading from Vienna over the whole world; the choir waltz, "The Blue Danube," is known as the second national anthem of Austria.

Frontiers

The Danube is a frontier stream which not only separates many states, but has also linked peoples at the river for thousands of years. The Danube basin was not only the stage for many wars in former centuries, but also for the two world wars in our time.

Figure 8. Meander of the Danube, Schlogen, Upper Austria

Since the last war people have lived peacefully together, and their countries have bloomed in new prosperity. New organisations promote economy and science, such as the international Danube Commission which cares for navigation and the International Study Group for Danube Research, which by its scientific studies makes possible a modern water economy for the whole Danube basin. Such a collaboration is necessary, for water knows no borders.

The Danube, designated correctly by Napoleon as the "King of the Streams," will no doubt continue to connect the nations in the future as it has done in the past.

Literature Cited

Commission du Danube. 1970. Annuaire Statistique de la Commission du Danube pour 1969, Budapest.

Liepolt, R. [ed.]. 1967. Limnologie der Donau. E. Schweizerbart' sche Verlags-buchhandlung, Stuttgart.

A COMMENTARY ON "USES OF A RIVER: PAST AND PRESENT"

E. D. Le Cren, Session Chairman

Men and rivers have been intimately associated with one another throughout most of the history of man as a species. The art galleries of Cro-Magnon man in Europe were caves in the gorges of the River Dordogne which produced the salmon, possibly a staple part of their diet. Modern western civilization originated in the ancient cultures of the Near East that were cradled in the great river valleys of the Nile (Hammerton, in these *Proceedings*) and the Tigris and Euphrates. The river systems of North America were the travel routes of the Indians and the means of first penetration, by canoe, of the early explorers. Finally, the power that lights and operates the building in which this meeting is held depends upon a river for its generation. The papers that describe seven different rivers provide a wealth of examples illustrating the importance of rivers to man and the powerful impacts that man has had on them.

The first section of the symposium considered rivers from a purely physical viewpoint; their part in the hydrological cycle and the natural laws that determine their courses and configurations. Secondly, rivers were considered as habitats for living organisms and features of their biota were described. Now, in this third part man is being introduced into the picture as a major component of the ecosystem.

It is as ecosystems that rivers should be studied—and managed. Systems that involve the physical environment, the whole community of plants, animals and microbes, and man as an exploiter of river resources. The following series of papers develop this approach to rivers and the implications it has in their rational management. They deal primarily with the rivers themselves, but, as is inevitable for ecological reasons, sometimes have to move away from the river bank into the catchment (or watershed). For the ecology of a river is often but a reflection of the ecology and use of its catchment, and the river cannot be understood without some knowledge of what happens on the land around it.

Most of the rest of this paper is devoted to a brief review of the range of uses which man makes of rivers, and an attempt to classify and list these uses and their impacts. It is hoped that this will serve as an appropriate overview of the series of case histories which provide illustrative examples of many of these uses

251

and impacts. The list of uses may serve to show the great range of ways in which the activities of man can be related to river ecology. Although in several of the rivers that are described man appears to have had his impact in only one or two ways, some of these examples illustrate how problems and conflicts may have arisen because of a failure to realise the wide range of multi-purpose potentialities provided by river resources.

A Classification of River Uses

There are several ways in which the uses of rivers could be classified. Further, if the impact of man on the ecology of rivers is being considered, a number of fringe human activities or interests must be included even though they cannot really be called a "use." For example, neither flood prevention nor the control of the vectors of human diseases are uses of rivers in the strict sense, but more accurately, attempts to control the effects of rivers on man.

In Table 1 the human uses and activities that affect rivers are listed under a classification based on human interests rather than on the impact they have. The paragraphs below amplify and discuss some of these.

Catchment Use. The importance of catchment use as an influence on river ecology (and indeed hydrology and physiography) has already been mentioned in the introduction to this paper. Though man may not be using a river itself, by his use of its catchment he can have a profound effect on the river indirectly. The effects of such changes as the removal of original climax forest vegetation, the cultivation of land previously forest or grassland, or the building of a town on previously undeveloped land would require a symposium on their own to discuss in detail and the present papers largely confine themselves to more direct human impacts on rivers themselves. Nevertheless, passing reference to catchment land use is made in several of the river case histories (e.g., Starrett for the Illinois River).

Navigation. The Illinois is an excellent example of a river that has been canalised, dredged and regulated by man to aid navigation and other examples are the Thames and Danube. The scale of impact varies largely with the size of the river and thus of the ships that will use it. Apart from the major engineering effects, ship traffic tends to disturb the water, causing higher concentrations of silt in suspension and bank erosion; both these have ecological effects. An incidental but nevertheless ecologically important effect of the building of canals is the way they can destroy barriers to migration between river systems and thus aid the dispersal of species. The introduction of the sea lamprey and the alewife to the Great Lakes of North America is a classic example, and the escape and spread of *Elodea* throughout England is another.

Table 1. A classification of human uses and influences on rivers

Main categories	Sub-categories	Some effect on rivers
Catchment use	Deforestation Cultivation Land drainage Urbanisation Mineral extraction	Changes in runoff pattern, changes in silt load, changes in temperature regime, chemical enrichment, toxic pollution
Navigation		Canalisation, dredging, flow regulation, species dispersal
Flood control		Dredging, canalisation, flood banks, weed control, abolition of floodplain
Power generation	Water mills Hydroelectric	Dams, diversions, obstructions to migration, changes in flow and temperature regimes
Abstractions	Domestic, Industrial Thermal power generation Agriculture	Reservoir building, diversions, reductions and changes in flow regime
Effluxions	Domestic sewage Agricultural sewage Industrial effluents Radionuclides Pesticides, Herbicides	Deoxygenation, biotoxins, increases in flow, canals and diversions
Biological exploitation	Commercial fisheries Fur-bearing mammals *Phragmites*, etc.	Sometimes weirs and traps. Over-exploitation
Hygiene	Control of disease vectors	Draining, dredging, bank management, pesticides
Recreation and culture	Angling, Boating Swimming Water skiing Shooting (hunting) Nature conservation Aesthetics	Creation of waters, habitat modification, disturbance, introduction of new species
Political frontiers	National, provincial and local	Defence works, problems in management

Flood Control. The very reasons which have enticed man to settle, cultivate, and eventually build cities and industries on the fertile floodplains of rivers subject these sites to periodic destructive floods and hence the need for engineering control works. These measures in their turn confine the energy of the flood waters and so increase the damage floods cause to the river itself and to its biota. The engineering works usually include regulatory dams, dredging and canalisation, flood banks (or levees) and weed control. All of these can have effects on the ecology of the river, usually tending to reduce the natural diversity of habitats and increase the seasonal variation in discharge, depth and temperature.

Power Generation. The energy source provided by gravity acting on water has long been utilised by man, and even in early historical times quite elaborate engineering works were constructed to harness this power. A major use of water power, particularly in Europe, was to grind corn, but from the time of the industrial revolution these water mills were increasingly used for other purposes. The last hundred years have seen the development of hydroelectric generators until they have become a major source of power. In their extreme development, as is well illustrated by the Columbia River, a major fraction of the potential energy has been harnessed and the whole nature of the river altered. As well as the series of barriers to migration provided by the dams the conversion of a continuously flowing river into a chain of lakes alters the whole ecology and has a profound effect upon the biota.

Abstraction. Man has many uses for water and rivers provide an obvious and convenient source of supply. The types of abstraction can be classified into four major categories according to the use that is made of the water.

Firstly, there is the domestic use for drinking, washing, waste flushing, etc. Such water needs to be nontoxic and pathogen-free and preferably clean and free of tastes and odors. The cost of treatment to render the water suitable for domestic supply is linked to the quality of the water at the point of abstraction, and this can have an influence on the other uses of the river upstream.

Secondly, there is the use of water for industrial processes. These are very varied, and so have varied water needs in terms both of quality and quantity. Many of the smaller industrial requirements are supplied from domestic distribution networks.

Thirdly, large quantities of water are often abstracted from rivers to cool thermal power stations. Cooling can be direct when the water is returned somewhat warmed but otherwise much the same in quantity and quality, or indirect, when cooling towers are used. These latter need much less water but evaporate some of it.

Fourthly, there is abstraction for agricultural irrigation. This is a major demand in some situations and the Nile is the classic example of this human use of a river.

These abstractions have varying effects upon a river depending largely upon the amount of water abstracted (which can range from an insignificant proportion to the whole flow) and what is its quantity and quality when it is returned to the river. Abstractions often involve storage dams, diversion canals and other engineering works that can influence river ecology. Finally, even if abstractions are taken from ground water rather than from rivers, they can have serious effects on river discharges.

Effluxion. The use of rivers for waste disposal (effluxion) is another example when the impact has increased rapidly over the last one hundred years or so, though, as the account of the River Thames shows, some of the problems date from the seventeenth century or earlier. The forms of waste received by rivers vary greatly according to their origin and can be classified on this basis.

Firstly, there is the excrement of man himself, nowadays usually in the form of sewage that has undergone a varying amount of treatment. Of a basically similar nature are the effluents from agriculture and especially those from intensive animal husbandry. The chief effect of these sewage effluents (and also many of those from industries based on biological materials) is to deoxygenate the river water by imposing a biochemical oxygen demand (BOD) and to enrich its fertility. The fertility of river water is also increased by the chemical fertilizers carried from the land by runoff.

Secondly, there is a large range of wastes derived from industry and mining. These can add suspended silts and other solids, but above all a wide range of substances toxic to organisms such as ammonia, cyanides, phenols and heavy metals, the toxicity of which can have profound ecological effects.

Some waste materials have special characteristics and these include radionuclides and the complex and often unknown wastes from modern organic chemical industries. Like pesticides and herbicides some of these may persist in the environment and tend to be accumulated by organisms and concentrated in the food chain.

Effluxions generally tend to increase silt loads and alter river ecology through their toxicities and oxygen demands, but engineering works associated with them may include canals and diversions (e.g., the Chicago sanitary canal on the Illinois River), and in some cases effluents provide the whole or a major part of the flow of a river. The River Danube and the Delaware River both provide examples of the effects of effluxion and the measures that may be taken to combat them.

Biological Exploitation. Fish provide the most valuable biological yield from rivers, and are the end result of the productivity of the ecosystem. Rivers are often very productive, perhaps because of the allochthonous contributions of organic matter from their catchments. Some of the more valuable commercial species of fish, however, are only spawned and reared in rivers and derive most of their growth from the sea (e.g., the salmon of the Columbia River).

Unless the fish are overexploited, commercial fishing does not usually have as great an impact on river ecology as other human activities, but some fisheries involve quite elaborate weirs and traps that may alter flow patterns and have effects similar to those of other engineering works.

Fish are not the only exploited river organisms. For example, fur-bearing mammals such as mink and beaver are trapped from northern rivers and in the Danube delta the reed *Phragmites* is harvested in large quantities as a source of cellulose.

Hygiene. Though not a *use* of rivers, man's attempts to control disease organisms and vectors have impact on river ecology and other uses and so must be included in this list. Notorious diseases such as typhoid and cholera are water-borne. *Schistosomum* (*Bilharzia*), whose alternate hosts are freshwater snails, is a common human parasite in some tropical countries. A large number of tropical diseases are transmitted by mosquitoes and other flies with aquatic larvae, the most typically fluviatile of which is *Simulium damnosum* the vector of onchocerciasis. The control of vectors may include the draining of riparian ponds and marshes, the control of bank vegetation and the use of molluscides and insecticides. All of these measures may have side effects on the ecology of rivers.

Recreation and Culture. As some later papers in this symposium emphasise, the political and economic values of rivers as centres for recreation are now being realised at least in the more developed countries. These recreations and sports associated with rivers are many and difficult to classify in any logical system but an attempt is made here (and in Table 1) to list the more important of them.

Angling is now a major participant sport and most rivers in Europe and North America are intensively used for recreational fishing. Boating, sailing, swimming, diving and water skiing are other popular water-based recreations. Shooting (hunting) ducks and other birds and mammals associated with rivers is another sport analogous to fishing.

Though few whole rivers, as such, have been designated nature reserves, parts of rivers and associated habitats are managed for nature conservation in several countries. The main aim of this management may be the preservation of rare species, community or geological phenomena, and ecological research or education. Some river nature reserves are used as tourist attractions (e.g., the Nile at the Murchison Falls).

Finally, rivers have exceptional value as scenic features of natural or man-made landscapes and provide much visual enjoyment and aesthetic refreshment. This, together with their significance to human survival and prosperity have assured them a place in art, culture and religion.

In general the recreational and cultural uses of rivers have tended to assist their conservation in a natural or semi-natural state; a policy that often conflicts with other uses, as several of the examples show. Management for angling, however, may involve "habitat improvement" engineering and the introduction of exotic species. Rivers have even been created or diverted for purely aesthetic reasons. The rising demand for aquatic recreation will tend to increase the ecological damage and conflict that too heavy a recreational pressure can cause.

Political Frontiers. An unusual—but frequent—use of rivers must be included in the list as it is one that has repercussions on river management. Man, like other territorial animals, has a need for devices to mark the boundaries between territories and rivers have been and still are used for this purpose. They are permanent, need no maintenance and are difficult to cross and so make rather effective boundaries.

Although occasionally their natural defensive character may be enhanced by military engineering usually their use as a frontier has little direct effect upon their ecology. It may even tend to delay the development of other uses, but, more often, causes political, economic and especially legal problems in management and control. In rare cases (e.g., the French *départments*) boundaries follow watersheds and more rational political areas result. Some of the problems of international rivers are discussed in the papers on the Danube and the Nile.

Effects of Human Manipulation and Use

In the preceding sections mention has been made of the manipulations or modifications of river features that are involved in human use. Many of these manipulations are engineering works on the physical structure of the river channel such as dams or weirs, embankments, canalisation, dredging and diversions. They obviously have a direct impact upon the course and bed of the river that they

257

aim to modify. Other manipulations such as abstractions are directed primarily to the quantity of river water and affect the pattern of discharge and its total amount. They need not, though often do, involve significant engineering works on the channel or bed. Effluents have some influence on the quantity of water but act more directly on its chemistry or what may be called water quality. The direct manipulations of the fauna and flora, e.g., exploitation, species introduction and pest control, will affect the biota directly but may have minor indirect effects on the physical and chemical features. Recreational and aesthetic use of a river will not normally have any direct influences unless it incidentally involves other manipulations or is very intensive.

Most of these various modifications to the physical, chemical, biotic or aesthetic aspects of a river act directly on one feature but at the same time indirectly on many of the others. Some of these effects are listed and crudely assessed in Table 2. This table is obviously incomplete and too simple and the judgment of benefit or harm is subjective. I have attempted to judge the latter from the viewpoint of the river itself rather than immediate human gain, hence the predominance of minuses (harm) over pluses (benefit). The presence in a section of both pluses and minuses indicates that the particular action can be either beneficial or harmful depending upon the circumstances or the way in which the manipulation is carried out.

For example, a dam will drown and destroy the river, as a river, upstream as far as the water is impounded. Downstream skillful manipulation of the sluices to regulate the flow might help to maintain a natural channel and protect it from the ravages of extreme floods and so be beneficial rather than deleterious. Similarly the introduction of an exotic water weed could help to stabilise the banks and bed, tend to regulate water depth and discharge, oxygenate the water, enrich the habitat diversity and, if it had an attractive flower, add to the beauty of the river. On the other hand, such a weed could become a pest and turn the river into an evil-smelling swamp choked with stagnation and decay! Thus any particular manipulation need not always be either good or bad.

The predominance of harmful effects in Table 2 probably arises partly because most human environmental manipulations attempt to simplify a complex natural feature or situation and in so doing stimulate reactions towards the original state, or disrupt a naturally stable dynamic equilibrium.

Interactions Between Different Uses

Many of the problems of river use derive not only from the disturbance of natural features but also from the interactions between different uses or different

Table 2. An approximate appraisal of the impacts of various human manipulations on river features. The numbers of pluses and minuses (from one to three) indicate the benefit or harm that can result from each manipulation to each river feature

Type of use or manipulation	River Channel	Water quantity and flow regime	Water quality	Biota	Aesthetics
Catchment land use	+ – –	+ – –	++ – –	+ – –	+ – –
Dams and weirs (impoundments)	+ – – –	+++ – –	++ – – –	++ – –	++ – – –
Boat traffic	– –		–	– –	– –
Canalisation	+ – – –	– –	– –	– – –	– – –
Dredging of bed	++ – – –		–	–	– –
Diversions	– – –	– – –	– –	– –	– – –
Bank modification	++ – – –	–	+ –	+ – –	– –
Abstraction	–	– – –	– –	– –	– –
Effluents	–	+	– – –	+ – – –	– – –
Biotic exploitation	–		–	– –	–
Species introductions	+ –	+ –	+ –	+ – – –	+ – –
Pest control	+ –		– –	– –	+ – –
Recreational uses	+ – –		–	+ – –	+ – – –

types of artificial manipulation. The case histories that follow show many instances of such conflicts; for example, the effects of power dams on the commercial fisheries of the Columbia River or the interactions between sewage disposal and water abstraction in the River Thames. In other instances some degree of balance between uses has been achieved and management for multiple use seems possible; the River Danube appears to be successfully avoiding the worst conflicts.

Perhaps multiple use of river resources will tend to ensure a balance between different manipulations so that natural features and systems will not be disturbed beyond the point where their built-in homeostatic mechanisms can adjust without disruption? Certainly in the examples discussed in the following papers those rivers in the Old World where a reasonably successful pattern of multiple use has been achieved (partly because of a longer history of more gently expanding exploitation) contrast with the problems of American rivers where two different dominant uses are in conflict.

The case histories also show that successful multiple use may involve sophisticated patterns of management (e.g., the model used to control pollution in the Delaware estuary) or require new and elaborate legal and administrative controls. Perhaps effective management for multiple use of such a complex and delicately balanced ecosystem as a river inevitably needs deep understanding and sophisticated methods of control?

Summary

Man uses rivers for a great many purposes such as navigation, power generation, water supply, waste disposal, biological exploitation, recreation and as political frontiers. He also affects rivers through his use of their catchments and attempts to control floods and eradicate pests. The engineering, physical, chemical and biological manipulations involved have profound effects on the river ecosystem and are more often harmful than beneficial. The different uses interact with one another, but successful multiple use is possible with careful management. These uses and their effects are listed and discussed with reference to the seven river case histories described in the symposium.

SESSION III

EFFECTS OF RIVER USES
Chairman: Theodore P. Vande Sande

REGULATED DISCHARGE AND THE STREAM ENVIRONMENT

J. C. Fraser

A river with a natural flow regimen is a rare phenomenon in this age of intensive resource development. For centuries man has been harnessing the power of falling water and diverting water from streams for irrigation and domestic purposes. By these activities man has long influenced river ecosystems, but until the present century his impact on the discharge or flow of most rivers was relatively insignificant.

Modern technology has made it possible to control the largest of rivers. Large dams and huge storage reservoirs are now commonplace. River discharges are being altered throughout the world. Man has now become the grand manipulator, if not the destroyer, of lotic ecosystems.

With a growing concern for the quality of life on this earth, the in-place value of rivers is taking on new meaning and new proportions. As a result we are focusing greater attention on the effects of our manipulations of the river environment. Our knowledge is still in its infancy, but it is growing. This report is intended to provide a review of some of the findings on the effects of controlled discharge on the river environment, with emphasis on fisheries resources. Fisheries biologists have been studying problems associated with controlled streamflow for many years in relation to the management of fisheries resources, and this paper is intended to provide a status report on where we stand in this field. There are, of course, many other aspects of discharge relating to the ecology of streams: a few of these will be touched upon briefly later in this report.

Discharge. A definition of discharge and some clarification of its components are in order. Discharge is frequently considered the outflow of a canal or stream where it enters another stream or body of water. However, for purposes of this report discharge will be considered as synonymous with streamflow since I will be largely considering water flows in natural channels. Discharge as used in this report is the volume of water flowing in a channel during a given time interval; it may be natural or controlled and may be considered in terms of a measured quantity at the point of release or as a measured quantity at any point in the stream.

Discharge in relation to its effect on the lotic environment is influenced by several component factors, sometimes referred to collectively as the "regimen" of the stream. The most important of these abiotic factors is velocity. The biotic communities of streams are highly oriented to currents of water. Most organisms of stream environments have rather narrow preferences for flow velocities, and therefore velocity plays an important role in the species composition and characteristics of the lotic biota. In this respect velocity exercises an influence on the "production" of a stream in terms of a particular species or group of desired species.

Depth is perhaps the next most significant physical factor in relation to discharge. The amount of discharge determines depth in a channel of given configuration. At least in some of their life habits most lotic species are limited in their preference of depth of water.

Width of a channel of given configuration is similarly determined by discharge and can be important in terms of area for spawning of fish, food production, temperature change, and satisfying spatial requirements of some species.

Timing of discharge is a major factor in assessing the effect of flow on the stream environment. The life cycles of lotic species are frequently related to natural seasonal variations in discharge.

Fluctuation of flow, or in other words, the short-term (e.g., hourly, daily, weekly) variation in discharge can have a marked influence on organisms with a narrow depth tolerance. In some situations fluctuation can result in temporary desiccation of part or all of the habitat for some aquatic organisms.

Quality is a component of discharge, especially in relation to a stream's capacity to transport sediment. Discharge can affect temperature, the transport of organic materials, nutrients, and other quality factors. Of special concern is the effect of discharge on salinity gradients in deltas and estuaries.

Some other features of streams having an interrelationship with discharge are channel configuration, stream-bed gradient, and stream-bed substrate. Each of these exercises both major and localized influences on velocities, depths, and to some extent width.

General Effects of Discharge. All of the foregoing factors of a stream-flow regimen are subject to measurement, but the relationship between a given quantity and the biotic community is imperfectly known. We know that lotic environments have been altered drastically by water development schemes throughout the world. The effects range from the obvious elimination of an aquatic environment by the complete diversion of a stream to an improvement or enlargement of the biota in terms of desired species or total biomass.

In recent years the biologist and other scientists and technologists have been seeking answers to the problems of predicting the effects of an altered stream-flow regimen on the aquatic environment and making recommendations for flows which would maintain specified or desired conditions. Studies along these lines have been primarily aimed at maintaining populations of desired species, such as the Pacific and Atlantic salmons, the sturgeons of Russia, and trout. Studies of estuaries and systems ecology are also shedding new light on the effects of altered stream flow.

As man's population increases, his demands for water increase. The construction of large dams and other water development projects will continue into the forseeable future. There is now more of an awareness of the need to assess the ecological impacts of water development projects than ever before, and many governments are now requiring such assessments before construction. Ideally, such biological assessment studies are conducted in advance of planning the physical features of the project, but unfortunately this is still a neglected process. In many cases little or no consideration is given to the ecological conse-quences of major water projects.

We are now aware of a number of effects of altered discharge on certain species. We know that food production, stimulus for migration, success of migration, spawning, survival of eggs and juveniles, spatial needs, species composition, water quality, and harvesting of fish can all be affected by stream-flow alteration.

The effects of regulated discharge are not confined to stream habitats, but extend to the biota of deltas, estuaries, lakes, and the ocean, as well as to adjoining terrestrial communities and to esthetic values. Riparian vegetation complexes are frequently dependent upon the flow regimen of a stream. Elimination of flooding by river storage facilities has had major adverse effects on many marshes, estuaries, and on stream-associated vegetation. The character of a stream bed can change markedly with an alteration in regimen, resulting in changes in the species of plants and animals capable of attachment on the new substratum. In some cases such changes have encouraged the production of species of insects troublesome or injurious to man. Concern is being expressed for many of the beaches on western North American ocean shorelines because of reduced sand replenishments resulting from continued reduction of river flows.

Effects on Fish Populations

The stream as a habitat for living organisms presents a highly variable and yet specialized set of conditions. The important consideration is that stream organisms have adapted to the various special conditions presented by a flowing-water environment. Many species have adapted to a rather limited range of velocities or depths of water, and their dependence upon sufficient flow to provide these conditions is usually complete.

The significance of water flow and the preference of lotic species for particular velocities is well illustrated in the literature. Ambuhl (1959) emphasized this factor and noted that the adaptation of forms to flowing water is not a simple orientation, but in many species has a very finite qualitative requirement for velocity.

The stream-flow requirements of aquatic organisms are still only vaguely known, but a number of studies have been made which shed considerable light on the needs of fishes, especially salmon and trout. From these studies we may gain an appreciation, if not an understanding, of the importance of discharge to the various stages in the life cycles of aquatic organisms. Subsequent discussion will center largely on salmon and trout. It must be recognized that these species are usually inhabitants of the higher velocity sections of streams, and their specializations for this environment are dissimilar in many cases to those specializations of species inhabiting streams of low velocities. It should also be recognized that the life habits of salmon and trout are geared to centuries of evolutionary adjustments to the seasonal variations of their flowing-water environment.

The literature gives ample evidence of major ecological changes throughout the world as the result of altering stream discharge. Total production of salmon, for example, has long been the subject of study and, although many variables other than flow exert an influence, certainly flow is one of the more important factors. Stream flow has been recognized as a limiting factor in the production of salmon by a number of investigators, including McKernon, Johnson, and Hodges (1950); Smoker (1953); Neave (1958); and Wickett (1958). Smoker calculated that a positive correlation existed in the period of 1935 through 1950 between the annual stream runoff in 21 western Washington watersheds and the annual total commercial landings of coho salmon (*Oncorhynchus kisutch*).

The effects of flows on species composition and distribution in streams have been noted by Curtis (1960) and Delisle and Wooster (1964), among others. Floods have been noted to have both beneficial and detrimental effects on populations, and fluctuations of flows and reduced flows have been cited as

266

influencing fish populations (Neave, 1958; Penaz et al., 1968; Pfitzer, 1954; and Lagler, 1969).

Many fish populations are dependent upon annual flooding for food and spawning. Ewer (1965) described the plight of a clam fishery of considerable importance to native people on the lower Volta River in Ghana. With stabilized flows there has been encroachment of tidal influences farther up-river, and the lower river fisheries are dependent upon annual flooding, which will rarely occur in the future because of upstream reservoir controls.

Butcher (1967) and Wharton (1969) cite the many changes brought about by discharge control on the river systems in Australia, where the future of the famous Murray cod hangs in doubt. In reference to the Murray River, Wharton noted:

> The teeming bird, fish, amphibian, reptile, mammal and invertebrate life that characterizes the river backwater systems after their periodic filling and which is almost beyond the belief of those who have not seen it at close quarters, is becoming a comparatively rare event and seems ultimately to be doomed.

The effects of discharge are not, of course, limited to the river and its flood plain—they extend to the estuaries and to the ocean. Diener (1963), Copeland (1966), and Odum (1970) cited changes in the ecology and productivity of estuaries brought about by reduced inflow of fresh water. The far-reaching influences of the Amazon River on the ecology of the ocean are commonly recognized. Andrew and Green (1960) noted that the Fraser River (British Columbia) discharge is known to have profound effects on the temperature, salinity, and other oceanographic features of Georgia Strait.

Migration. Migrations of fish are affected by the amount of discharge in a number of ways. Discharge can cause migrations to commence, create barriers at high or low flows, cause delays, disrupt normal routing, and change the speed of travel. The role of discharge in inducing migrations of fish is important in many species and may vary between species and between streams for the same species. Most, but not all, salmon migrations occur at times of the year when increasing or seasonally high discharge can be expected. In some rivers, such as the Sacramento River in California, there are peaks of migrations, but some salmon enter the river in almost every month of the year.

Davidson et al. (1943) observed different reactions of adult pink salmon (*O. gorbuscha*) to flows in three streams in the Pacific Northwest. In one case the salmon milled about in the bay until sexually mature before attempting to enter the stream even though flows were suitable for migration at an earlier date. There was indication that they responded to changes of flow after they collected at the mouth. In another stream which also had adequate flows the salmon moved directly into the lower reaches of the stream, where they remained until sexually mature. The flow of the third stream fluctuated in close relationship to daily rainfall, and here the salmon moved into the stream when the flow was high due to rain. They moved upstream rapidly, except during extreme freshets, and stopped in deep holes, where they remained until sexually mature. If no rain freshets occurred in this third stream, the salmon remained in the bay until sexually mature and then attempted to enter the stream regardless of its flow. These observations indicate that in some streams migrations of salmon can be delayed by reduced flows or elimination of natural freshets until the fish are sexually mature, and then they may migrate regardless of flow.

In reporting on studies of pink salmon in McClinton Creek, British Columbia, Pritchard (1936) demonstrated a positive significant correlation between numbers of fish migrating from the sea to the stream each day and the maximum daily water height in the stream and the daily rainfall in the area.

It has also been observed that reductions of river flow may in some cases be a stimulus to the migration of Atlantic salmon (*Salmo salar*). Hayes (1953) noted that in comparison with natural rains the effect of artificial freshets produced by release of stored water should be considered as small. He believed it impossible to produce a flow large enough to inhibit salmon ascent in the La Have River, Nova Scotia. Generally, however, extreme freshets have been observed to reduce or even stop salmon migrations (e.g., Pritchard, 1936).

The upstream movement of salmon in several rivers in England also suggests a triggering of migration more by natural freshets than by high flows as such (Alabaster, 1970).

Andrew and Green (1960) observed that discharge influences the timing of the Adams River sockeye salmon (*O. nerka*) run. In years of low spring and summer discharge, the spawning migration is delayed and takes place over an extended period of time and at a later average date. They noted that in such years the fish in the last part of the run often fail to arrive on the spawning grounds or arrive too late for efficient spawning.

Reduced flows can cause undesirable delays in migration. Brett (1957) noted that delays in the upstream migration of adult salmonids cause premature use of energy reserves and develop stresses causing death or reduced reproductive success. The low tolerance of salmon to delays was also noted by Hourston (1958).

Points of difficult passage or complete barriers to migration can develop with reduced discharge, despite the appearance of adequate flows for passage in most of the stream (Andrew and Green, 1960; Hourston, 1958; Deschamps, Wright, and Magee, 1966).

The effects of delay at the Hell's Gate velocity barrier on the Fraser River is a classic example of a major disruption of the ecology of a river by a change in velocity in one location. Delays caused by high or low flows can be just as important in their effects on migrating salmon as a physical barrier.

Upstream migrants become confused and trapped, or at least delayed, by a discharge leading them into incorrect paths of migration. Hydroelectric power station discharges to a river from a tributary channel can cause such off-route delays, as noted by Andrew and Green (1960). Ganssle and Kelley (1963) described a case of actual flow reversal in the delta of the San Joaquin River in California caused by heavy pumping for irrigation at one location. They predicted that in the future such flow reversals will increase in duration and intensity, making it difficult, if not impossible, to maintain or restore salmon runs in the San Joaquin River system.

Downstream migrant salmonids usually have a strong orientation to current. The dependence on this orientation varies with the species, and among the salmon it is particularly pronounced. Mackinnon and Hoar (1953) demonstrated that chum and coho salmon smolts responded positively to changes in water flow. Both species were found to respond to the stronger of two parallel laminar currents, but the chum salmon were better able to distinguish very small differences of flow.

The rate of downstream migration of juvenile fish can be greatly influenced by the discharge and its velocity. The migration rate of young chinook salmon (*O. tshawytscha*) in the Snake and Columbia rivers was found by Raymond (1968) to be in general directly related to stream flow. Low river flows increased the travel time and contributed to delays in the migration of young salmon to the ocean.

Raymond (1969) compared the migration rate of juvenile chinook salmon from Ice Harbor Dam to the Dalles Dam before and after operation of the new John Day Reservoir located between them on the Columbia River. Before John Day Reservoir was formed, fish traveled the distance in an average of 14 days at a rate of 18 km per day. After the formation of John Day Reservoir the average time was 22 days, or 11 km per day. Delays in the downstream migration of sockeye smolts resulting from reduced stream velocities were also cited by

Andrew and Green (1960) as a possible reason for reduced survival of the smolts. Similar concern was expressed by Gordon (1965).

Again, as with upstream migrants, the stimulus provided by natural freshets can be a significant factor in the migration of juveniles. Freshets were noted by Chapman (1965) as appearing to cause large downstream movements of juvenile coho salmon.

The role of discharge is thus extremely important to anadromous species. The stimulus and orientation provided by a current appears essential to the migration of many species. A current-oriented downstream migrant is placed in a time-delay and stress situation in the absence or curtailment of current and certainly cannot be expected to adjust to a flow reversal situation as described by Ganssle and Kelley (1963).

Spawning. Virtually all stream-dwelling fishes have adapted their spawning activities to some range of water velocity and depth. Knowledge of this range is one of the key factors in the management of salmon and steelhead populations and is probably a key factor in the ecology of most, if not all, streams. Considerable attention has been given by many investigators to discharge in relation to those species which commonly spawn in rapidly flowing water, but discharge can be equally important to many species which spawn in slow-moving sections or in the still waters of seasonally flooded areas.

Velocity of stream flow is a factor in salmon and trout redd construction, in the fertilization of the eggs, and in supplying oxygen to the eggs. Subsurface flow is essential in maintaining high oxygen levels for salmon (Alderdice, Wickett, and Brett, 1958; Wickett, 1954 and 1958). Generally, salmon appear to select gravels with an adequate oxygen supply by sensing a current or upwelling of flow through the gravels. Andrew and Green (1960) suggest such a selection process. Surface flow plays a major role in the subsurface flows which keep the eggs oxygenated.

Work has been done by a number of fisheries biologists to quantify the water velocity and depth preferences of spawning salmon and trout. These studies have demonstrated that salmon and steelhead have a rather narrow tolerance to velocity and depth when choosing spawning areas. These velocities and depths have been measured on many rivers, and although there are some variations between rivers, the velocity and depth tolerances for salmon and steelhead in west coast American rivers have been fairly well determined. It has been a major step in quantifying the elements of a stream's ecology.

Dan Slater and Jack Savage of the U.S. Fish and Wildlife Service and George Warner of California pioneered some of the work done in measuring the depth, velocity and gravel size preferences of chinook salmon. The methods were later described by Warner (1953 and 1955) and further developed by Kier (1964), Puckett (1969), and others.

As a result of these and other studies of the depth and velocity preferences of spawning salmon and steelhead, it was shown that the discharge of a river can determine the amount of spawning area available and that spawning does not necessarily increase proportionately with increased flow. Warner (1953) found two peaks of available spawning area for chinook salmon in the American River, one at a discharge of 2,700 c.f.s. and a larger one at 500 c.f.s. The available spawning area increased as the flows dropped from 4,800 c.f.s. to 2,700 c.f.s. This resulted from reduction of excessive velocities and depths over the lateral or floodplain gravels. As the discharge dropped further from 2,700 c.f.s. to 1,300 c.f.s., the amount of suitable spawning area decreased. This was attributed to the exposure of the lateral beds as the water level dropped. The velocities in the main or central stream bed were still too high. As the flow dropped below 1,300 c.f.s., conditions of velocities and depths improved in the center portion of the stream, making more of these gravels suitable for spawning until another peak was reached at 500 c.f.s. Below this flow the amount of gravel suitable for spawning diminished rapidly.

McKinley (1956) observed that flows of 5,000 c.f.s. and above created unfavorable depths and velocities and would virtually eliminate spawning area for chinook salmon in the Cowlitz River. Similarly, Deschamps et al. (1966) noted that a flow in excess of 400 c.f.s. in the Wynoochee River was detrimental to the spawning of fall chinook, coho, and chum salmon.

Thus the discharge for a particular stream can be a determining factor in the spawning success or failure of a salmon or steelhead population. Velocity and depth are the most important features of the discharge for spawning. Too high or too low a discharge can result in velocities or depths outside the tolerance range for spawning. Optimum discharge for salmon or steelhead spawning depends upon the configuration and gradient of the stream bed and the presence of usable gravel. Although increased spawning area might be expected with increased flow above that which provides at least some spawning area, it is not always a steady increase. There may be two or more peaks of usable spawning area, depending upon configuration of the stream bed. Suitable conditions for spawning decrease with excessively high or low discharge.

It has also been noted that freshets may be essential in some streams to stir and clean gravel beds used by salmonids for spawning. In the absence of such freshets the gravels may become compacted and the interstices filled with fines and sediment to the point where they are unusable by the fish.

Rearing Capacity (including embryo development and food production.) Surface flow influences subsurface flow and is therefore important to the survival of salmon and steelhead eggs. However, excessive or widely fluctuating flows can be detrimental to salmonid egg survival. Lister and Walker (1966) found that winter discharge stability increased the egg-to-fry survival of chum salmon and that the magnitude of coho salmon fry emigration was also related to winter discharge stability. This study further demonstrated the very strong relationship between discharge and reproductive success in salmon.

Salmon eggs require a flow of well oxygenated water through the gravels in which they are incubating (Wickett, 1954, 1958; Vaux, 1968). The main influence on subsurface flow is the surface stream; that is, the surface discharge of a stream exercises a major influence on the intragravel water flow (Sheridan, 1962).

The velocity of intragravel water flow is important in itself. Shumway, Warren, and Doudoroff (1964) reared eggs of coho salmon and steelhead trout (*Salmo gairdnerii gairdnerii*) at different concentrations of dissolved oxygen and at different water velocities. They found that irrespective of the oxygen concentration tested, reduced water velocities resulted in reduced size of the hatching fry. This effect of velocity was nearly as pronounced at high oxygen concentration as at low concentrations. These studies seem to confirm the suggestion of Wickett (1954) that the high mortalities in chum salmon eggs experienced in some areas could be explained on the basis that oxygen demands of the eggs were not being met due to very low oxygen content of the water or very low velocity of the intragravel water.

Egg survival and the success of the juvenile fish after hatching determine the rearing capacity of the stream, and food and shelter are essential elements for this survival and growth. Discharge influences the fish-food species composition (Phillipson, 1954) and total production, as well as the availability of shelter. In salmon and trout streams the production of fish-food organisms is highest in the riffle or rapid-flow areas, and the turbulent and "white-water" areas provide a significant part of the shelter needed by the young fish.

Pearson et al. (unpublished) found that the peak aquatic insect production on the riffles studied occurred at velocities of about 2.0 ft/sec. Therefore, based solely on fish-food production, a suitable discharge for a stream would be the flow covering the largest area of riffle with water velocities of about 2.0 ft/sec.

Studies have shown that low discharge affects food production in winter by encouraging the formation of anchor ice, which can also be directly damaging to fish and fish eggs present at the time (Somme, 1960). Low flows can increase predation, particularly on downstream migrant juvenile salmonids, until the predator populations stabilize in relation to the new flow regime (Hourston, 1958). Fluctuations of flow, such as those below some hydroelectric plants, can cause the stranding and reduction of fish-food organisms and young fish, thus reducing the rearing capacity (Green and Andrew, 1961; Phillips, 1969; Powell, 1958). Disturbance of the flow regimen may alter the natural drift of food organisms and affect the feeding habits and possibly the production of fish. Freshets have been noted to contribute to food supplies by dislodging bottom organisms.

Turning to the rearing capacity of streams, some recent studies on juvenile salmon production are very interesting. Two recent unpublished papers point up the significance of discharge and velocity to the rearing capacity of a stream. Erho et al. (unpublished) found that coho salmon smolt production was in direct proportion to the July, August, and September discharge in Speelyai Creek in Washington. The data support the thesis advanced by Smoker (1953) that stream flow is a limiting factor in the production of coho salmon.

The studies of Pearson, Conover, and Sams (unpublished) further support the contention that the most important factor determining the juvenile coho carrying capacity of a stream is the summer discharge. They studied the summer rearing requirements of juvenile coho salmon and the effects of diminished flows in nineteen streams in four river systems in Oregon. They concluded that production of juvenile coho was directly related to summer stream flows and that increases in populations associated with increased flows were also velocity-related. Their studies indicate clearly that reduction of summer flows would result in a reduced production of coho salmon. They found that each stream and each pool in a stream has a definite rearing capacity which is influenced by food production and spatial requirements of the fish—in turn affected greatly by velocity of the stream flow.

The spatial requirements of stream fishes have a relationship to discharge which should be given greater attention. Most stream fish exhibit a strong territorial orientation (Onodera, 1962; Allen, 1969). In many cases this territoriality is influenced greatly by the velocity of the current. The studies of Kalleberg (1958) are of special interest here. He found that juvenile salmon and brown trout occupy and defend well-defined territories which become smaller with increased velocities. This depression in the area of individual territories with increased velocities enabled juvenile fish, which previously were without territories, to occupy and defend an area of their own. Conversely, reduced velocities cause individual fish to enlarge the area of their territories, and the smaller and less aggressive fish are often displaced in the process; the more aggressive fish take up more space in the stream and consequently less space is available for the remaining fish.

Kalleberg noted that with reduced flows increasing numbers of fish were forced to select less desirable feeding stations because of the expanded territories of the more aggressive individuals. Thus the competition for space becomes a competition for food. The less aggressive individuals remain in the smaller size units over longer periods of time and are thus exposed more to predators because of both their smaller size and their less desirable location in the stream. As a result, more fish are eliminated from the population simply because of this territorial influence. Thus, within some kind of limits, the resident salmonid carrying capacity of a unit of stream bed can be reduced or increased by velocity alone.

Obviously, the effect of current velocity on territoriality, shelter, and carrying capacity needs quantifying if we are better to define the discharge needs of salmonids and other stream fish. Shelter is, of course, an important factor in the carrying capacity of trout streams and is influenced by depth, the characteristics of the stream bed and banks, riparian and aquatic vegetation, clarity or turbidity of the stream, and velocity of the water creating turbulent and white-water areas. The role of velocity in creating shelter is often overlooked in considering the effects of reduced or increased discharge on stream fish populations.

Harvesting of Fish. Harvest factors are important in man's consideration of the stream environment. Discharge can affect the vulnerability of aquatic organisms to various types of capture methods and gear. Reduced flows or excessive flows can negate the use of boats, nets, and other means which have commonly been used. Sport fishermen have long taken an interest in discharge of streams from the standpoints of fish production, wadeability, visibility, and esthetics.

The harvestability of fish in a river can be influenced by discharge, and this in turn can affect the ecology in terms of size, species, and composition factors of the fish populations.

Water Quality

Discharge can influence the quality of water by its dilution of contaminants. Increased discharge can alleviate pollution problems which are upsetting the normal environment. Reduced flow can, of course, intensify the adverse effects of downstream pollutional loadings. It has become common practice in the United States to consider the value of reducing water quality problems through flow augmentation by water storage projects. One cannot help but wonder if this process should be encouraged in lieu of correcting the basic problem, the source of pollution itself. In any event, the effect of flow augmentation on the ecology of the river, as well as its value for alleviating pollution, should be assessed. In many instances the ecology of the river (at least in terms of species directly valuable to man) could benefit from flow augmentation.

Solar radiation, winds, bank storage, spring seepages, and tributary streams usually have a greater effect on river water temperature at lower flows. Increased water temperatures with reduced flows are of concern to the salmon and trout manager and have been reported by a number of workers including Gordon (1965), Hourston (1958), Parsons (1955), Stober (1963), and Kent (1963).

Increased temperatures do not always occur with low flows. For example, Curtis (1960) found that the temperatures of water in the Pitt River, California, remained essentially the same at greatly reduced flows because of the cooling effect of springs in the stream bed. Conversely, low flows in the winter can be a contributory factor to lower-than-normal water temperatures and in some cases the formation of anchor ice.

The absence of high flows can eliminate warm floodplain waters upon which some stream species are dependent for spawning, food production, and nursery areas.

Vegetation

Heavy discharge can retard the encroachment of terrestrial vegetation onto the stream bed and inhibit aquatic plant growths. In the case of salmon and trout streams, the removal of rooted plants from the spawning areas is a normal factor in the ecology. In other streams the dislodging effects of high discharge can be detrimental to aquatic plant growth of importance to the normal ecology.

I have observed the encroachment of willows and other vegetation on chinook salmon spawning beds in several California rivers where flood flows have been controlled. The effect of such encroachment is to render the spawning areas unusable by salmon even though stream flow at the time of spawning may be entirely satisfactory. The same effect has been noted on a number of trout streams.

Reduced discharge can create conditions favorable to floating or rooted aquatic plants and thus contribute to marked changes in the river's ecology (Hynes, 1970). Stabilized low flows in the lower Tuolumne River in California undoubtedly contributed to favorable conditions for luxurious growths of water hyacinth in some years. The plants were considered an impairment, if not a total block, to salmon migration despite sufficient water discharge for passage of salmon in other parts of the river.

Conversely, high flows can result in reduction of riparian vegetation and thereby reduce cover and food production for fish. Riparian vegetation often plays a part in the ecology of a stream, and its character and composition can be altered by an increase or decrease in discharge. The esthetics of a river are greatly influenced by the riparian vegetation. Some terrestrial plant species respond favorably to periodic flooding, and some entire forests are dependent upon periodic flooding (Lagler, 1969).

Although many species of aquatic plants can tolerate only low-current velocities, Whitford and Schumacher (1962) carried out experiments indicating that certain species of plants grow only in a current greater than 15 cm/sec; lotic species have a higher respiratory rate and mineral uptake than related limnetic species, and these requirements can only be satisfied by a current above 15 cm/sec.

Solids Transport

The fact that high discharge moves solids in streams is contributory to the environmental characteristics of most streams. High stream velocities generally result in a substratum of rock or gravel, whereas low velocity areas are usually characterized by a substratum composed of finer materials. Many species of plants and animals have adapted at least part of their life cycle to the results of such movements and are more or less limited in their distribution according to the bottom material (Hynes, 1970).

Salmon and trout, for example, are dependent upon gravels which are relatively free of fines. The silt-cleansing actions of freshets are therefore important to the life cycle of these and many other species. In other cases, lower river species may be dependent upon the fines deposited from upstream areas by freshets. The production of fish-food organisms can be greatly affected by the absence or occurrence of freshets and the resultant effects on the substrate in the stream.

Reduced flows below large dams have had a profound effect on the ecology of the Trinity River in California. Prior to reservoir development, annual freshets washed away accumulated sediments, cleansing the gravel beds and retarding the growth of stream-bottom and stream-bank vegetation. The river below the dams now serves as a delta for the deposition of sediment from uncontrolled tributaries (Anonymous, 1970). Because the river lost its ability to flush sediments, the pools were filled with sand, and the salmon-spawning gravels were being covered over. It was estimated that the spawning area for 14,000 chinook salmon had already been destroyed by this process. Eustis and Hillen (1954) noted similar problems on the Colorado River.

In drainages with steep slopes and unstable soils landslides will occur and introduce large volumes of fines to the rivers. Under natural conditions where most rivers have seasonal freshets, such landslides are carried away, but in controlled-flow streams they can contribute to siltation for many years.

Determining Discharge Recommendations

The task of determining the discharge needs of a stream for maintaining a fisheries resource or for maintaining the ecology of the stream is being pursued by many government and private agencies. As more water is diverted from streams or stored in reservoirs, the ecological changes become more apparent and the values of the remaining streams become greater. As a result, increasing attention is being given to the ecological impact of water development projects and other activities on watersheds.

A primary target of consideration in pre-water development project studies is usually the question of stream discharge. How much water should by-pass project structures at various times for downstream purposes? The project developer, of course, is usually concerned with maximum development of the water supply for off-stream purposes. Such purposes are frequently incompatible with maintaining a natural discharge regimen. Naturally, the project engineer wants to devote as much of the stream's water as possible to the purposes which are helping to pay for the project. As a consequence of this atmosphere, the biologist is under considerable pressure to determine the minimum discharge of water which will barely maintain the stream resource. Unfortunately, this approach frequently results in maintaining a mere vestige of the former aquatic resources.

Discharge recommendations are often based more on a biologist's or engineer's guess of what the water development project might be able to provide in the way of discharge than on a quantified evaluation of the relationship between discharge and the ecology of the stream, its esthetics, and other in-place uses. Such guesswork on the part of biologists and resource administrators has not been all bad because without these efforts there would surely be fewer rivers and fewer river-associated resources than there are today. What is wrong is that we should continue such unenlightened procedures when we have the ability to develop and use more definitive processes.

The key to adequate determination of stream-flow recommendations is to quantify the water-flow needs of the organisms involved. Some significant advances have been made in this regard with salmon, steelhead, and trout resources. This approach requires considerable knowledge of the life processes of aquatic organisms and will necessitate continuing research. But we should quantify more of the known water-flow needs of fish and the more important aquatic organisms without waiting for the needed basic research on all aspects of the ecology of aquatic environments.

A number of workers have quantified such factors as spawning-flow velocity, rearing-pool velocity, and riffle-food productivity, and others have attempted to place numerical values on flow in relation to food production, shelter, and spawning (Pearson et al., unpublished; Warner, 1953 and 1955; Delisle and Eliason, 1961; Delisle and Wooster, 1964; Deschamps et al., 1966; Kier, 1964; Hutchinson and Fortune, 1967; Chambers, Allen, and Pressey, 1955). These are meaningful steps toward determining a realistic discharge recommendation.

Several studies have unfortunately relied on judgmental interpretation of photographs or measurements of the wetted area of a stream at various flows. This is not to say that photographs and measurements of wetted areas are not helpful in gaining an understanding of the effects of various discharges on the physical characteristics and appearance of a stream; but they are no substitute for quantifying the water needs of the organisms and the environment to be maintained. Photographs are helpful in visualizing various flows and evaluating the esthetic and recreation aspects of discharge.

Particularly annoying is the process by which some biologists and some engineers have attempted to dignify arbitrary and purely guesswork decisions in connection with flow recommendations by converting them to impressive-sounding formulae or arithmetic "rules." In the absence of thorough studies, judgmental decisions of biologists and other professional scientists will have to suffice in many cases, but they should be recognized as judgmental and should not be disguised by a mathematical formula. Especially disconcerting is the formula which assumes that one-third, one-half, or some other arbitrarily chosen portion of the mean or low flow of the stream is the fundamental governing factor for maintenance of a particular species or the stream's ecology as a whole. Another frequently used approach is simply to assume that the stream's ecology will be maintained by the low flow. Such approaches have no justification except expediency.

Statutory Considerations

We often hear a professional colleague deploring the thoughtless disruption of the ecology of some segment of the environment and then turning his wrath upon public officials who he considers are not doing their job. Unfortunately, the public officials may be fulfilling their responsibilities quite well under the limitations of their statutory guidance. This is often true in relation to protecting water discharge to maintain a river's values.

Few areas of the world have laws which give any status or protection to water supplies for the purpose of maintaining a stream, its ecology, or its other in-place values. Water laws throughout the world, and especially in the United States, were formulated to stimulate development of the land and frequently encourage the diversion of water away from the stream.

Provision for in-stream water rights or protection of a basic discharge, except possibly for purposes of navigation, are nonexistent in most water laws. Many states in the United States have no legal machinery for maintaining discharge in a stream to protect the stream's ecology or other in-place public values. The well established doctrine of diversion as the basis for water rights and the use of water remains as an invidious force in the water laws of most states. It will continue to prevent appropriate consideration and protection of stream flows until rescinded by new legislation which clearly gives weight and protection to water for in-place stream values.

One of the tragedies of our times is the legacy of private ownership of rivers or of water diverted from them in the United States. True, such ownership stimulated and facilitated settling of the land in the pioneering frontier days, but it now forms an effective block to environmental management of water resources in the public interest. The use and values of a river are no longer tied solely to the individual landowner or holder of a water right—they frequently extend more importantly to the public as a whole.

A number of state laws in the U.S. provide language to effect some consideration of "minimum" flows for protection of fish life, but all too frequently these have been interpreted so conservatively as to constitute a legislative state-wide impairment of the ecology and esthetics of rivers. The legal situation in some states is so weak or so protective of water for diversion that despite the availability of water and money to correct or prevent an ecological mistake, it cannot be accomplished.

Fortunately, the rivers and streams of Australia have been retained in public ownership. The environmental water needs of Australia will certainly be more easily considered as a result of this farsighted action. The wild rivers legislation in the United States is a major step in the direction of protecting some river values, but much more needs to be done.

There is need for a worldwide review and revision of water laws if the ecological and environmental values of streams are to receive effective attention. Where no legal foundation has yet been developed to recognize and protect the interests of the general public in the ecology and in-place uses of streams, the legislatures should be asked to take corrective action. The ability to preserve adequate streamflows in relation to future developments should be given statutory implementation wherever it does not presently exist. It may well be necessary to provide the power and the funds to condemn and purchase water rights, including riparian rights, to prevent or to correct serious ecological mistakes caused by inappropriately regulated river discharges.

Acknowledgments

I am grateful to Dr. J. A. R. Hamilton for permission to use an unpublished paper on the Speelyai Creek studies (Erho et al.) and to Lincoln S. Pearson for permission to make reference to an unpublished report on stream rearing of juvenile coho salmon in Oregon. Special thanks go to Dr. W. D. Williams (Monash University), T. P. Vande Sande, and G. McCammon for their review and helpful comments on the draft.

Literature Cited

Alabaster, J. S. 1970. River flow and upstream movement and catch of migratory salmonids. J. Fish. Biol. 2:1-13.

Alderdice, D. F., W. P. Wickett, and J. R. Brett. 1958. Some effects of temporary exposure to low dissolved oxygen levels on Pacific salmon eggs. J. Fish. Res. Bd. Canada 15(2):229-249.

Allen, K. R. 1969. Distinctive aspects of the ecology of stream fishes: a review. J. Fish. Res. Bd. Canada 26(6):1429-1438.

Ambuhl, Von Heinz. 1959. The significance of flow as an ecological factor. Schweizerische Zeitschrift fuer Hydrologie, Vol. xxi Fasc. 2:133-270.

Andrew, F. J., and G. H. Green. 1960. Sockeye and pink salmon production in relation to proposed dams in the Fraser River system. Inter. Pac. Salmon Fish. Comm. Bull. 11, 259 p.

Anonymous. 1970. Task force findings and recommendations on sediment problems in the Trinity River near Lewiston and a summary of the watershed investigation. State of California, The Resources Agency, 32 p.

Brett, J. R. 1957. Salmon research and hydro-electric power development. Bull. Fish Res. Bd. Canada 114:1-26.

Butcher, A. D. 1967. A changing aquatic fauna in a changing environment. Proc. and Papers IUCN, 10th Tech. Mtg. (Lucerne, 1966), IUCN Pub. New Series 9:197-218.

Chambers, J. S., G. H. Allen, and R. T. Pressey. 1955. Research relating to study of spawning grounds in natural areas. Annual Rept. Wash. Dept. Fisheries to U.S. Army Corps Engrs., 175 p.

Chapman, D. W. 1965. Net production of juvenile coho salmon in three Oregon streams. Trans. Am. Fish. Soc. 94(1):40-52.

Copeland, B. J. 1966. Effects of decreased river flow on estuarine ecology. Water J. Poll. Control Fed. 38(11):1831-1839.

Curtis, Brian. 1960. Observed changes in a river's physical characteristics under substantial reductions in flow due to hydro-electric diversion, p. 164-174. In Seventh Tech. Mtg., Inter. Union Cons. Nature Nat. Resources, Athens, Greece, 1968. Vol. IV, Nat. Aquatic Res.

Davidson, F. A., Elizabeth Vaughn. S. J. Hutchinson, and A. L. Pritchard. 1943. Factors influencing the upstream migration of the pink salmon (Oncorhynchus gorbuscha). Ecology 24(2):149-168.

Delisle, G. E., and B. E. Eliason. 1961. Effects on fish and wildlife resources of proposed water development on Middle Fork Feather River. Calif. Dept. Fish Game, Water Proj. Branch. Rept. No. 2, 36 p.

——————, and T. W. Wooster. 1964. Changing the flow regime and its possible effects upon aquatic life and fishing—Middle Fork Feather River. Calif. Dept. Fish Game Water Projects Br. Admin. Report, 36 p.

Deschamps, Gene, Sam Wright, and J. K. Magee. 1966. Biological and engineering fishery studies Wynoochee Reservoir, Washington. Wash. Dept. Fisheries Rept., 40 p.

Diener, R. A. 1963. Effects of engineering projects on estuaries. U.S. Fish Wildl. Serv. Circ. 161:52-54.

Erho, M. W., J. D. Remington, L. D. Rothfus, and J. A. R. Hamilton, Experimental rearing of coho salmon in Speelyai Creek, Washington. Coop. project Wash. Dept. Fisheries, Cowlitz Co. Pub. Utility Dist. and Pacific Power and Light Co. (Unpublished.)

Eustis, A. B., and R. H. Hillen. 1954. Stream sediment removal by controlled reservoir releases. Progr. Fish-Cult. 16(1):30-35.

Ewer, D. W. 1966. Biological investigations on the Volta Lake, May 1964 to May 1965, p. 21-31. *In* Man-made lakes. Symposia of the Inst. of Biol. 1966.

Ganssle, David, and D. W. Kelley. 1963. The effect of flow reversal on salmon. Annual Rept. (1962-63) Delta Fish and Wildlife Protection Study, Rept. No. 2, Appendix A:1-15.

Green, G. H., and F. J. Andrew. 1961. Limnological changes in Seton Lake resulting from hydroelectric diversions. Inter. Pacific Salmon Fish. Comm. Progr. Rept. No. 8, 76 p.

Gordon, R. N. 1965. Fisheries problems associated with hydro-electric development. Can. Fish Cult. 35:17-36.

Hayes, F. R. 1953. Artificial freshets and other factors controlling the ascent and population of Atlantic salmon in the La Have River, Nova Scotia. Bull. Fish. Res. Bd. Canada, 99, 47 p.

Hourston, W. R. 1958. Power developments and anadromous fish in British Columbia, p. 15-23. *In* The investigation of fish-power problems, H. R. MacMillan Lectures in Fisheries, Univ. Br. Columbia.

Hutchinson, J. M., and J. D. Fortune, Jr. 1967. The fish and wildlife resources of the Powder Basin and their water requirements. Oregon State Game Comm., Fed. Aid Fish Rest. Proj. F-69-R-5. Progr. Rept., 10 p.

Hynes, H. B. N. 1970. The ecology of flowing waters in relation to management. J. Water Poll. Control Fed. 42(3) I:418-424.

Kalleberg, Harry. 1958. Observations in a stream tank of territoriality and competition in juvenile salmon and trout. Inst. Freshwater Res. Drottningholm, Sweden No. 39:55-98.

Kent, Ronald. 1963. North Platte River fish loss investigations—1963. Wyoming Game Fish Comm. Fed. Aid Proj. 763-6-1, 62 p.

Kier, W. M. 1964. The stream flow requirements of salmon in the lower Feather River, Butte County, California. Calif. Dept. Fish Game, Water Proj. Br. Office Rept., 17 p.

Lagler, K. F. [Ed.] 1969. Man-made lakes—planning and development. Food Agric. Org. United Nations, Rome, Italy, 71 p.

Lister, D. B., and C. E. Walker. 1966. The effect of flow control on freshwater survival of chum, coho and chinook salmon in the Big Qualicum River. Can. Fish Cult. 37:3-25.

Mackinnon, Dixon, and W. S. Hoar. 1953. Response of coho and chum salmon fry to current. J. Fish. Res. Bd. Canada, 10(8):523-538.

McKernon, D. L., D. R. Johnson, and J. I. Hodges. 1950. Some factors influencing the trends of salmon populations in Oregon. Trans. Am. Wildl. Conf. 15:427-449.

McKinley, W. R. 1956. Relative abundance of optimum spawning area in the lower Cowlitz River for anadromous fish as a function of river discharge. Wash. Dept. Fisheries Admin. Rept.

Neave, Ferris, 1958. Stream ecology and production of anadromous fish. p. 43-48. In The investigations of fish—power problems. H. R. MacMillan Lectures in Fisheries, Univ. of Br. Columbia.

Odum, W. E. 1970. Insidious alteration of the estuarine environment. Trans. Am. Fisheries Soc. 99(4):836-847.

Onodera, Kosi. 1962. Carrying capacity in a trout stream. Bull. Freshwater Fisheries Research Lab. (Tokyo, Japan) 12(1):1-41.

Parsons, J. W. 1955. The trout fishery of the tailwater below Dale Hollow Reservoir. Trans. Am. Fish. Soc. 85:75-92.

Pearson, L. S., K. R. Conover, and R. E. Sams. Factors affecting the natural rearing of juvenile coho salmon during the summer low flow season. Oregon Fish Commission. (Unpublished), 64 p.

Penaz, Milan, Frantisek Kubicek, Petre Maruan, and Milos Zelinka. 1968. Influence of the Vir River Valley Reservoir on the hydro-biological and ichthyological conditions in the River Svratka. Prirodoved Pr. Ustavo Cesk Akad. Brne. 2(1):3-60.

Pfitzer, D. W. 1954. Investigations of waters below storage reservoirs in Tennessee. Trans. N. Am. Wildl. Conf. 19:271-282.

Phillips, R. W. 1969. Effect of unusually low discharge from Pelton Regulating Reservoir, Deschutes River, on fish and other aquatic organisms. Oregon State Game Comm., Basin Investigations Sect., Spec. Rept. No. 1, 39 p.

Phillipson, G. N. 1954. The effect of water flow and oxygen concentration on six species of caddis fly (trichoptera) larvae. Proc. Zool. Soc. London. 124: 547-564.

Powell, G. C. 1958. Evaluation of the effects of a power dam water release pattern upon the downstream fishery. M. S. Thesis, Colorado State University, 149 p.

Pritchard, A. L. 1936. Factors influencing the upstream spawning migration of the pink salmon (*Oncorhynchus gorbuscha* Walbaum). J. Biol. Bd. Canada, 2(4):383-389.

Puckett, L. K. 1969. Fisheries surveys on Thomes and Stony Creeks, Glenn and Tehama Counties with special emphasis on their potentials for king salmon spawning. Calif. Dept. Fish Game, Water Proj. Branch Admin. Rept. No. 69-3.

Raymond, H. L. 1968. Migration rates of yearling chinook salmon in relation to flows and impoundments in the Columbia and Snake Rivers. Trans. Am. Fish. Soc. 97(4):356-359.

—————. 1969. Effect of John Day Reservoir on the migration rate of juvenile chinook salmon in the Columbia River. Trans. Am. Fish. Soc. 98(3): 513-514.

Sheridan, William. 1962. Waterflow through a salmon spawning riffle in south-eastern Alaska. U.S. Fish Wildl. Serv. Spec. Sci. Rept. — Fisheries 407:1-20.

Shumway, D. L., C. E. Warren, and P. Doudoroff. 1964. Influence of oxygen concentration and water movement on the growth of steelhead trout and coho salmon embryos. Trans. Am. Fish. Soc. 93(4):342-356.

Smoker, W. A. 1953. Streamflow and silver salmon production in western Washington. Wash. Dept. Fisheries, Fisheries Res. Papers 1(1):5-12.

Somme, Sven. 1960. The effects of impoundment on salmon and sea trout rivers, p. 77-80. *In* Seventh Technical Meeting, Inter. Union for Cons. Nature and Nat. Res., Greece, 1958. Vol. IV, Nat. Ag. Res.

Stober, Q. J. 1963. Some limnological effects of Tiber Reservoir on the Marias River, Montana. Proc. Montana Acad. Sciences, 23:111-137.

Vaux, W. G. 1968. Intergravel flow and interchange of water in a streambed. U.S. Fish Wildl. Serv. Fishery Bull. 66(3):479-489.

Warner, G. H. 1953. The relationship between flow and available salmon spawning gravel on the American River below Nimbus Dam. Calif. Dept. Fish Game Admin. Rept., 5 p.

————. 1955. The relationship between flow and available salmon spawning gravel on the Feather River below Sutter Butte Dam. Calif. Dept. Fish Game Admin. Rept., 5 p.

Wharton, J. C. 1969. Recreation and conservation in the aquatic environment. Proc. Roy. Soc. Victoria, 83(1):47-54.

Whitford, L. A., and G. J. Schumacher. 1962. The current effect, an important factor in stream limnology. *In* Congress of the Sociatas Internationalis Limnologiae. Madison, Wisconsin, 1962.

Wickett, W. P. 1954. The oxygen supply to salmon eggs in spawning beds. J. Fish. Res. Bd. Canada, 11(6):933-953.

————. 1958. Review of certain environmental factors affecting the production of pink and chum salmon. J. Fish. Res. Bd. Canada, 15(5):1103-1126.

MORPHOMETRIC CHANGES

T. Blench

River morphometrics is literally the routine of measuring quantities deemed relevant to river regime. The term "regime," like "climate," is appreciated intuitively but is difficult to define. Both are visualised in terms of the joint existence of many erratically fluctuating quantities whose averages over sufficient time change slowly, if at all. The long-term averages receive most attention in quantitative studies because they yield useful correlations easily, but the patterns of fluctuation of these averages give mental pictures their detail and character despite relative mathematical intractability. Metaphoric reference to the regime of a river as its climate would be apt. A regime change implies a change in "morphometric factors."

The scientific, quantitative study of river regime depends on morphometrics and gives the means of predicting the hydraulic consequences of interferences with rivers by engineers and nature. For two reasons all parties interested in arresting and reversing the deterioration of the environment should appreciate the rudiments of regime behaviour. The first is that, although engineers hold a vital role in the planning and construction of water resources projects, few have received college instruction in the hydraulic consequences of interfering with regime. The second is that the hydraulic consequences are likely to set off damaging and irreversible reactions in the environment; expert, exhaustive knowledge of these reactions cannot be imparted in engineering education. Accordingly, this paper will attempt a nontechnical presentation of the state of the science, the effects of interference with rivers—on themselves and on the environment— and the difficulties in disseminating knowledge and achieving interdisciplinary cooperation.

History of Applied Morphometrics

At the end of last century irrigation engineers in India had begun to realise that the artificial rivers (canals) they created could not be made to retain their designed and constructed breadths, depths, and slopes; instead, they would keep adjusting themselves, by erosion or deposition, till they had attained dimensions decided by unknown natural laws. The patient collection of measurements of finally self-adjusted channels resulted, in the 1920's, in a centrally directed, scientific analysis from which the basic principles and equations of self-adjustment were formulated for duned sand beds. Their functional forms are still valid for rivers of the proper type; research by other specialised groups continues into details, new aspects, and all types of bed conditions. Also, at the end of last century geologists, mainly of the U.S. Geological Survey, started similar river studies for different reasons. In cooperation with engineering researchers, these have continued to date, supporting canal findings in general and covering almost every type of channel. The net result is that there is now available among specialists an effective, developing science of the regime of rivers that form themselves from their own transported material. Recent publications (Blench, 1969*a*, *b*, 1970; Leopold, Wolman, and Miller, 1964; U.S. Geological Survey, undated) present its essentials and contain key references of the considerable literature.

UNESCO (1962) recognised the importance of the subject by listing "evolution of river bed, and sedimentation" as the third of nine "major scientific problems . . . in scientific hydrology" for the International Hydrologic Decade (IHD); a subhead was "Elaboration of a general theory of a river bed development and sediment transport in a stream." The ninth subject, which implies the third, is: "influence of man on continental waters." These topics were emphasised again after the IHD had started (UNESCO, 1964).

The Basic Principle of River Regime

Only rivers that form their channels from their transported sediment will be considered hereafter in this paper since they alone react rapidly to interference by man or by nature. The basic principle behind their self-formation, established from observation, can be stated:

A channel carrying a definite long-term pattern of water and sediment discharge, and subjected to a definite set of constraints that can adjust to the action of the flow, will acquire a definite regime.

That is, the channel will acquire a definite set of measures of its forms—a definite "morphometry." From a primary engineering viewpoint the common measures are breadth, depth, slope, meander length, and meander belt width, averaged over suitable times and distances. Useful, but more specialised, measures include grain size distribution of the materials of the bed sides and floodplain, height and length of dunes that form on a bed, and limits of variation of various quantities. Note that usable measures are normally time or space or space-time averages.

288

Figure 1 illustrates a powerful, simple test for definiteness of regime. For each of a sequence of years, starting a short time before a barrage was built across the Indus River, the downstream gauge heights for each of four different flow discharges were plotted for summer and winter. Had the regime been definite—or, in usual technical language, had the river been "in regime" or "in equilibrium"— the gauge records would have shown erratic fluctuations about horizontal lines. However, the completion of the barrage in 1924 halted the natural flow of sediment downstream, resulting in a change of downstream regime shown by a sudden abnormal gauge drop. The regime then tended toward normal till 1933, when modifications of the barrage produced another sudden drop followed by a gradual recovery, which by 1945 appeared to be aiming at a new regime of gauges higher than before the barrage was built. In fact, the ultimate regime would be different from the old because the canals upstream of the barrage withdraw flow and are operated to draw less than the share of sediment appropriate to their discharges.[1]

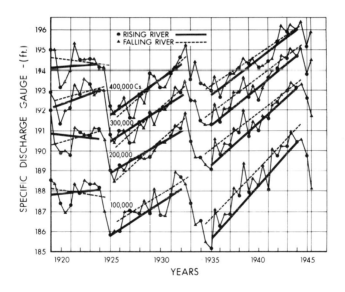

Figure. 1. Specific gauges, Indus River

[1]Gauges plotted to correspond to a specified discharge are called "specific gauges"; if the specific gauge record shows a trend, the channel is changing its regime.

Simple Morphometric Relations

It follows from the basic principle above that morphometric quantities for channels in regime must be related to each other, for they are related to the same two primary causes which are definite for in-regime conditions. The following figures relate various practically important quantities and provide the means for estimating the probable ultimate consequences of most major interferences with river regime. However, they are fitting-curved to field data which yield points scattered relatively widely about any fitting-curve; reasons for scatter include secondary factors, irregularity of measured quantities, and impossibility of deciding exactly how measurements should be made for in-regime conditions. Therefore, although they show trends reliably, they should not be trusted to find exact values for any individual case.

Figure 2 shows that over the range of sizes from tiny models to the largest rivers there is direct proportionality of channel breadth to meander length; thus rivers of the same meander pattern—sinusoidal, serpentine, D-shaped, etc.—tend to be geometrically similar in plan. Because of this peculiarity, it is not possible to tell the size of a river from an air photo without information from the topography; for example, a comparison with buildings, trees, and field sizes.

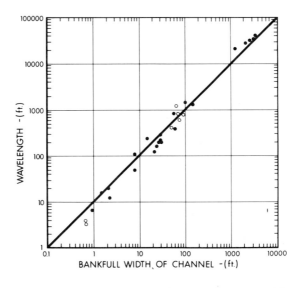

Figure 2. Meander length proportional to channel breadth

In Figure 3 each line of the topmost graph is for a particular river system and was drawn to fit points recording surface breadth of channel against discharge (at a standard frequency of excedance) at various discharge measurement sites within the system. The graph below replaces breadth with the corresponding depth. The last graph, for velocity, is irrelevant here. The systems were selected from different physiographic settings; in fact, one system was a man-made one— an irrigation system in India. These plots and many others suggest that, if all causes except the discharge are constant, the standard measure for breadth varies as about the square root of the corresponding discharge, and the measure for depth varies as about the cube root of the discharge. Detailed research shows that the rules are somewhat more complex, but the simple ones are adequate to prevent major blunders in design and planning.

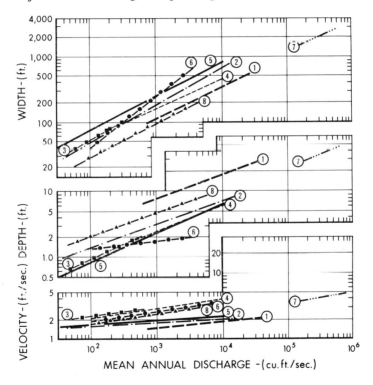

Figure 3. Breadth and depth related to formative discharge

Figure 4 shows how slope depends on discharge in rivers that either have duned sand beds or have gravel beds combined with depth much larger than the gravel size. Slope S does not depend so simply on discharge Q as do breadth b and depth d, so a quantity Z is plotted instead of S. Z is found by multiplying S

291

by the sixth root of b and dividing by a quantity that virtually measures the size of the bed-material. The bottom line of the figure relates Z to Q for straight canals that have very small sediment loads (indicated by $f^{111}[C] = 1.0$ where C measures the proportion of total flow that is bed sediment). The intermediate line is for straight canals that have very small load but have been neglected so long that they have developed into rivers with marked meandering. This degree of meandering is denoted by writing $k = 2.0$ on the line; $k = 1.0$ is for straightness. The same line would do for a straight canal with a considerable load, which is denoted on the top line by $f^{111}(C) = 2.0$, since Z is proportional to the product of k and $f^{111}(C)$. The top line is for this considerable load and considerable meandering. As in Figure 2, verification is from tiny models up to the world's largest rivers. For gravel rivers where the depth is small compared with gravel size the chart is inapplicable.

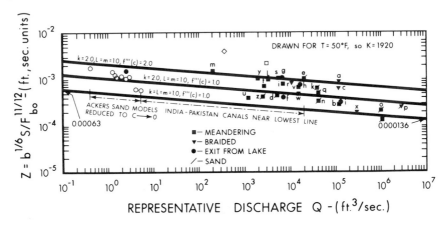

Figure 4. Specific slope, Z, related to discharge, load and tortuosity

Figure 5 condenses mechanical analyses of active riverbed sands of different mean sizes into one dimensionless plot. It emphasises that Nature imposes simple laws on river dimensions, even on the distribution of particle sizes in the material found on a duned sand bed.

Applications of the Morphometric Relations

The natural scientist wishing to know how rivers will change after they have undergone engineering action, direct or remote (for example, trapping their sediment in a reservoir at any distance upstream or downstream), must be in a position to estimate the quantitative consequences. Only then can he undertake the task of estimating the likely effects on the environment (e.g., drop of water

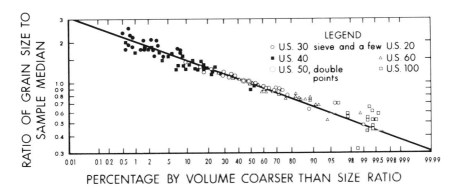

Figure 5. Particle size distribution in active river-bed sand

table due to river degradation). He should not accept assurances of river stability from planners till he has formed his independent judgment and then had his objections irrefutably dispelled. Because the figures of this paper show the observed natural, broad interconnections among ruling dimensions of most rivers, they can be used nontechnically, as well as technically, to indicate whether, and to what extent, a river plan may be defective or physically impossible. The following paragraphs will show, by easy steps, how to apply the figures to some hypothetical cases

Case 1 — Diversion — Figure 6. An unprotected sand-bed stream, A, is diverted into another, B, of similar type, at a point C 15 river miles upstream of B's exit E into a large river; assume that the diversion doubles the discharge of BE at C. CE meanders widely and relatively inactively through a cultivated floodplain which is narrow enough for the meanders to strike the erodible valley sides in a few places. Periodic floods spill over the floodplain without doing much damage and with some mildly beneficial effects. Ranch houses are on the top of the valley walls. Channel slope along the winding stream is 2 ft per mile. Engineers assess the effect of the diversion to be an approximate doubling of the "formative discharge," that is, the particular average discharge that corresponds to the river's self-formed dimensions according to accepted formulas or graphs like Figures 2 - 4. What effects should be considered likely in terms of general principles?

The ultimate effect on channel breadth can be forecast from Figure 3. If a point for the original breadth b and formative discharge Q were plotted on that figure, the new ultimate breadth b^1 and the new formative discharge, assumed to be $2Q$, would define a second point that would have to lie on a line, through the first point, having the general slope of the other lines since both points concern

293

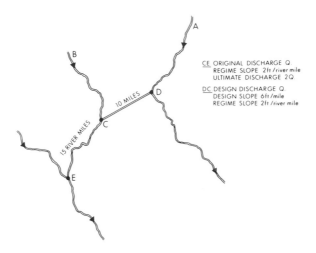

CE ORIGINAL DISCHARGE Q
 REGIME SLOPE 2ft /river mile
 ULTIMATE DISCHARGE 2Q

DC DESIGN DISCHARGE Q.
 DESIGN SLOPE 6ft /mile
 REGIME SLOPE 2ft /river mile

Figure 6. Diagram for Cases 1, 2

a river of one type. As this slope is about 1/2, the ratio b^1/b would be about $2^{1/2}$ (the plot is to double-log scale). That is, the channel would broaden about 40% eventually. The ultimate loss of land by cultivators would be 0.40 times b times river length.

The effect on meandering follows from Figure 2. The form of the meander pattern will not ultimately change because boundary nature has not changed. Therefore meander belt width will increase in the same ratio as meander length, which increases in the same ratio as channel breadth, namely, $2^{1/2}$ times. Accordingly, as cliffs are hit harder and more often, they are liable to suffer more frequent slides and possible loss of an occasional ranch house. During the change to the ultimate state there will be increased nuisance from bends chewing out land from one side of the river and depositing it on the other.

The ultimate effect on depth also follows from Figure 3. As the depth lines slope at about 1/3, the ultimate depth would be about $2^{1/3}$ times the original, or 12.5% more. However, recourse must be made to Figure 4 to find whether the bed goes down or the water level goes up to suit this new depth; in this case, the bed goes down at the river mouth since the water level is fixed there by the major river into which B spills.

Figure 4 is used exactly like Figure 3. That is, if (Z, Q) is plotted, then $(Z^1, 2Q)$ must lie at the end of a line drawn from it at the band slope (which is 1/12 downwards to the right). So Z^1 is Z divided by the twelfth root of 2, which

294

is approximately 1.06. Then as F_{bo} in Z measures the size of bed-material which does not change, it follows that Z varies only as $b^{1/6}S$, and therefore $S^1/S = (Z^1/Z)(b/b^1)^{1/6} = (1/1.06)(1/1.06)$ since b^1/b has been found previously to be $2^{1/2}$. Thus the ultimate slope will be about 12%, or about 1/8th, flatter than the original. As the river reach is 15 miles and the original slope 2 ft per mile, the original rise from mouth to head of reach was 30 ft. Therefore the river at the head of the reach should be expected ultimately to run a little under 4 ft lower than originally.

However, this ultimate drop in river levels, which would reduce flooding and might not be sufficient to drop water table objectionably, could be preceded by decades of deleterious actions while the enhanced discharge was carving out its new channel. In fact, the doubled discharge into the initially inadequate channel would cause enhanced flow levels both downstream and upstream of C. Downstream of C the high floods would dump sterile sand on the floodplain during floods; upstream the ponding of the unaltered discharges would cause the bed to rise to maintain the regime depth. In due course the water and bed levels at C would return to normal, on their way down to being 4 ft lower (see above), and the scouring bed material would pass into CE, lengthening the time required for it to degrade below old conditions. When ultimate stability was achieved, the river upstream of C would have to be 4 ft lower in the whole reach upstream of C which was in erodible material; and it would be somewhat broader because of bank collapse to suit the 4 ft of bed degradation. In estimating the time for these transitional occurrences, it should be remembered that most rivers are fairly inactive except when in flood.

It is emphasised that the gloomy possibilities might never occur since the upper river might be in bedrock or be isolated by a waterfall, and the lower river might have bedrock just under its visible noncohesive bed. However, the planner should check all possibilities; the ones described are normal expectation.

Case 2 — Straightening a River — Figure 6. Suppose that in Case 1 the diversion from A to B, at C, is designed as a practically straight canal DC, 10 miles long, sloping 6 ft per mile and running through the same alluvium as A and B; there would be a regulator at D to maintain A's water levels. Because the original A and B and, of course, the diversion DC are assumed for this illustration to have the same kind and degree of sediment load and to run through the same kind of material, and the original discharge of CE has been stated to be the same as that of the artificial channel DC, it follows that DC must aim to acquire the state of the old CE. So it will acquire meandering, making its "river mileage" an estimated 15 miles, and a slope of 2 ft per river mile against its original 10 per mile on the straight. The drop to which it was designed is 10 x 6 = 60 ft, and the drop at which it aims is 15 river miles x 2 ft per mile = 30 ft. So the bed level of DC at D will eventually degrade by 30 ft relative to C, while C will eventually degrade by the 4 ft computed in the preceding case. That is, the bed level in DC at D will finish 34 ft below design. The design blunder here used to be popular, on a small scale, in highway ditches; the author has seen 70 feet of retrogression in an irrigation escape. The mistake continues.

The process of attaining the final state is along the lines of Case 1. However, the tremendous load of eroded sediment moving along DC into CE would exaggerate the transitional behaviour there. In fact, CE would aim to acquire a slope somewhere between its original 2 ft per mile and 12 ft per mile before it started to return through normal to its ultimate lower level of 4 ft at C; the 12 ft, instead of 6, is because a meandering channel needs a greater slope than a straight one to carry its load.

Figure 4 merely adds quantitative detail to this commonsense picture. To illustrate with minimum calculation, suppose the original discharge of B is 1,000 cusecs, breadth b is 64 ft so that $b^{1/6}$ is 2, the slope is 2/5,000 = 0.0004, and $F_{bo}^{11/12}$ is 1.2 (corresponding to about 0.4 mm sand). Then Z for original B and ultimate CD is about 0.00067; plotting this against Q = 1,000 cusecs in Figure 4 gives a "state point" just above the band center line. This indicates that if $k = 2.0$, then $f^{111}(C)$ would be a little larger than 1.0, and the value of C would be (from diagrams not given in this paper) 1 or 2 parts per hundred thousand by weight of the water flow. For the straight CD, as designed, Z would be $6/2 = 3$ times greater than 0.00067; that is $k \cdot f^{111}(C)$ would be three times greater than for B. But k would be half of the original 2.0 because CD is straight, so $f^{111}(C)$ would be enhanced six times compared with B (and would stay at this till CD started to develop tortuosity or degraded). This enhancement would correspond (by diagrams not given in this paper) to C of the order of 100 for CD, which is 50 to 100 times that of its parent and 10 times an amount considered abnormally large for most rivers. As practically all this load would be picked up by boundary erosion, it follows that erosion would be violent. Without the preceding calculation the position of the "state point" of CD, which would be about quarter of a band width above the top band, shows that CD would have an abnormally large load even were it crooked enough to have $k = 2.0$. This would warn even a layman that the design would lead to violent erosion and flattening.

Case 3 — Reservoirs. Downstream of a reservoir the river must degrade since the trapping of bed load reduces $f^{111}(C)$ downstream, and thence Z and S (see Figure 4). In most rivers $f^{111}(C)$ ranges between 1.0 and 2.0. This implies that a reservoir selected at random is as likely as not to cause an ultimate downstream reduction of slope by about one-third. There are rivers with mobile sand beds where the reduction of slope could be zero because the reduction of formative discharge, caused by removal of peak floods by storage, could counteract the effect of reduced sediment load. Also, gravel rivers are likely to have enough large stones in the bed, or from the collapse of banks that contain large stones, to form a paving that will refuse to erode. The importance of a reduction of even one-third can be seen by imagining (i) a distance of 50 miles to the first control point downstream of a dam and (ii) a sand bed with slope 2 ft per mile. The reduction of the drop of 100 ft by one-third implies an ultimate abrupt drop of 33 feet just downstream of the dam. This might take about a century to develop fully (barring natural obstacles interfering), as the volume of earth to be removed would be at least that in a prism 50 miles long with a base measuring 33 ft vertically and a meander belt breadth (about 10 river-channel breadths) horizontally. To this would be added the material from tributary degradation.

Tributaries would degrade for two reasons. First, even before the degradation of the main stream reached them, the removal of flood peaks from the main stream would leave them "up in the air" when they flooded. Second, of course, the main stream degradation, when it reached them, would add to the effect.

Upstream of a reservoir the river has to aim at regaining its regime slope, drawn back from wherever the delta has reached. As all reservoirs must fill solid eventually, the new river would be the old one displaced downstream to pass through the top of the dam. This would require the whole river to be lifted by the height of the dam; thus centuries or millenia might be required for the whole task if the dam were high. Nevertheless, deltas are formed mainly from suspended load which may exceed bed load a hundred times in quantity. So, with heavily laden rivers and even very high dams, effects near the original reservoir head can usually be noticed some 10 years after construction and can be causing flooding and waterlogging at the lake-head after 20 years. It is fortunate that deltaic advance is likely to be proportional to the cube root of time. Small reservoirs can fill solid with sediment in a few years, whereas the very large ones might lose capacity at about 1/2 to 3/4% per year.

Canada is fortunate not to have rivers with large suspended loads.

Fundamental Causes of an Ecological Emergency

From a hydraulic viewpoint, the case history used here is merely one of governmental ignorance of regime principles and hydraulics mentioned above in Case 3, so needs no further comment. However, I see little hope of arresting and reversing the senseless destruction of our environment in the short time available unless we appreciate fallible man's general misdirection and mismanagement of the science and technology available to him; with that appreciation the natural cooperativeness of the individuals of our society, plus the considerable group cooperativeness that has been achieved already in various contexts, gives us a reasonable chance of success. Therefore I shall outline, for discussion and possible correction, my personal picture of the root causes of the ignorance of both hydraulics and ecology that caused the occurrence.

The case itself was presented recently to governments in an emergency brief, entitled "Death of a Delta," by an interdisciplinary group at the University of Alberta (Anon., 1970b); it was released thereafter to the press and appeared in the June 24, 1970, Canadian issue of *Time* magazine (Anon., 1970a). In the following, quotes and Figures A to D are from this brief. In short, a dam on the Peace River in British Columbia was closed by late 1967 and filling, tempered by progressive releases through five of the ten planned turbines, continued to November 1970. The river downstream runs first through British Columbia and thence across Alberta into Wood Buffalo National Park (Federal) where it joins with the Athabasca River to form the Slave River.

The 1,000 square mile delta of the Peace and Athabasca Rivers . . . comprises a unique ecological system of lakes and rivers (Figures A and B). . . .

The productivity and uniqueness of the Peace-Athabasca delta is entirely the result of its hydrological regime. Before construction of the dam, the Peace River had an extremely variable flow, characterized by spring flows in the order of 350,000 cubic feet per second (cfs), followed by a gradual decline during summer and fall to winter flows of about 15,000 cfs. Lake Athabasca and the major lakes of the delta are connected to the Peace River through several outlet channels (Figures B and C). During the spring flood, the water level in the Peace River at the delta generally exceeded the level of Lake Athabasca and the other delta lakes. Consequently, the flow in the outlet channels changed direction and large inflows from the Peace River caused the lakes to rise rapidly by 6 to 8 feet, to levels between 688 and 692 feet. At such high stages of water almost the entire area was under water, but such yearly flooding is essential to maintain the ecology of the delta.

The Federal Inland Waters Branch has produced the only published study of flows and water levels in the delta (Coulson, A., and R. J. Adamcyk—The effects of the W. A. C. Bennett Dam on downstream levels and flows. Technical Bulletin No. 18, Department of Energy, Mines and Resources, Inland Waters Branch, Ottawa, 1969), a study which is unfortunately inadequate and incorrect in several respects. It asserts that Lake Athabasca levels should not drop below 682 feet in winter, yet during February of 1970 a level of

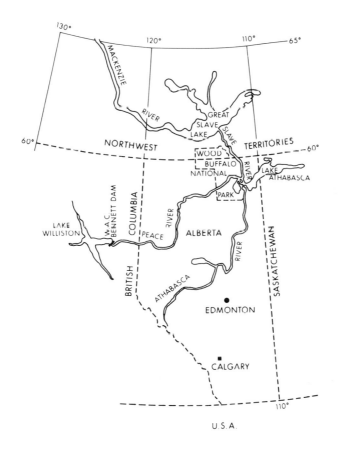

Figure A. Location of the Peace–Athabasca Delta

675 feet was measured (Figure D). Also in a sweeping final statement the downstream effects of Bennett Dam are called "beneficial because of reduction in flood peaks and increases in winter flows!" Yet, there have never been any significant flood problems along the Peace River, and the increased winter flows of 1969-70 were unable to prevent the catastrophic drop in the level of the delta lakes. All the other adverse effects of the dam, such as interrupted navigation, deterioration of national park values, and the elimination of subsistence trapping and commercial fishing in Lake Athabasca, are entirely ignored. . . .

299

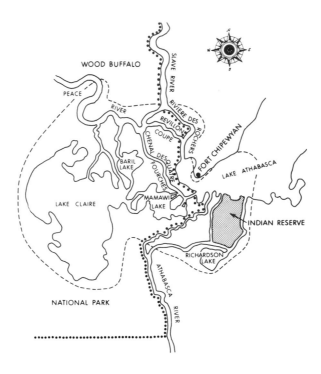

Figure B. The Peace-Athabasca Delta

The National Parks Policy states that "The basic purpose of the National Park system is to preserve for all time areas which contain significant geographical, geological, biological or historic features as a national heritage for the benefit, education and enjoyment of the people of Canada." Despite that statement of dedication to future generations of Canadians, extensive and significant changes have been allowed to occur in Wood Buffalo National Park, as a consequence of the lowering of the water levels in the vast Peace-Athabasca delta.

What once was a fascinating and varied natural complex of lake and marsh is fast becoming a succession of isolated mud flats whose communicating streams are drying up. The Park was established originally for the protection of the wood bison but the replacement of their preferred sedge meadows by unpalatable grasses and dense thickets of willows may seriously restrict the supply of natural forage available to them.

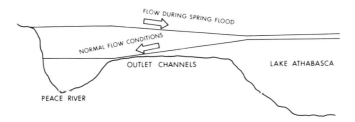

Figure C. Flow of water in the outlet channels

The lakes and marshes of the Peace-Athabasca delta are one of the most important areas for migratory waterfowl in all of Alberta. They are, in fact, unusual because they receive birds from each of the four major flyways in North America. Twenty-two species of ducks, five species of geese, swans, pelicans, grebes, gulls and great flocks of shorebirds have always appeared there each year on migration. Even the rare and spectacular whooping cranes feed and rest in the delta enroute to their nesting grounds only a few miles northward . . .

If the Peace-Athabasca delta is sacrificed as a major waterfowl area, the vast populations wanting to go there would have to depend upon the Hay-Zama Lakes complex, which already is in grave danger of pollution from petroleum extraction. Present conditions on the Peace-Athabasca delta pose a threat to the elimination by direct and indirect effects, of one of the largest waterfowl hunting potentials in North America. . . .

It is expected that the commercial fishery in Lake Athabasca will collapse within three to five years as recruitment to the harvestable fish populations of walleye, pike, and lake trout declines, and as stocks of cisco are eliminated as the basic food supply for the carnivorous species listed above. Aside from the commercial harvestable fish, the availability of various species for native domestic consumption will be drastically reduced. . . .

Insofar as the Government of Canada has duties to safeguard the rights of the Crown with respect to Wood Buffalo National Park, the Migratory Birds Treaty, and Acts in behalf of the Treaty Indians residing in the region affected, it should institute similar action to safeguard the rights being threatened.

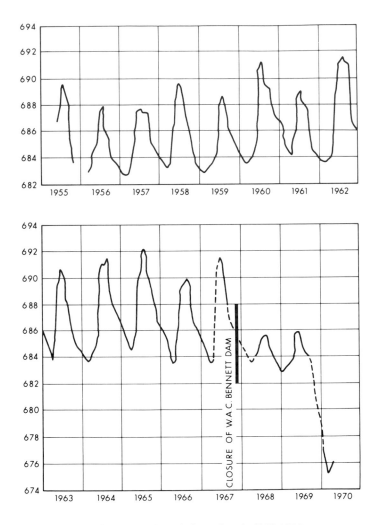

Figure D. Lake Athabasca Levels 1955-1970

The immediate practical problem after the brief was issued was to produce an engineering remedy at maximum speed. Luckily, Alberta possesses a Cooperative Highways and River Research Programme whose members are the Civil Engineering Department of the University of Alberta, the Research Council of Alberta, the Alberta Highways Department, and the Alberta Water Resources Department. Personnel of all these, from the top down, have taken courses in River Engineer–

ing that have been taught at the University since 1948 (Blench, 1959, 1966). The organisation has been conducting field and laboratory research and collecting data on rivers for the last decade; one practical aim is to have dossiers on all rivers likely to be included in projects, so that all the material for scientific planning and design will be available in advance. The organisation has good international liaison. The appropriate members dispensed with protocol and had a provisional analysis of causes, since published (Kellerhals, 1971), and a tentative emergent plan ready by the time the Governments of Canada and British Columbia had arranged to cooperate; a final plan will take longer, as the technical problems of such an involved delta are considerable. Actually, protocol could have delayed matters since the Federal Government alone has interprovincial jurisdiction, is responsible for maintaining navigation down the Athabasca, Slave, and MacKenzie Rivers to the Arctic Ocean, and is responsible for the National Parks. The British Columbia Government built the dam with Federal acquiescence; no engineering body had foreseen damage to lake fishing, of which Alberta had the minor part and Saskatchewan the major.

With action in hand to remedy matters, the practical question now is how such a situation can arise. I do not think any one person could answer, so merely record my own views for criticism.

First, the Dam project was designed, like others everywhere, with no thought of ecological consequences. Two reasons are: (a) experience shows that most mistakes can be patched up eventually without public reaction and at a cost that is far less than the benefits from the project, and (b) engineers and ecologists live in separate worlds. The former reason has been politically acceptable because the public has regarded the environment as expendable, cannot discover the long-term dollar cost of the usually slowly developing consequences of omission in planning, and does not realise that engineers can be taught river mechanics in colleges just as they are taught soil mechanics as part of civil engineering. The enormity of (b) is being realised now in some circles, and one good reaction is the establishment of schools of environmental engineering and, in the USA, the general establishment of federally-funded water resources centers at universities. The general remedy seems to be for colleges, works agencies, and governments to adjust their systems to compel interdisciplinary cooperation.

Second, Canada has only just started to construct a Science Policy and its federal works departments have had nothing approaching the works and research experience of their U.S. counterparts. Comparison with India and the USA is useful here. The Provinces of India developed their enormous irrigation projects separately, starting before 1900, and had acquired good research facilities by the 1920's. Research was coordinated through facilities provided by the federal government which also maintained its own central research station but never dictated to provinces. Thus, after the war when large dam projects were in view, India and Pakistan were able to create efficient organisations, similar to the U.S. Bureau of Reclamation, out of provincial resources and still leave provincial agencies active and important. In the USA major water resources projects started several decades ago under federal agencies which have now achieved relatively good cooperation, even among different disciplines, and have probably left major mistakes far behind; state agencies, once relatively unimportant, are now devising their own State Water Plans. Research, as in Indian provinces, started early within the various federal agencies and was accompanied by appreciation of the fact that future leaders in new ideas, methods, and knowledge are to be found in the graduate schools of universities. Therefore federal researchers of various classes were assigned to universities, and research facilities were established there in such a way that an outside observer could have difficulty in finding which was university and which was federal staff or facilities. Further, the federal government has published statements of operations its agencies have dropped, or decided not to pursue themselves, on the grounds that the requisite abilities existed elsewhere. This concentration on utilising all the talents of the country to the fullest advantage greatly benefits the U.S. in research and development. Its giving of grants, at the start of the IHD, to all Land Grant colleges to start Water Resources Centers according to their own detailed ideas is a new example of the policy.

In Canada the spirit of research has been kept alive in the Universities by the National Research Council (NRC), which parallels the U.S. National Science Foundation and operates largely through Associate Committees drawn from volunteers from the universities; even its annual awards to university staff are made by rotating committees from the universities. Awards are to individual professors according to research ability and originality; records of their work are kept on file. Basic and applied research are equally acceptable, and needless repititions are not supported. The system has operated efficiently and amicably, but has never had the funds for the major project-oriented engineering research at colleges that the U.S. Government regards as essential for the development of the nation's technology.

In terms of this background of different policies, I think the following events, relevant to the case history under discussion, show the great importance of a "democratic" science policy that rests on the effort of all scientists and technologists and reflects their joint opinions, as against an "autocratic" one dictated by a central government body.

The advent of the IHD, starting 1st January, 1964, and the first move toward a Canadian Science Policy roughly coincided. In 1963 the Hydrology Subcommittee (founded 1958) of the Associate Committee on Geodesy and Geophysics of the NRC issued "Proposals for Canadian Participation in the International Hydrologic Decade," in which appeared:

5.5 Evolution of River Beds and Sedimentation

Tests of general theories of the self-adjustment of rivers with sand and gravel beds should be undertaken by a number of countries for various types of rivers. Quantitative regime surveys in sand and gravel rivers and laboratory, and large scale field model investigations, should be undertaken to develop transport laws valid in rivers. Analyses of effects of human interference with river regimes, and the causes and prevention of local erosion should be undertaken along with studies of sources and quantities of river sediments.

This item was intended to include the UNESCO (1962, 1964) topics mentioned above which, apart from including the whole of applied morphometrics (or river regime), opened the gate wide to cooperation with ecologists. NRC influence to implement this ceased with the start of the IHD when a National Committee for the IHD (NC/IHD) was created under the reorganised Department of Energy, Mines and Resources (EMR), whose Inland Waters Branch (IWB) replaced the older Water Resources Branch of another department; the expertise of that branch had been, essentially, the measurement of river discharges. The NC/IHD did not profess technical expertness and so decided to accept technical advice, for all Canadian hydrology, from a tiny subcommittee of the Hydrology Subcommittee. It had no funds but could give its published "approval" to research projects for the IHD, A couple of years later the government gave funds to the Department of EMR for distribution to assist project-oriented research by various bodies, including universities and, of course, itself. I feel that these arrangements had to restrict research support to topics comprehended by a small, specialised group—a group utterly inadequate, through no fault of its own, to deal with hydrology as a whole and lacking the NRC's knowledge of researchers, and research funding.

It soon became clear that EMR funds would not be given to universities for the purposes of item 5.5 quoted above. The first major example I know is the rejection, in 1965, of the proposed first edition of a book (Blench, 1969b) offered to the NC/IHD as a Canadian project in response to UNESCO's call for "Elaboration of a general theory of a river bed development and sediment transport in a stream." The rejection said: " . . . the preparation of text books is not part of the Decade program, this project in particular as it constitutes com-

305

mercial enterprise." The commercialism consisted of three professors risking a few thousand dollars of their own money because there was no other source available at the time. For the second, more expensive edition, through the University of Alberta Press, the never-failing NRC came to the rescue with a guarantee to absorb most of the possible loss. It is noteworthy that, if the technical branches of the Department of EMR had studied the text, they would have known that, barring special circumstances, Lake Athabasca should be expected to drop and the Athabasca River upstream of it to degrade slowly; alternatively, they could have consulted the Alberta Cooperative Highways and River Research Organisation (above).

The second example is the denigration of the UNESCO topics in a Special Study (Bruce and Maasland, 1968), one of whose authors was from the IWB, as follows:

Sub-Category 210 — Erosion and Sedimentation

Erosion of valuable soils is due mainly to moving water. It is estimated that 25 per cent of Canada's usable land has been affected more or less seriously by erosion. An understanding of this process as it applies to Canadian soil types is necessary to devise suitable protection measures. The streams in the Prairies and in agricultural areas, carry sediment loads and deposit the sedimentary materials in reservoirs, thus reducing the useful storage. Erection of structures such as bridges and piers in rivers and lakes changes the patterns of scour and deposition of sediment. If a good knowledge of the process is not available, this can result in undermining of structures and/or need for costly dredging. The Advisory Committee considered this an important but not crucial field for water research in Canada and recommended a somewhat less rapid than average increase to bring the present $250,000 per year expenditure to about $570,000 in 1972-73. Research directed specifically towards control of erision and sedimentation is considered under Sub-Category 507—water quality control.

Finally, an interdisciplinary group at the University of Alberta requested, from the Department of EMR, a grant to establish an interdisciplinary Water Resources Research Center. The group insisted that funds for this purpose were more important than for specific projects by parts of the group, and that it should control detailed disposal of the money according to certain specifically stated objectives. After two rejections the group gave up hope of aid from this source and obtained official University status without immediate budget support. The group still aims to raise additional governmental support and has already held its first public symposium to discuss the interdisciplinary aspects of the

Peace-Athabasca Delta case whose existence is due to interdisciplinary plus intradisciplinary noncooperation.

In trying to draw a simple constructive moral from all this, I should not forget my own inglorious part in the case history. Several years ago I visited the Delta while disposing of a navigation problem on the Athabasca River and, for some time thereafter, used to mention casually to engineering acquaintances that the proposed Peace River Dam was likely to drop Lake Athabasca considerably and that they ought to visit the place during the hunting season. It seems unbelievable now that I did not link the ducks and the inevitable drop in the lake. I just had not the remotest idea of the extent of the particular ecological damage that would occur. Had an ecologist been along on the trip, or had there been an interdisciplinary center at the University, there would have been an immediate interchange of information that would have allowed the alarm to be raised and passed to the Federal Government at top level by authorities to whom attention would have been paid. So, when I see the mistakes of limited agencies charged with an impossibly wide task, I see my own. And I think that a major constituent of necessary action to develop a science policy is that universities form interdisciplinary organisations—no matter what the financial discouragement—aimed at apprising themselves and the public of all the actions required in a unified fight against environmental destruction. This kind of action ensures that governments will receive the information essential for devising a science policy and, in a democratic society, will eventually ensure that governments will be compelled to use it properly.

Literature Cited

Anonymous. 1970a. Death of a delta. *Time*, Can. Ed.(26):8-9.

—————. 1970b. Death of a delta. Brief dated Oct. 2nd. Univ. Alberta Water Resour. Center.

Blench, T. 1959. River engineering as a college course for civil engineers. Eng. J., Eng. Inst. Can., Montreal, P. 86-88, 95.

—————. 1966. River hydrology, or fluviology, for engineers. Civil Eng. p. 56-57.

—————. 1969a. Co-ordination in mobile-bed hydraulics. J. Hydraul. Div. Amer. Soil Conserv. Serv. 95 (HY6), 1871-1898 (closure 96[HY6]: 341-347.

—————————. 1969*b*. Mobile-bed fluviology. 2nd rev. ed. Univ. Alberta Press, Edmonton, Alberta.

—————————. 1970. Mobile-bed hydraulics. J. Hydraul. Res. 8(2):159-188.

Bruce, J. P., and D. E. L. Maasland. 1968. Water resources research in Canada, Spec. Study No. 5 for Science Secretariat, Privy Council Office, Ottawa. Queen's Printer, p. 72.

Kellerhals, Rolf. 1971. Factors controlling the level of Lake Athabasca. Proc. Peace-Athabasca Delta Symp., Jan. (From Director, Univ. of Alberta, Water Resour. Center, Edmonton, Alberta).

Leopold, L. B., M. G. Wolman, and J. P. Miller. 1964. Fluvial processes. W. H. Freeman & Co., San Francisco.

NRC. 1963. Proposals for Canadian participation in the International Hydrologic Decade. Hydrol. Subcomm., Ass. Comm. Geod. Geophys., Nat. Res. Counc., Ottawa, Canada.

UNESCO. 1962. Note NS/NR/17 dated Paris, 15th Oct., 1962.

—————————. 1964. Int. Hydrol. Decade Intergovernmental meeting of experts of 7-17 April 1964. Annex II p. 9, 11, 13. Reference NS/188 dated Paris, 5 June 1964.

U.S. Geological Survey. Undated. Sediment transport in alluvial channels 1963-65. U.S.G.S. Prof. Pap. 462.

SEDIMENTATION (SUSPENDED SOLIDS)

H. A. Einstein

All parts of the earth experience precipitation; but the total annual amount of precipitation and its distribution over the various seasons vary widely from place to place. The precipitation falling on oceans or lakes immediately mixes with and becomes a part of these bodies. The precipitation falling on solid ground in part evaporates immediately, in part percolates into the ground if sufficiently porous, and in part runs off at the surface. It is this last part of the precipitation which contributes mostly to the river flow, carrying suspended sediments with it.

The water and its sediment follow the path of greatest slope, collecting in little draws, in little valleys, and finally in larger valleys, and increasing more and more the concentration of the flow until it finally may be called a river. By that time the flow is able to shape for itself a permanent channel in which the duration of the flows becomes increasingly longer. It is this kind of river channel which constitutes the subject of this conference, with particular emphasis on its ability to sustain all kinds of life. The influences sustaining or hindering the free development of life in these channels are studied.

As the precipitation hits the ground and begins to concentrate in rivulets, it always meets some loose particles or rock or other solid materials along its course which, if the flow concentration is sufficient, it picks up and carries downhill as sediment. Sediment has accompanied the water flows through all the geologic periods and must be considered an integral part of the river flow.

Sediment, as it accompanies the flow downstream, may have a wide variety of sizes and shapes, densities, and resistance against physical and chemical attack. The size has probably the greatest influence on its ability to be moved by a flow. Sediment may include sizes all the way from boulders down to the most minute clay particles; it may constitute a gel or even go in solution. In all these forms it contributes to the solids load of the stream (Anderson, 1954). But all these different sizes and forms of the load do not move equally fast or in the same way. The coarser particles have usually a sufficiently high settling velocity in the water of the stream to concentrate predominately or exclusively near the bed. Very often they actually deposit on the bed, more or less permanently, and thus become part of the bed. When a significant part of the bed consists of such sediment particles this material is called the bed material of the stream.

309

When the bed material of a stream is analyzed for sizes, it is usually found that all sizes down to a certain limit are well represented, but that the sizes smaller than that limit are found only in very insignificant amounts. This is most interesting because in most streams these same fine sediment sizes constitute a large, often predominant part of the sediment load—of the particulate matter passing a given river section (Einstein and Anderson, 1940). Engineers have introduced the term "wash load" for this fine part of the sediment load, while the part strongly represented in the bed is called "bed-material load." This division is made for practical reasons only. First, there exists a great difference in the travel speed of the two categories. The wash load moves in the river channel with a velocity of the same order as that of the water. The bed-material load, on the other hand, has a travel velocity which is one or several orders of magnitude smaller than that of the flow. The wash load may easily travel during one flood (most of the sediment moves during flood flows) from its point of origin to the point of its final deposition. The bed-material load travels during the flood only a short distance, and even this motion is continuously interrupted by periods of rest in the bed. It is not uncommon that bed particles require 50 or 100 years to travel the distance covered by the wash-load particles in one flood.

Another important difference between the two parts of the load is directly connected with mode of motion. The determination whether or not a particle is to move at a specific instance is made at the point from which its move originates. For the wash load this determination is made in the watershed at the point of original scour, after which the load moves continuously. The rate at which the wash-load particles move in a stream is, therefore, determined by the conditions in the watershed (Einstein and Chien, 1953; Gottschalk and Brune, 1950). The bed-material load, on the other hand, travels in the river in a series of moves and stops, each new move depending on the flow condition at the specific time and place of rest. The rate of transport of the bed particles depends, therefore, on the availability of the particles in the bed and on the local flow conditions. With a given sediment composition of the bed, the rate of bed-material loading is a direct function of the flow rate in the river. The bed-material load, unlike the wash load, can be predicted as a function of the discharge in a given stream channel.

Dr. Blench has just been telling you about some of the further relationships that can be derived as a function of the bed-material load in cases where the entire bed of the stream is composed of movable sediment. We shall see where and when the wash load may have an influence on the environment. The most obvious location is in the areas where the sediment is being deposited.

Well known and most important of these areas is the deposition of wash load in the over-bank or the stream. Most alluvial streams have well defined channels which can accommodate low and medium flows. At high flows they often overflow their banks and inundate parts of the valley outside the channel proper. One must realize that the entire valley is the flood channel or the river. During the high flood flows the largest wash loads are carried by the stream. The part of the flow which goes over-bank carries with it a proportionate part of the wash load, often with flow velocities so much reduced that it becomes unable to maintain all the wash load in suspension. A part of the load is deposited in the over-bank areas, gradually aggrading it. These deposits usually contain a wide variety of sizes and often significant amounts of organic material, making these over-bank areas most desirable for agriculture. In many parts of the world agriculture has been able to maintain itself for long periods of time in a rather primitive, but effective, way on the basis of the continuous soil renewal by sedimentation. An actual water ecology will usually not develop in these areas because of the short duration of the inundated periods and the long dry periods between them. However, because of the fineness of the deposited sediment and the high rate of channel aggradation in many alluvial rivers, there are many such valleys with drainage difficulties in which large areas of the over-bank become waterlogged, develop swamps, and attract all sorts of wildlife by their luxurious vegetation. During the last century many such valleys in Europe were drained and developed agriculturally, for health reasons (mosquito abatement) and because of the high fertility of the soil. One may debate whether the destruction of the original ecology is desirable or not. It is a fact, however, that the engineers and politicians of those days who performed and organized these works of "melioration" were and still are honored as heroes.

It is much less well known and appreciated that wash load is continuously being deposited on the banks of the active channels of alluvial rivers. Most such banks consist of much finer material than the riverbed, i.e., of wash-load material. As long as the river is neither widening nor narrowing its channel, bank material is being scoured and deposited. Those who live or work near the bank of such a river know that the river is only statistically stable; scour and deposition occur constantly. Since the banks consist of wash-load material, one must draw the conclusion that any accretion of bank material represents a deposit of wash load. Very little is known about the rates of bank material deposition.

If a stream maintains an equilibrium width, this status must be interpreted as a statistical equality of the rate of scour and the rate of deposition. To determine the statistical rate of bank scour is also to obtain the equal rate of wash-load deposit in this same stream bank. Immediately downstream from a dam of sufficient size to hold back the wash load, scour must be expected to continue at the same rate as before operation of the dam, while there is no wash load available to be deposited. Thus in many channels downstream from such dams a considerable amount of bank scour is observed. Only few observations exist from which the rate of bank scour can be derived quantitatively. One such stream is the Red River below Denison Reservoir (unpublished information). Before construction of Denison Reservoir the Red River downstream from the dam maintained a constant channel width and a rather constant alignment. After construction of the dam practically all the sediment supply was cut off, including the wash load. Repeated resurveys of a set of cross sections downstream from the dam indicated bank scour widening the channel between 5 and 10 ft per year. For years before the dam was constructed the suspended load of the river had been monitored at various points near the dam site and downstream. Over a length of several hundred miles the total wash-load transport was roughly the same, with only insignificant tributary contributions in that reach. After construction of the dam the suspended load near the dam was practically zero, but increased systematically in the downstream direction to reach the pre-reservoir rate about 200 miles downstream. Assuming that the full load was approached exponentially, the rate of scour just about explained the increase of the load.

From this evidence one may conclude that in a river such as the Red River the entire wash load will exchange against similar material of the banks. An equilibrium seems to exist between the bank material and the wash load similar to that between the bed and the bed-material load. It appears that this relationship exists independent of changing meander patterns, but is extremely important for the maintenance of an efficient cross section. One may conclude that structural or other artificial means must be applied in the river channels downstream from dams to prevent excessive bank scour. In some cases, vegetation may be sufficient to achieve stability. Very little is known today about how to choose and establish such bank-stabilizing vegetation (and how to prevent adjacent landowners from interfering with it!).

A heavy wash load in a river interferes with many types of plant and animal life. If reservoirs are established of sufficient size to eliminate dangerous floods, and if at the same time the banks are sufficiently stabilized to prevent any large-scale bank erosion, the river downstream should clear up, thereby providing an acceptable habitat for many forms of life, much more closely resembling the streams which the first settlers of this country encountered before agriculture, mining, and various other human endeavors changed the rivers into their present form. This is possible, as exemplified by the Missouri River, the "Big Muddy," which has today sufficiently cleared up that trout have come in spontaneously.

Reservoirs and lakes have always been a preferred habitat for various kinds of life. These same water bodies were described above as the permanent resting place for large parts of the suspended load of streams. It is of interest, therefore, to specify more accurately where and how the various parts of the suspended load are deposited. There are so many combinations of river size, hydrology, chemistry and size distribution of the sediment, salinity and temperature of the water, etc., that one can only give some very generalized rules which may or may not apply to a particular case.

The bed load and the coarser part of the wash load deposit predominately in the form of a delta at the upstream end of the lake or reservoir (Harrison, 1953). After this delta has advanced a certain distance into the lake, the extension of the upstream channel causes its slope to be reduced, and deposition begins to fill the channel itself. The channel capacity is reduced, and overflow deposits sediment of the same kind on both sides of the channel. These over-bank deposits are coarsest near the stream channel and become increasingly finer as one moves away from the channel proper. As in the over-bank of the alluvial valleys, swamping sometimes results. But the bulk of the wash load, as well as the greater part of the bed load, travels down the channel into the reservoir or lake proper. The river flow mixes with the lake water after deposition of the coarser parts of its load in the delta. It is still carrying with it the finest of the solid particles—the clays. In the eastern part of this country, where kaolinites predominate, the individual clay particles remain separate and eventually settle out into rather dense deposits covering the entire area of the reservoir bottom. In the western part, where montmorillonites predominate, the clay fractions have the tendency to flocculate, particularly where the water has a considerable salt content. The resulting deposits are fluffy and have a much higher water content. In cases of high concentration inflows into reservoirs (e.g., Lake Mead) (Einstein, 1941; Grover and Howard, 1938; Knapp, 1943; Sherman, 1953) density currents have been observed to develop which may carry the sediment-laden water into the deepest part of the reservoir, where the wash load is deposited with a nearly horizontal surface—a lake in the lake. These deposits may make an acceptable habitat for certain forms of life, but are generally undesirable because of the irregular and rather sudden addition of new layers of deposit.

Another place of deposition for the wash load is in the ocean. The rules of deposition in the ocean are similar to those in reservoirs or lakes with two important differences: First, the ocean is much larger and has currents and wave action of larger intensity which disturb the rather simple and smooth deposition pattern of the lakes and permit the sediments to follow much more complicated and farther-extended deposition patterns. Second, the ocean water and usually also that of the lower ends of the river channels have a much higher salinity than

313

that of most lakes and reservoirs, causing practically all clays to flocculate. This basically changes their settling velocity and thus the distance over which the sediment can move in suspension.

For the latter reason most clays never reach the ocean proper if the river develops into an estuary at its mouth (Einstein and Krone, 1961); most of it is deposited in the estuary, where it is in most cases undesirable. Flocculated clay deposits are usually very fluffy and with the low density goes low strength, long-duration compaction, and a large displacement. Many estuaries are used as harbors, and with the continuously increasing size of the commercial vessels such deposition is very undesirable. Also, the dredging of this material is disagreeable and often ineffective. Because of the high water content one must move large volumes. Upon agitation the sediment-water mixture becomes fluid and cannot easily be stored on dry land. If it is stored in the waterway itself, it may be moved by tidal and wave motion back into the same locations from where it had been dredged. This rather unfortunate development has been observed in several estuaries.

A strictly ecological problem is the survival of the eggs and of the young of anadromous fish such as the salmon. The fish lay their eggs in the pores of gravel beds in freshwater streams. The eggs need oxygen for survival, which they obtain by percolation of river water through the gravel deposits. As long as the river water is clear, the process goes on without difficulties. The eggs and the young fish after hatching find ideal living conditions in the gravel pores and develop normally. If the stream carries wash load, some of the particles deposit in the pores of the gravel and prevent the necessary passage of water. The supply of oxygen is reduced and many of the eggs die. Under natural conditions this is just one of the hazards counteracted by the large number of eggs provided per grown fish. The fine sediment deposited in the pores of the gravel bed during medium flow periods is periodically washed out by flood flows which disturb and clean the bed.

With the development of various water resources many of these streams have lost so much of their low water flows that the fish are unable to reach their historic spawning grounds. It has become necessary to replace the natural spawning ground with artificially prepared deposits of gravel. To attract the fish a flow is provided over the gravel which must continue into the same stream in which the historic spawning grounds are located. This can best be achieved if the artificial spawning grounds are located in the diversion canals. These canals carry with their flow an equal amount of wash load as the river used to carry, and this sediment must be expected to deposit in the artificial gravel bed just as it used to deposit in the natural gravel bed. But the diversion canal has no flood flows to clean the gravel; this must be done artificially. To design equipment and to organize the operation of cleaning the gravel, it is essential to know the frequency with which the cleaning operation must be repeated. On the basis of the results of flume experiments which duplicated the gravel bed and the wash load in full scale (Einstein, 1968), it is possible today to predict the rate of pore filling if the flow condition over the gravel bed and the concentration and composition of the wash load are known.

The results of the flume experiments showed that the rate of deposition for each grain size is a function of the local sediment concentration and the settling velocity of the particular sediment size. Starting from the upstream end of the gravel bed and filling the pores completely, the deposition proceeds gradually downstream. Only unflocculated sediment was studied for which surface forces are negligible compared with the particle weight. Because of their much higher settling velocity, silt particles are the greatest part of these deposits while the deposition of clay particles is unimportant.

Many geologic gravel deposits have been found in which the pores are not filled with fine particles. One can hardly assume that the particular streams were not carrying fine sediment in addition to the gravel. Judging from the behavior of our contemporary streams, one must rather assume that those streams carried, in addition to the gravel, the fines and all the intermediate sizes. The gravel was moved only at flood flow while all smaller grains moved through as wash load without depositing in significant number. Then, as the high flow stage gradually ended, layers of gradually reducing grain size were deposited on top of the gravel, effectively building a filter which prevented the fines from entering the gravel during the long-duration lower flows. At the beginning of the next high stage this filter was scoured and removed by the flow.

Another very interesting relationship between wash load and the ecosystem of the river was pointed out by Dr. R. B. Krone, Professor of Hydraulic Engineering, University of California at Davis. He noted that the inflow to the Bay of San Francisco carries continuously a small concentration of clay as wash load. This clay is mostly montmorillonite and as such has a very large surface and is electrically quite active, as indicated by its strong tendency to flocculate. He argues that such a clay is able to absorb rather large amounts of dissolved and very fine particulate matter of different chemical composition, even at the low clay concentrations. He thinks that this clay may be relied upon for removing large quantities of metallic and other poisons from the water as it deposits in the Bay. He warns that although the progressive reduction of clay inflow into the Bay will reduce the amount of dredging, it will also deprive the system of this natural cleaning agent. He expects the toxicity of the Bay water to increase, even without a significant increase in the inflow of toxic materials.

315

The presence of this suspended clay also significantly reduces the ability of sunlight to penetrate into greater depths, effectively limiting growth of plants such as algae. In general, one may look at the growth of algae as the first step of a food chain, and thus necessary and desirable. However, there are conditions under which the food chain is interrupted and the algae, based on an oversupply of nutrients, will overpopulate the area. During the night they consume all the oxygen dissolved in the water, leaving none for the other forms of life dependent on it. Such algae "blooms" have been observed in quiet parts of bays and seas and have resulted in massive fish kills. Dr. Krone thinks that these undesirable events may be prevented by a moderate concentration of suspended clay.

All this may indicate that the ecology of a water area, be it a river, a lake, or an ocean, cannot be fully assessed without due consideration of the various effects of its suspended sediment. Predictions of future developments of life in a water course must in part be based on the sediment conditions expected to prevail. At the beginning of the century one could safely predict in most cases that the sediment load was increasing. Today there are still many areas where new agricultural or housing developments will increase the sediment supply from the watersheds, but there are also many streams in which the construction of reservoirs and other water-supply developments drastically reduce sediment supplies; each case must be analyzed for its own particular conditions.

This meeting has as its theme "River Ecology and the Impact of Man," and let me emphasize that we engineers invited to take part realize the importance of the various forms of life in the river. But I must emphasize with equal strength that the various forms of life in a river are purely incidental, compared with the main task of the river, which is to conduct the water run-off from an area toward the oceans. Some people think the only purpose of the river is to provide them with fishing on Sunday; some people see it only as a place to crank their boats up to full speed and beyond; some people look at the river only as a cheap way of disposing of their old tires; and some people think the only purpose of a river is to support life and its ecology. Even if the latter attitude is today fashionable, we must remember that the main purpose of the river remains that of conducting water. All the other activities are only permitted as long as they do not interfere with the main purpose. It is imperative, therefore, that every student of river ecology familiarize himself thoroughly with the rules and laws of river hydraulics, of which the motion of its sediment is an important factor.

Selected Bibliography

Anonymous. 1940-1942. A study of methods used in measurement and analysis of sediment loads in streams. Hydraulic Laboratory, Univ. of Iowa.

Anonymous. 1954. Interim report—distribution of sediment in reservoirs. Sedimentation Sect., Hydrol. Br., Proj. Invest. Div., Bur. Reclamation.

Anderson, H. W. 1954. Suspended sediment discharge as related to stream flow, topography, soil, and land use. Amer. Geophys. Union. Trans. 35(2):268-281.

Brune, G. M. 1953. Trap efficiency of reservoirs. Amer. Geophys. Union Trans. 34(3):407-418.

Einstein, H. A. 1941. The viscosity of highly concentrated underflows and its influence on mixing. Trans. Amer. Geophys. Union Pap. Hydrol., p. 597-603.

——————. 1968. Deposition of suspended particles in a gravel bed. Hydr. Div. Proc. Amer. Soc. Civil Eng.

——————, and A. G. Anderson. 1940. A distinction between bedload and suspended load in natural streams. Amer. Geophys. Union Trans.

——————, and N. Chien. 1953. Can the rate of wash-load be predicted from the bed-load function? Amer. Geophys. Union 34(6):876-882.

——————, and R. B. Krone. 1961. Estuarial sediment transport patterns. J. Hydr. Div. Proc. Amer. Soc. Civil Eng.

Gottschalk, L. C., and G. M. Brune. 1950. Sediment design criteria for the Missouri Basin loess hills. Soil Conserv. Serv. TP-97.

Grover, N. C., and C. S. Howard. 1938. The passage of turbid water through Lake Mead. Amer. Soc. Civil Eng. 103:720-732.

Harrison, A. S. 1953. Deposition at the heads of reservoirs. Iowa 5th Hydr. Conf. Proc., p. 199-226.

Heinemann, H. G. 1962. Volume-weight of reservoir sediment. J. Hydr. Div. Proc. Amer. Soc. Civil Eng.

Knapp, R. T. 1943. Density currents—their mixing characteristics and their effect in the turbulence structure of the associated flow. Proc. 2nd Hyd. Conf. Iowa City.

Sherman, I. 1953. Flocculant structure of sediment suspended in Lake Mead. Amer. Geophys. Union Trans. 34(3):394-406.

Vanoni, V. A. 1944. Transportation of suspended sediment bv water. Amer. Soc. Civil Eng.

A COMPARATIVE ASSESSMENT OF THERMAL EFFECTS IN SOME BRITISH AND NORTH AMERICAN RIVERS

T. E. Langford

The importance of rivers as agents for the dissipation of waste heat has increased markedly during the past two decades in many countries. The majority of this heat originates from cooling water used in the generation of electricity, and in both the United States and Great Britain the demand for electricity has been doubling every ten years.

Modern British generating stations require about 120 cfs of cooling water per 100 MW generated (Hawes, 1970), and some nuclear stations in the U.S. may require up to 50% more. Thermal efficiencies vary over a similar range in both the U.S. and Britain, i.e., 20-25% in older plants and 30-35% in the latest plant. At best, a power station has to dispose of some 50-65% of the energy it generates as heat, via the cooling water to the environment. For the average plant in the U.S., the loss is about 5,300 BTU/KWH (F.P.C., 1969).

In Britain a 400-MW unit has to dispose of some 2,000 million BTU/hr. This is sufficient to raise the temperature of a flow of water of 15 mg/hr (or twice the dry-weather flow of the Thames at Teddington) through 5°C (Ross, 1959). Today power stations of 2000-MW capacity are in operation in Britain, and larger ones are under construction. The size of the heat disposal problem is, therefore, considerably greater than in 1959.

In comparison, the U.S. power industry is constructing units of up to 1100-MW capacity for very large installations with total productions up to 8000 MW (Adams, 1969).

The ecological consequences of heated discharges have caused great concern in the U.S., and there is, as a result, a vast amount of literature, both academic and emotional, available on the subject of temperature and aquatic ecosystems. Much of this literature has been very well reviewed by a number of authors (Cairns, 1955; Raney and Menzel, 1967; Hawkes, 1969), and it is not my intention to attempt a further, full review in this paper. There are also some 300 programmes of research either completed, planned, or in progress dealing with the physical, chemical, and biological aspects of power developments and heated discharge in lakes, rivers, estuaries, and coastal waters (E.E.I., 1969).

In Britain public concern about heated discharges has not been as obvious, though interest has been increasing steadily from conservationists (Mellanby, 1969), the authorities (Bottomley, 1971), and the press (Tinker, 1970). For some fifteen years, however, the Central Electricity Generating Board, responsible for the production of electricity in the whole of England and Wales, has carried out, or provided facilities for, research into the physical, chemical, and biological aspects of heated discharges on British rivers, lakes, estuaries, and coastal waters (Alabaster, 1962, 1969; Hawes, 1970; Ross, 1970).

The aim of this paper is to summarize briefly the most relevant data on the effects of heated discharges in British and American rivers. From a comparison, it is possible to draw up a number of general rules for minimising ecological effects on a rational rather than emotional basis. I hope this will be of use in forecasting effects and planning discharges from future power developments.

Use and Availability of Cooling Water

The pattern of demand for electricity differs considerably in the U.S. and Britain. In Britain peak demand usually occurs in winter when water temperatures are low and river flows high. In the U.S., particularly in the eastern and southern states, air-conditioning machinery creates a peak demand period in the height of summer, when natural water temperatures are high, humidity is high, and river flows are at their lowest.

In 1968-69, 10% of the total fresh water runoff in the U.S. was used for cooling purposes (Lof and Ward, 1969). Most of this water was available for reuse. In Britain the total amount of fresh water circulated in power stations exceeds 40% of the mean annual runoff. Further, although new power stations in Britain are of 2000-MW capacity, there is not sufficient fresh water in any river at summer flows to cool even one 500-MW unit directly (Hawes, 1971). As a result of this acute water shortage, all power stations on river cooling are now built with full cooling-tower capacity. Thus a 2000-MW installation (four 500-MW units), requiring some 50 million gallons of water per hour (2,400 cfs), need only abstract 3% of this water from the river. One percent is lost by evaporation, along with a large amount of heat (Ross, 1970). Two percent is returned to the river to allow for make-up, preventing overconcentration of solids in the cooling system.

In the U.S. water is much more plentiful. There are a number of rivers with dry-weather flows of over 3,000 cfs, and most major river systems are controlled by large numbers of dams. Cooling towers are used only in certain regions or localities where necessary (F.P.C., 1969; Churchill and Wojtalik, 1969).

There are in both countries, however, power stations which abstract and return 70 to 90% of dry weather flows in some rivers (Langford, 1970). At such stations there is often a reversal of current in the river as outlet water is drawn back towards the intake. This reversal may also occur as a result of upstream winds in some British rivers (Langford, 1971a).

Water Temperature

Natural Temperatures. River temperatures in Britain rarely exceed 24°C under natural conditions (Macan, 1963). The upper limit is normally 16-22°C, depending upon latitude and altitude. Over the ten years from 1957 to 1966 the Severn at Ironbridge exceeded 22°C only once (Langford, 1970).

In the United States, particularly in the East and South, 30°C is the normal upper natural limit for many rivers. Temperatures of up to 37°C have been recorded in shallow backwaters (Profitt, 1969), and Wurtz (1969) quotes temperatures of 33.2 to 34.2°C in parts of the Mississippi. In the Pacific Northwest water temperatures in the Columbia River have a natural range similar to that of the Severn (Nakatani, 1969).

The temperature pattern of larger rivers is more stable than that of smaller streams. Detailed studies of the Severn in Britain (Langford, 1970) showed very small diurnal fluctuations, i.e., rarely more than 2°C, even in summer. Figure 1 shows the daily records for part of June 1966 during varied weather and river conditions. The normal daily fluctuation was depressed at the height of the spate and was reinstated as the river fell to normal summer level. The greatest daily fall in river temperature was 3.5°C during a heavy rain and hailstorm.

In small streams daily fluctuations of 6°C were recorded by Macan (1958), and Sprules (1947) recorded a diurnal fluctuations of 9.5°C. Maximum rates of change in large rivers are generally low, averaging less than 0.2°C per hour. In smaller streams rates of 0.8°C to 1.7°C have been recorded (Wurtz, 1969).

Animals and plants in larger rivers are therefore accustomed to more stable temperature patterns than those in smaller streams, though overall ranges are larger.

Effects of Power Stations on River Temperature. It is impossible to generalize about the effects on river temperature of operating power stations, as there are so many factors involved, including river flow and velocity, design of outfall, shape of river channel, power output, type of cooling, dilution factor, and natural temperature.

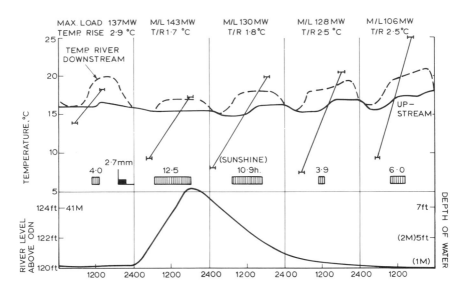

Figure 1. Effects of summer spate on the temperature of the river water
(After Langford, 1970)

The usual temperature rise across the condensers in British power stations is 8-10°C, and in the U.S., 8-15°C. Maximum discharge temperatures in Britain rarely exceed 30°C, though in some recirculating rivers temperatures up to 32°C have been recorded in hot summers. In 1970 the River Trent reached 30°C at Nottingham for a period, but this temperature was unusual (Bottomley, 1971). Usually, these high temperatures last for only a day or two each year.

In the U.S. discharge temperatures reach 35-42°C, depending upon conditions (Coutant, 1962; Churchill and Wojtalik, 1969; Profitt, 1969), though again the highest temperatures last for only a few days.

The total effect on river temperature varies widely. In some rivers a positive thermal stratification occurs over part or all of the river (Churchill and Wojtalik, 1969; Goubet, 1969). In others the discharge flows along one side of the river channel (Mann, 1965; Churchill and Wojtalik, 1969). And in yet others thorough mixing may occur as a result of weirs or the shape of the river (Langford, 1970). Total rises in the Severn downstream of Ironbridge power station after complete mixing varied from 0.5°C during spates to 8°C during lower river flows (Langford, 1970). Localised increases of 11-12°C have been recorded in stratified or poorly mixed British rivers (Langford, 1971a). On the other hand, effluents from "fully closed cooling" stations, even those of 2000-MW capacity, are so small that it is difficult to detect temperature effects even 100-150 m downstream.

322

In the U.S. the effects also vary widely but, due to high ambients, temperature effects are mainly on a higher level. Thus Coutant (1962) recorded temperatures of 37-41°C in the Delaware, where cooling water was thoroughly mixed vertically over part of the cross section of the river. In the Green River temperatures of 32°C were recorded downstream of a power station using a large proportion of the river flow at 25°C ambient. At the Colbert Stream Plant the effects of the discharge were only detectable in a very small area of the Tennessee River, and the main river temperature was not altered; and Profitt (1969) showed that in the White River the mixed water two miles downstream of the Indiana Power and Light Company plant was never more than 3°C above ambient, and the average elevation was about 0.3°C, though the maximum discharge temperature reached about 42°C at one time.

Power station discharges may also affect the rate of temperature change, exaggerate the diurnal variation, and alter the long-term patterns. For example, at Ironbridge the power station operated for sixteen to twenty hours each day. Thus when operation began, at 0600 hours, there was a sharp rise in temperature downstream between 0600 and 0800 hours. During this two-hour period, the rate of rise was 2.5°C at low river flows. At Peterborough even more marked fluctuations occurred in the outfall channel each morning, when rapid changes of 10-15°C were recorded over a period of one to two hours (Cragg-Hine, in preparation). At Ironbridge the summer diurnal increase downstream of the power station was 100% greater (Figure 1)—i.e., 4.7°C—during some periods. In winter, increases of 3-4°C occurred downstream, but not upstream. In the short term, therefore, the general effect of the power station was to destabilize the river temperature, for up to 3 km downstream of the outfalls.

In the longer term mean weekly spring temperatures were three to five weeks advanced downstream, and autumn temperatures, one to two weeks retarded. Also, degree-hours over 0°C were accumulated at a faster rate downstream, and these totals were three to five weeks in advance during various seasons. In a more extreme case, Alabaster (1969) showed that mean temperatures in the Trent downstream of some power stations was above normal summer levels in November.

Chemical Effects of Power Station Discharges

Apart from the rise in temperature, power station discharges are often altered chemically during the cooling process. Davies (1966) showed that cooling tower discharges have lower ammonia levels, higher concentrations of nitrate, and lower levels of organic nitrogen when the cooling water is abstracted from a polluted river. However, there are chlorine by-products in very small quantities and an increased concentration of dissolved solids.

It has been adequately established that little, if any, deoxygenation occurs in rivers below direct-cooled power stations (Hawes, 1970; Alabaster, Garland, and Hart, 1971). Further, Aston (in press) has shown that in very polluted rivers rapid aeration occurs during passage of the river water through cooling tower systems, and oxygen concentrations in cooling water can be considerably raised, even though temperatures are higher than ambient (Figures 2 and 3). Aston estimated that up to 3.9 metric tons of oxygen per day are injected into the River Trent from the cooling water at Drakelow Power Station, resulting in both chemical and biological improvement. Recent work by Alabaster et al. (1971) has shown that river water extracted from below power stations on the Trent is less toxic to fish than that from upstream.

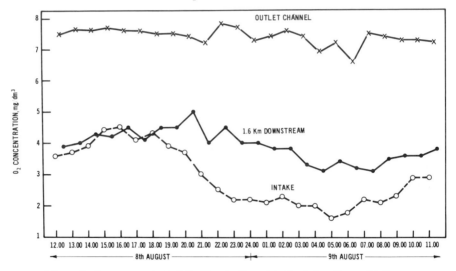

Figure 2. Oxygen content of the Trent adjacent to Drakelow Power Stations, 8th and 9th August, 1967

In the U.S., where temperatures may be higher, some deoxygenation may occur below direct-cooled stations. Profitt (1969) showed decreases from 12 to 9 mg/l, 15 to 10½ mg/l, and 13 to 10 mg/l between intake water and effluent at the Indiana Power and Light Company plant on the White River, but only on one occasion—in July 1965—was a drop from above saturation level to below saturation level recorded. Dissolved oxygen records from cooling tower discharges in the U.S. are not available, but it would be expected that oxygenation would occur as in Britain.

There has been relatively little research on the effects of chlorine by-products on river ecosystems, and in some fish mortalities in the U.S. it has not always been possible to separate chemical and thermal effects. It is obvious, however,

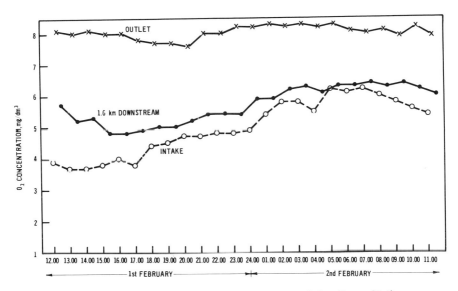

Figure 3. Oxygen content of the Trent adjacent to Drakelow Power Stations, 1st and 2nd February, 1968

that when attempting to assess the effects of heated discharges on rivers, careful consideration must be given to these chemical changes.

Biological Effects of Cooling Water Discharges

There are very few uncomplicated cases of thermal loading in rivers, either from natural or industrial sources. Either the discharge is chemically altered or the receiving river changes physically or chemically between intake and outfall (Churchill and Wojtalik, 1969; Langford, 1971a). Even in natural thermal areas there is often a high concentration of minerals in the spring water (Winterbourn, 1968). Thus the actual effects of thermal loading need to be assessed from field studies of a number of sites, as well as from laboratory experiments.

In Britain biological studies have been made on a variety of rivers—both clean and polluted—around some twenty power stations (Figure 4) with various cooling systems. Invertebrate surveys have been carried out on long- and short-term bases (Mann, 1965; Langford, 1971a), and life histories of benthic invertebrates, emergent insects, and fish have been studied (Steel, 1961; Negus, 1956; Langford, 1971a; Cragg-Hine, in press).

In the United States there are many biological studies in progress but data, particularly concerning invertebrates, are scarce (Coutant, 1962, 1969).

325

Figure 4. Map showing power station sites used for ecological surveys

In the following section I will summarize the more relevant data on various factions of the river ecosystem in relation to heated discharges.

The Flora — Algae and Micro-Flora. Patrick (1969) has reviewed the literature on temperature and freshwater algae, and this review needs no expansion. Few studies on algae below heated effluents have been published.

In the U.S. Trembley (1965) found that species diversity was reduced where temperatures reached 40°C in the Delaware. Also, above 34.5°C increases in blue-green algae were noted. Chlorination decreased the number of individual cells slightly, but not the species diversity. Similar results were found on the Susquehanna and Potomac rivers (Patrick, 1969).

In most temperate waters temperatures of 30°C caused increases in species diversity though some diatom species were favoured (Strangenberg and Pawlaczyk, 1961). Swale (1964) studied the phytoplankton of the River Lea around Rye Meads power station in Britain and found that temperature increments of 3-12°C had no effect on species diversity or production.

On the other hand, at a power station on an oligotrophic lake, Fleming (unpublished) showed that colonisation rate and growth of periphyton (expressed as chlorophyll a per unit area) was higher in the heated effluent than in the rest of the lake.

The effects of passage of phytoplankton through conditions of thermal shock and power station condensers were observed by Savage (1969) and Griffith (op. cit., Patrick, 1969). They concluded that there were no significant effects on the algal cells under conditions which would be experienced in a power station cooling system at those temperatures.

Rooted Plants. Studies of rooted plants around heated discharges are rare in both Britain and the U.S. Recently, Anderson (1969) concluded that the effluent from a power station on the Patuxent River (Maryland) suppressed the vetegative growth of *Ruppia maritima*, though it was not lethal to the species. *Potomageton perfoliatus* was tolerant of temperatures up to 35°C and was replacing *Ruppia* in some areas. Also, *Spartina alterniflora* and *Phragmites communis* could tolerate temperatures up to 35°C. In Britain most of the common indigenous aquatic plants have been found downstream of power station effluents at temperatures as high as 32°C. At Lincoln, on the River Witham, growths of *Cladophera* were found below the outfall in January, but none were found upstream. Temperature at the time was 18-20°C just downstream and 60°C upstream. At Peterborough casual observations have shown that the under-

327

water leaves of *Nuphar lutea* were present all year round in the effluent channel, but not in the river upstream. Also, surface leaves and flowers seemed to appear earlier. *Nuphar lutea, Myriophyllum spicatum, Callitriche* spp., *Phragmites* sp., *Carex* spp., and *Potomageton pectinatus* have all been found in reaches which exceed $30^{\circ}C$. It is quite evident, however, that the effects of heated discharges on rooted plants need much more investigation.

Invertebrate Fauna — The *Protozoa*. In Britain the fresh-water Protozoa in artificially heated waters have not been studied. In the U.S., however, Cairns (1969) has shown that acute thermal shock, where temperatures reached $50^{\circ}C$, caused a drastic reduction in species diversity in an experimental community. This community recovered to almost its normal level in about 120 hr. Exposure to temperatures of $31^{\circ}C$ for 24 hr caused a less marked reduction in diversity, and the system recovered in about 32 hr. In the Potomac and Savannah rivers no marked changes in the protozoan communities were observed below heated discharges.

Macro-Invertebrate Communities and Species Diversity. The change in diversity of the macro-invertebrate fauna of a river, expressed in one of any number of ways, is a reasonable indication of any dramatic effects of an effluent. Figure 5 shows the diversity of the fauna in various reaches of the River Trent, expressed as mean number of species present, in relation to dissolved oxygen, ammonia, and temperature (Langford and Aston, in press). There was a slight drop in the mean number of species below three power stations, but the major changes in the river fauna were obviously related to changes in dissolved-oxygen levels, irrespective of ammonia or temperature.

In a clean river, the Severn, species diversity was also unrelated to temperature up to $28^{\circ}C$ (Figure 6). Figure 7 shows the degree of similarity between invertebrate communities upstream and downstream of twenty power stations in Britain, in relation to the maximum river temperature downstream. Degree of similarity is expressed simply as

$$\frac{\text{Number of families common to both reaches}}{\text{Total number of families recorded in both reaches}} \times 100$$

There was no relationship between the "degree of similarity" and temperature up to $32^{\circ}C$. Therefore, although there was some variation of the fauna upstream and downstream of outfalls, other factors were mainly responsible. Longer term studies have also shown that in British rivers thermal effects on the composition of the invertebrate fauna are small (Mann, 1965; Langford, 1971a; Aston, in press).

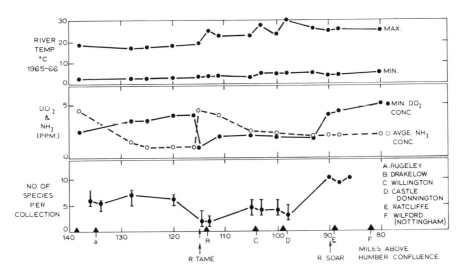

Figure 5. Average number of invertebrate species collected from reaches of the River Trent in relation to temperature and chemical factors in 1965

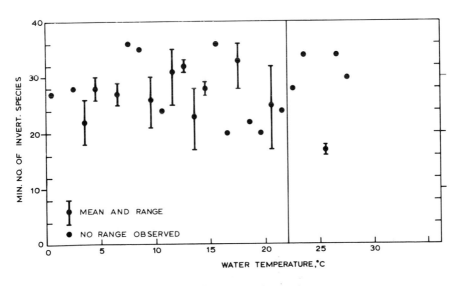

Figure 6. Species diversity at River Severn sampling stations relative to temperature

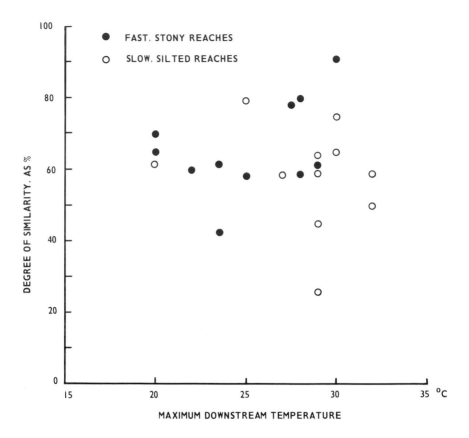

Figure 7. Similarity of the invertebrate faunas upstream and downstream of
20 power station discharges, in relation to the maximum recorded temperature
Similarity = % families common in both sampling areas

In contrast, Coutant (1962) found a drastic reduction in species diversity, numbers of animals, and standing crop where river temperatures exceeded 37°C. At 40-42°C few organisms survived, though recolonisation was reasonably rapid as temperatures fell. At peak temperatures the number of species upstream was 40, downstream 4. In the Schuylkill River, however, Wurtz and Dolan (1960) found very little reduction in the number of species collected 0.25 mile below a heated discharge even though temperatures exceeded 37°C for some days. Merriman (1970) and Mann (1965) suggested that changes in substrate and river flow were more responsible for changes in river communities below Haddam Neck (U.S.) and Earley (U.K.) power stations than were increased temperatures. At Haddam Neck the fauna of areas adjacent to the discharge became more diverse even though discharge temperatures exceeded 38°C at times.

In the Tennessee River Churchill and Wojtalik (1969) found no significant changes in bottom fauna below the Widows Creek plant, and in Cane Creek, below the Colbert Steam Plant, changes in the bottom fauna were most likely due to changes in substrate.

Below the Paradise Plant on the Green River, Kentucky, reductions in the invertebrate fauna were recorded on some five miles of shoreline where river temperatures exceeded 35°C for long periods. Cooling towers were eventually constructed to keep river temperature down to 32°C. Profitt (1969) also found no living invertebrates where outfall temperatures exceeded 40°C.

Tolerance of Indigenous Species. The tolerance of certain invertebrate species to ecological disturbance varies with the *type* of change. For example, stoneflies (Plecoptera) and mayflies (Ephemeroptera) are intolerant of organic pollution (Hynes, 1960), but reasonably tolerant of inorganic pollution by metal salts (Carpenter, 1924, 1925); the reverse is true for Oligochaeta, leeches, crustaceans and molluscs. In Britain many of the pollution-intolerant species were found to be tolerant to heating in the Severn at least to 28°C (Langford, 1971*b*).

Coutant (1962) showed that certain species were first to disappear as temperatures rose; *Chironomidae, Berosus, Argia,* and *Physa* sp. were most tolerant.

At a temperature of 34°C Profitt (1969) collected planarians, Hyalella, and two species of mayflies. Both Coutant and Profitt suggested that some *Chironomidae* were favoured by temperatures up to 35°C, as were larvae of *Psychomia.* The most tolerant indigenous British invertebrates are probably *Limnodrilus hoffmeisteri, Asellus aquaticus, Limnea pereger, Bithynia tentaculata,* and the damselfly *Ischnura elegans.* All these species have been observed regularly in cooling ponds and effluent channels.

Very few species are restricted by warm water discharges in Britain. Mann (1965) suggested that some flatworms, leeches, and shrimps were less abundant below Earley power station. Langford (1971*a*) and Cragg-Hine (personal communication) also suggested that leeches and molluscs were less abundant below some outfalls. Other factors may have been acting to restrict their distribution, however, as these species were collected at higher temperatures below other outfalls. Both Mann (1965) and Langford (1971*a*) indicate that the planarian *Dugesia tigrina* may be restricted by warm water discharges.

Introduction of Exotic Species. Records of exotic invertebrates in rivers warmed by power station discharges have been noted by Mann (1965) and Langford (1971*a*). The exotic gastropod *Physa acuta* has been recorded in the Rivers

331

Trent, Ouse, and Witham and in the cooling-tower ponds of four British power stations. Also, the exotic oligochaete *Branchiura sowerbyi* has been recorded below Earley (Mann, 1965) and Coventry power stations (Aston, unpublished).

There are no records of exotic species in American rivers, though it may be significant that *Physa* spp, were among the most tolerant species recorded by Coutant (1962).

Life Histories. Steel (1961) found that breeding activity of the isopod *Asellus aquaticus* was advanced by at least one month below Earley power station. Also, Negus (1966) showed that there was a similar advance in the freshwater mussel *Anodonta anatina* in the same reach of the Thames. Temperatures varied from 1-12°C above ambient.

Aston (in press) found that the peak of cocoon production in some oligochaete species in the River Trent occurred in May upstream of Drakelow Power Station, but not until October downstream (Figure 8). Also, there were always higher numbers of sexually mature specimens of *Limnodrilus hoffmeisteri* downstream (Figure 9). The average temperature increment downstream was 10°C. Subsequent experiments (Aston, in press) show that *Limnodrilus hoffmeisteri* can reproduce successfully over a temperature range of 5-30°C, and *Tubifex tubifex* over a range of 10-25°C (Figure 10).

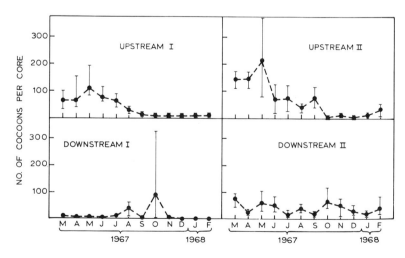

Figure 8. The numbers of tubificid cocoons in core sample taken from the River Trent above and below Drakelow Power Stations showing averages and the total spread of results Reproduced with permission of Dr. R. J. Aston (Aston, in press)

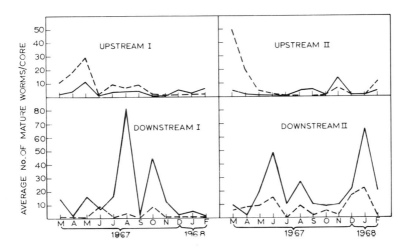

Figure 9. The average number of sexually mature worms of L. hoffmeisteri (———) and
L. profundicola (– – –) per core sample taken from the River Trent
upstream and downstream from Drakelow Power Stations
Reproduced with permission of Dr. R. J. Aston (Aston, in press)

Studies on the insects of the River Severn (Langford, 1971a) have also shown
that some mayfly and stonefly species hatch, grow, and emerge over wide
temperature ranges. For example, the winter-growing stonefly, *Taeniopteryx
nebulosa*, hatched from the egg over a range of 17°C to 5°C, though in the
downstream reaches tiny nymphs were absent during a longer period of summer
than upstream (Figure 11). Growth of nymphs after September was similar in
upstream and downstream reaches. Nymphs of the summer mayfly, *Ephemerella
ignita*, hatched from the egg over a temperature range of 6°C to 25°C in the
Severn (Figure 12). In each of these two species hatching in 1966 occurred
simultaneously upstream and downstream even though mean temperatures and
degree-hours were three to five weeks in advance downstream after the winter.
In 1967 both species hatched earlier downstream.

Emergence trapping of adults during 1970 showed that the mayfly and caddisfly species continued to emerge even when water temperatures were 26-28°C
downstream (Figure 13). The emergence period of the mayfly *Heptagenia
sulphurea* was longer upstream than downstream (Figure 14)—the reverse of
what might be expected. The emergence patterns of the caddisfly *Psychomia
pusilla* were very similar in heated and unheated reaches (Figure 15). It is obvious
that emergence of some species is not dependent on temperature after a certain
date. In 1971 catches have shown patterns similar to those in 1970 even though
water temperatures were 8°C lower in June this year than in June last year.

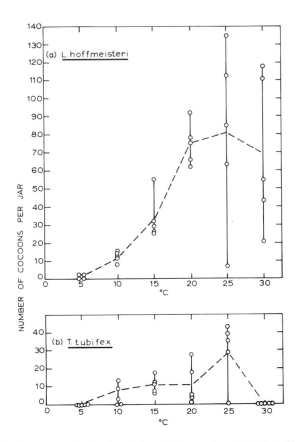

Figure 10. Cocoon production in relation to temperature in (a) L. hoffmeisteri and (b) T. tubifex. Each point represents the number of cocoons laid by 10 worms over a period of 5 weeks

The influence of temperature is not fully clear. Macan (1964) showed that emergence of some Odonata was related to the earliest date on which certain temperatures occurred. Gledhill (1960) and Sprules (1947) showed that the emergence of some Plecoptera, Ephemeroptera, and Trichoptera in small streams was not related to water temperature.

Fish — Mortalities. It is evident that free-living fishes will detect and avoid uncomfortable temperatures before they are killed (Alabaster, 1962). Where mortalities have been recorded in both the U.S. and Britain, fish have usually been trapped in an effluent channel or exposed to a sudden discharge of hot water.

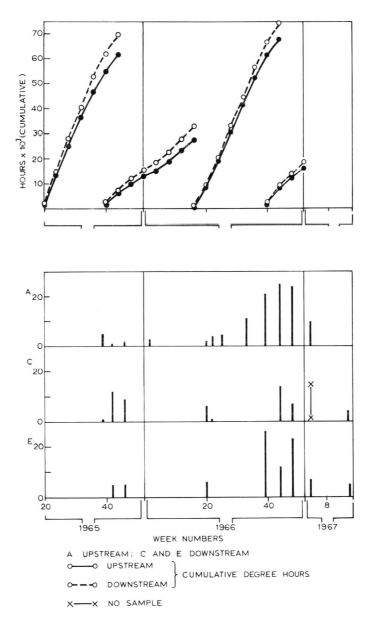

Figure 11. Numbers of <u>Taeniopteryx</u> <u>nebulosa</u> nymphs collect from River Severn at Ironbridge

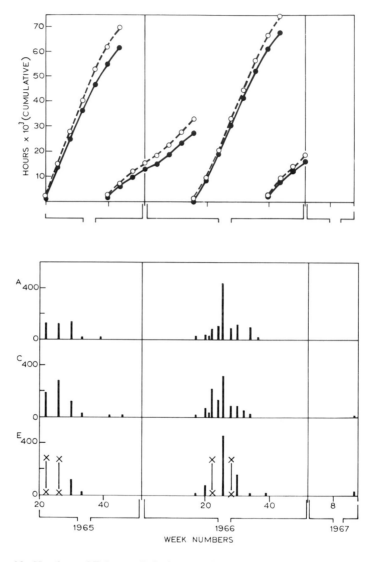

Figure 12. Numbers of Ephemerella ignita nymphs collected from R. Severn at Ironbridge

In the heated channel at Peterborough, Cragg-Hine (in press) showed that coarse fishes of several species, for example, roach, bream, silver bream, perch, and pike, were not killed by violent temperature fluctuations of 10-15°C occurring over one to two hours each morning.

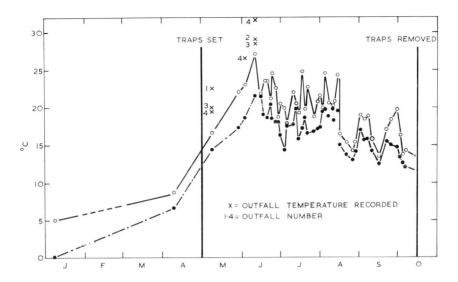

Figure 13. Water temperature at emergence traps upstream and downstream of
Ironbridge Power Station, 1970

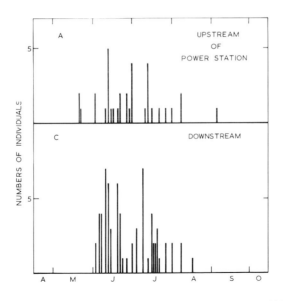

Figure 14. Catches of adult Heptagenia sulphurea, River Severn, 1970

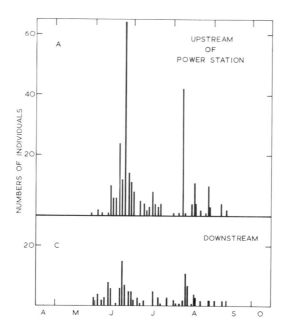

Figure 15. Catches of adult Psychomyid pusilla, River Severn, 1970

Hawkes (1969) gives a comprehensive table of upper tolerance limits for some U.S. and British fishes. Except for a few species, these limits do not, in fact, differ very widely for fish common to both countries or those peculiar to one country. Thus the species in the U.S. are living much nearer their upper temperature tolerance limits, as temperatures under natural conditions are higher in the U.S. than in Britain.

Distribution Around Heated Discharges. Alabaster (1962, 1969), in his excellent work on fish distribution in relation to heated discharges showed that up to 26°C most coarse fish are attracted by the discharges. At 30°C, however, they tend to move away. Also *sudden* exposure to temperatures of 8°C above summer ambients are lethal to trout and the more sensitive coarse (rough) fishes.

Where stratification occurs, there is less effect on fish distribution than where effluents are fully mixed (Alabaster, 1969). In places where these discharges affect only part of the river, fish are able to move quickly to colder areas if necessary.

Very little is known about the effects of heated discharges on the migrations of Atlantic salmon and sea-trout in Britain. Ross (1970) quotes a case where migrating salmon were temporarily halted by warm water from Carlisle power station, but in the Severn there is no evidence that salmon have been halted by the effluent from Ironbridge Power Station during the past 38 years.

In the Columbia River there has been little evidence so far to show that the distribution or migration of the anadromous fish has been altered by the normal operating conditions at Hanford. It is probable, however, that most migrations take place before midsummer (Nakatani, 1969).

Sonic tagging programmes on the Columbia and Connecticut rivers have shown that migrating shad will pass through, around, or under heated discharges with little hesitation (Coutant, 1969; Merriman, 1970), but there have been few studies of migration on a river which has a fully mixed heated discharge. In both Britain and the U.S. there is considerable room for further work on migratory fishes and power station intakes and outfalls.

A number of authors have shown that in the U.S. there tend to be accumulations of fish around heated discharges during the colder seasons, but that both the numbers of fish and species diversity are reduced when temperatures reach 35-37°C (Profitt, 1969; Merriman, 1970).

Growth. The direct effects of temperature on the growth of fish have been described by various authors (Brown, 1946; Brett, 1956; Fry, 1957; Nakatani, 1969) and discussed by de Sylva (1969). Few studies have been made in the field around actual heated effluents.

Cragg-Hine (in press) showed that at Peterborough roach (*Rutilus rutilus*) and bream (*Abramis brama*) began to grow rapidly some two or three months earlier than normal, and scale check formation also occurred earlier (Figures 16 and 17). The overall growth rate, however, was similar to that of fish in the main river (Figure 18). Although there was no evidence that growth slowed down at peak temperatures, overcrowding of fish in the channel and competition for food probably depressed the overall growth rate.

Nakatani (1969) showed that salmonids reared in warm water below Hanford grew faster than those in the colder water. Cairns (1956), on the other hand, demonstrated a lack of growth in fishes subjected to sublethal temperatures, even when food intake was increased.

Under natural conditions, however, fish would be subjected to wide temperature fluctuations below any effluent due to variations in river conditions and their own movements.

Feeding. Under experimental conditions the feeding rate of fishes increases with temperature (de Sylva, 1969). Hathaway (1927) found that at 20°C the feeding rate of some Centrachid fishes was three times that at 10°C. Brown found that over 19°C feeding rate in brown trout declined sharply.

339

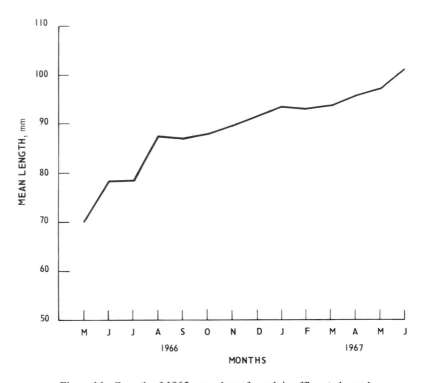

Figure 16. Growth of 1965 year class of roach in effluent channel
from May 1966 to June 1967

Dandy (unpublished) found that there was no change in the seasonal diet of trout in the warm water of Cavendish Dock, the cooling pond for Roosecote Power Station. Cragg-Hine (in press) found that in the warm water of the Peterborough channel all the indigenous river fishes were present and feeding for much of the year. Some species also fed during winter instead of fasting, as they would in colder waters. Fish in the warm channel ingested more vegetable material than did river fish. This high proportion of vegetable material may also have led to a reduction in overall growth rate.

Becker (1969) found little evidence of significant changes in insect production or feeding rate and diet of young chinook salmon below Hanford on the Columbia. It was noted, however, that in small areas receiving heated water via intragravel seepage on the shoreline, food intake was temporarily restricted during summer flows. The fish also consumed fewer insects.

In Ukrainian rivers, where temperatures reached up to 30°C, feeding intensity of carp was reduced at 29-30°C, though *Coregonus* spp. were found to be feeding actively at only 3°C below their lethal temperature (E.I.F.A.C., 1968).

340

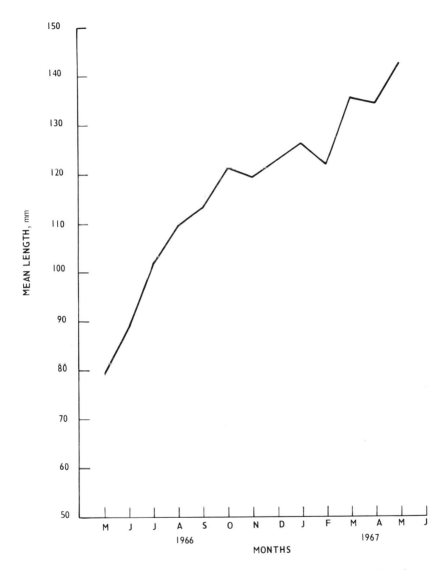

Figure 17. Growth of 1965 year class of bronze bream in effluent channel from May 1966 to May 1967

Spawning.   Temperature obviously plays an important role in controlling spawning time and development rates in fishes (E.I.F.A.C., 1968; Clark, 1969). In Lake Lichen (Poland), warmed by a power station discharge, advances in the spawning time of three species—*Blicca bjoernka* (silver bream), *Tinca tinca* (tench), and *Lucioperca luciperca* (pikeperch)—were observed. Other species— *Perca fluviatilis* (perch) and *Esox lucius* (pike)—were unaffected.

341

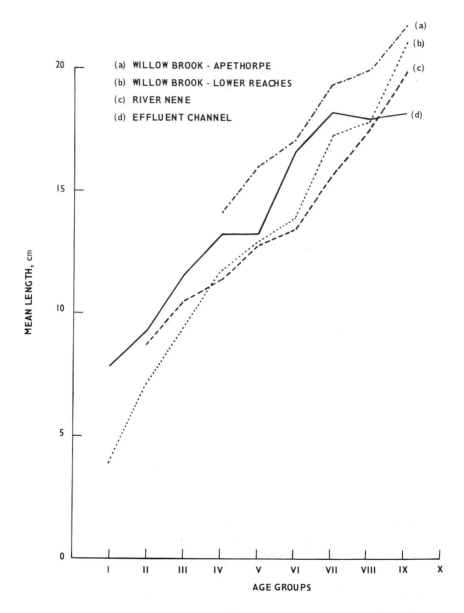

Figure 18. Comparison of growth of roach in River Nene and effluent channel with growth in Willow Brook

At Peterborough all the species in the heated channel spawned at the same time as in the river, though recent work by Bray (personal communication) suggests that roach spawned six to eight weeks earlier than normal in the heated part of the River Witham at Lincoln. Temperatures during the spawning time of roach and bream at Peterborough were about 10-15°C in the river and 20-26°C in the heated channel.

In the United States, as well as in Britain, field data on spawning times are sparse. In the Connecticut there was no change in the numbers of shad entering to spawn after the start-up of Haddam Neck (Merriman, 1970) and no change in spawning time. Eggs were collected at temperatures up to 25°C, and young *Alosa* larvae were collected from the power station effluent channel during temperatures of 35-37°C.

In the Columbia River the distribution of autumn chinook nests was related not to the thermal plumes from the reactors, but to gravel size, current velocity, and water depth (Nakatani, 1969). Young salmonids could be reared in warm-water reactor effluent, though at temperatures over 4°F above ambient, mortalities increased in fish hatched from "October" eggs. "December" spawn was less affected and could tolerate increments up to 8°C without apparent damage or increased mortality. Warming the colder waters of early winter may, in fact, be beneficial for egg incubation in fall chinook.

Parasites and Disease. De Sylva (1969) considers that heated discharges may tend to improve conditions for the growth and development of fish parasites. There are very few data, however, to support this theory. Nakatani (1969) suggested that overcrowding in warm waters may cause increases in columnaris mortalities. Under hatchery conditions, however, columnaris incidence fell off after early July over temperatures of 17-18°C. Incidence of columnaris in coarse fishes was similar in heated and unheated areas of the Columbia.

In Europe the parasite *Eimenia carpelli* in the gut of common carp had its development time reduced from 17 days at 17°C to 7 days at 19.9°C. Also, the absence of ice cover on heated waters attracts more waterfowl, which may lead to higher parasite densities in fish attracted to the outfalls (E.I.F.A.C., 1968).

Future Developments

There have been many forecasts of the possible effects of expansion of power in Britain and the U.S. if the rate of increase continues at 8-10% per annum. Estimates of future water use in the U.S. vary from one-sixth of the average run-off to the whole of the dry-weather flow by the year 2000.

There are a number of rivers in the U.S. which are still able to stand considerable increases in direct-cooled power plant discharges. For example, it is estimated that the Ohio has cooling capacity for a further 60,000 MW of fossil-fueled capacity or 40,000 MW of nuclear power plant capacity, based on one station per 25 miles.

In Britain, although total generating capacity is only 50,000 MW, compared with 300,000 MW in the U.S., there are nine times more megawatts produced per square mile (Hawes, 1970). Water shortage will not allow the use of rivers for direct cooling of future plant. In fact, as older, direct-cooled facilities are replaced by stations with fully closed systems, the amount of heat discharged to rivers will be reduced considerably, at least over the next 20 to 25 years.

Criteria for Future Discharges

It is obvious, even from the fairly limited data, that the extent of the ecological effects of a heated discharge depends mainly on the maximum temperature and the area affected by this temperature.

In Britain and the Pacific Northwest, where natural maxima are around $20\text{-}22^{\circ}C$, it is evident that temperatures of $30^{\circ}C$, though not likely to kill fish, may affect their distribution and growth. Most invertebrates common in big rivers may be able to survive $28\text{-}30^{\circ}C$, at least for short periods. In the eastern and southern U.S. temperatures of $40\text{-}42^{\circ}C$ may be reached in effluent channels, and these are without doubt lethal to most fishes and invertebrates. Temperatures of $35\text{-}37^{\circ}C$ do not appear to cause significant ecological damage, though if persisting for long periods, they may do so.

From the evidence so far, it must be concluded that in a fairly clean British river, with a mixed fishery and diverse fauna, temperatures over the whole width of the river should not exceed $30^{\circ}C$ for more than very short periods, and never more than $32^{\circ}C$. When this does happen, however, recovery is fairly rapid, and fish are not normally killed unless trapped.

In the southern and eastern rivers of the U.S., fully mixed temperatures should not exceed $37^{\circ}C$ except for very short periods, though effluents may be allowed to reach $38^{\circ}C$ under certain conditions.

Experience shows that it is ecologically better to disperse an effluent over part of the cross section of a river, rather than spreading the plume across the whole surface, particularly where emergent insects and plankton are involved.

Obviously, in a river where all or the major part of the flow is used and returned at temperatures up to $10^{\circ}C$ above ambient, there is a need for some kind of cooling system which reduces the amount of effluent, if not its temperature. On the other hand, where only a portion of the river will be affected, ecological and economic considerations must be carefully balanced. Mount (1970) suggests that a mixing zone should not exceed 30-40% of a stream width, Thus a zone of escape would still exist for both resident and anadromous fishes. If this type of design is possible, effluent temperature need not be as tightly controlled.

The merits of plume dispersal over the whole surface of a river or complete mixing at the point of discharge will depend entirely on the resulting temperature increment. If migratory fish are the main consideration, they can still pass below the plume spread on the surface, though resident surface-feeding fishes and invertebrates with aerial stages may still be affected in the immediate area of the outfall.

Cooling towers of various designs have advantages as far as the river is concerned—that is, small effluents and oxygenation and chemical improvement—but may not be beneficial to the atmospheric or terrestrial environment in the U.S. The design, operation, and relative costs of various cooling systems have been adequately described and discussed by Parker and Krenkel (1969). There is no doubt that, where adequate water is available for direct cooling and river dilution factor is high, it is unnecessary to construct cooling towers.

In the past the discourse about heat from both conservationists and power companies has been too emotional, considering the small amount of actual recorded ecological damage caused by cooling-water discharges. It is time for a more objective and rational approach.

Fetterholf (1970), in a recent paper, summarized the advantages and disadvantages of heated discharges into Lake Michigan from an objective, biological point of view. He showed that, even with the calculated expansion of the power industry, the daily energy input from cooling water would be only two trillion BTU/day, compared with 597 trillion BTU from the sun. Under these conditions, the policy statement by the Federal Water Quality Administration (op. cit., Letterhof, 1970), that "in no event will heat discharges to Lake Michigan be permitted to exceed $1^{\circ}F$ rise over natural to the point of discharge," is particularly unrealistic.

345

Future Research

There is a need for further research on a number of aspects of heated discharges. Even though many programmes are being carried out, published biological data from actual research are still relatively scarce. More work on the aquatic flora, both algae and macrophytes, is needed, together with long-term field studies of invertebrates around operating power plants. The migratory fishes also merit more investigation.

It has always seemed to me that predictive ecological data on the effects of heat may be better obtained from studies of operating power plants over periods of three to five years, than from the prolonged "before-and-after studies" which both conservationists and planners revere so much.

The aim of biologists in this field should be the ability to predict with some accuracy the effects of any proposed development discharging a heated effluent. This is obviously a long-term view but need not be too long if a plant in operation is studied.

Already, in Britain the Central Electricity Generating Board has had to go some way toward alleviating thermal effects in rivers, and there is no doubt that future policy and acute water shortage will dictate that it continue to do so. In the U.S. I do not believe that, except for localised areas, there are many examples of real "thermal damage" in rivers, and I would expect that reasonable ecological and economic pressures applied with objectivity will ensure that a "thermal crisis" will not arise. I believe that "thermal effluents" have a great deal of expanding to do before they cause drastic ecological changes of the magnitude of those caused by other industries and by the impoundment of rivers.

Acknowledgements

I would like to thank my colleagues, Dr. R. J. Aston, Dr. David Cragg-Hine,* and John W. Whitehouse for their assistance and permission to reproduce data and figures from their papers in press or in preparation. Also, my sincere thanks to John R. Daffern for his patient work on figures, drawings, and bibliography. For editing the manuscript, my thanks go to Theodore Vande-Sande, session chairman of this conference.

The work is published by permission of the Central Electricity Generating Board.

---

*Dr. Cragg-Hine is now with the Northern Ireland Inland Fisheries Trust, Coleraine.

Literature Cited

Adams, J. R. 1969. Ecological investigations related to thermal discharges. Pacific Coast Elec. Ass. Eng. Oper. Sect. Annu. Meet., Los Angeles.

Alabaster, J. S. 1962. The effects of heated effluents on fish. Int. J. Air. Water Pollut. 7:541-563.

——————. 1969. Effects of heated discharges of freshwater fishes in Britain, p. 354-381. In P. A. Krenkel and F. L. Parker [eds.], Biological aspects of thermal pollution.

——————, J. H. N. Garland, and I. C. Hart. 1971. Fisheries, cooling-water discharges and sewage and industrial wastes. C.E.R.L. Symp. Freshwater Biol. Elec. Power Generat.

Anderson, R. R. 1969. Temperature and rooted aquatic plants. Chesapeake Sci. 10(3,4):157.

Aston, R. J. In press. Field and experimental studies of the effects of temperature on Tubificidae - Oligochae.

Becker, C. D. 1969. The food and feeding of juvenile chinook salmon in the Columbia River at Hanford, p. 27-31. In Biological effects of thermal discharges. Annu. Progr. Rep. for 1968, AEC Res. Develop. Rep. December 1969.

Bottomley, P. 1971. CERL Symposium on Freshwater Biology and Electrical Power Generation - Leatherhead, Surrey.

Brett, J. R. 1956. Some principles in the thermal requirements of fishes. Quart. Rev. Biol. 31(2):75-87.

Brown, M. E. 1946. The growth of brown trout (*Salmo trutta* : Linn). III. The effect of temperature on the growth of two year old trout. J. Exp. Biol. 22:145-155.

Cairns, J., Jr. 1955. The effects of increased temperatures upon aquatic organisms. 10th Purdue, Ind. Wastes Conf.

——————. 1956. Effects of increased temperatures on aquatic organisms. Ind. Wastes 1(4):150-152.

Cairns, J., Jr. 1969. The Response of Fresh-water Protozoan Communities to Heated Waste Waters - Chesapeake Science 10(3,4):177-185.

Carpenter, K. E. 1924. A study of the fauna of rivers polluted by lead mining in the Aberyswyth district of Cardiganshire. Ann. Appl. Biol. 9:1-23.

——————. 1925. On the biological factors involved in the destruction of river fisheries by pollution due to lead mining. Ann. Appl. Biol. 12:1-13.

Churchill, M. A., and T. A. Wojtalik. 1969. Effects of heated discharges on the aquatic environment: The TVA experience. Amer. Power Conf., Chicago.

Clark, J. R. 1969. Thermal pollution and aquatic life. Sci. Amer. 220(3):19-27.

Coutant, C. C. 1962. The effect of a heated water effluent upon the macroinvertebrate riffle fauna of the Delaware River. Proc. Amer. Acad. Sci. 36:58-71.

——————. 1969. Behaviour of sonic-tagged chinook salmon and steelhead trout migrating past Hanford thermal discharges, p. 21-26. In Biological effects of thermal discharges. Annu. Progr. Rep. for 1968, AEC Res. Develop. Rep. December 1969.

Cragg-Hine, D. In press. Proc. 5th Coarse Fish Conference. University of Liverpool.

Davies, I. 1966. Chemical changes in cooling water towers. Int. J. Air Water Pollut. 10:853.

E.E.I. 1969. Edision Elec. Inst. Rep.

E.I.F.A.C. 1968. Water quality criteria: Report on water temperature and inland fisheries. Eur. Inland Fish. Adv. Comm. Rep.

Fetterholf, C. 1970. Are we in hot water with thermal discharges? 45th Annu. Conf. Mich. Water Pollut. Contr. Ass.

F.P.C. 1969. Problems in disposal of waste heat from steam electric plants. Fed. Power Comm. Staff Rep.

Fry, F. E. J. 1957. The aquatic respiration of fish, p. 1-63. In The Physiology of Fishes. 1. M. E. Brown, ed. Academic Press, New York.

Gledhill, T. 1960. The Ephemeroptera, Plecoptera and Trichoptera caught by emergence traps in two streams in 1958. Hydrobiol. 15:179-188.

Goubet, A. 1969. The cooling of riverside thermal power plants. *In* Engineering Aspects of Thermal Pollution. F. L. Parker and P. A. Krenkel [eds.]. Vanderbilt Univ. Press.

Hathaway, E. C. 1927. The relation of temperature to the quantity of food consumed by fishes. Ecol. 8:318-434.

Hawes, F. B. 1970. Thermal problems "old hat" in Britain. Elec. World, April 1970.

—————. 1971. Water in electricity generation. C.E.R.L. Symp. Freshwater Biol. Elec. Power Generat.

Hawkes, H. A. 1969. Ecological changes from waste heat, p. 15-53. *In* F. L. Parker and P. A. Krenkel [eds.], Engineering aspects of thermal pollution, Vanderbilt Univ. Press.

Hynes, H. B. N. 1960. The biology of polluted waters. Liverpool Univ. Press.

Langford, T. E. 1970. The temperature of a British river upstream and downstream of a heated discharge from a power station. Hydrobiol. 35(3-4):353-75.

—————. 1971a. The distribution, abundance and life histories of stoneflies (Plecoptera) and mayflies (Ephemeroptera) in a British river warmed by cooling-water from a power station. Hydrobiol. 38(2):339-377.

—————. 1971b. The biological assessment of thermal effects in some British rivers. C.E.R.L. Symp. Freshwater Biol. Elec. Power Generat.

—————, and R. J. Aston. In press. The ecology of some British rivers in relation to warm water discharges from power stations. Proc. Roy. Soc.

Lester, F. 1970. Op. cit., Jon Tinker: The Daily Telegraph Colour Supplement, September 1970.

Letterholf, C. 1970. Are we in hot water with thermal discharges. Pres. to 45th Conf. of Mich. Water Poll. Control. Assoc.

Lof, G. O. G., and J. C. Ward. 1969. Economic considerations in thermal discharges to streams, p. 282-301. *In* F. L. Parker and P. A. Krenkel [eds.], Engineering aspects of thermal pollution. Vanderbilt Univ. Press.

Macan, T. T. 1958. Methods of sampling the bottom in stony streams. Mitt. inf. Ver. Limnol. 9:21.

——————. 1963. Freshwater ecology, Longmans, London.

——————. 1964. The Odonata of a moorland fishpond. Int. Rev. Ges. Hydrobiol. 49:325-360.

Mann, K. H. 1965. Heated effluents and their effects on the invertebrate fauna of rivers. Proc. Soc. Water Treat. Exam. 14:45.

Mellanby, K. 1969. Can Britain afford to be clean? New Sci. 43:648-650.

Merriman, D. 1970. The calefaction of the Connecitcut River. Sci. Amer. 222(5):42-52.

Mount, D. I. 1970. Technical statement of Michigan State Water Quality Hearings, March 1970.

Nakatani, R. E. 1969. Effects of heated discharges on anadromous fishes, p. 294-353. In P. A. Krenkel and F. L. Parker [eds.], Biological aspects of thermal pollution. Vanderbilt Univ. Press.

Negus, C. L. 1966. A quantitative study of the growth and production of unionid mussels in the River Thames at Reading. J. Anim. Ecol. 35:513-532.

Parker, F. L., and P. A. Krenkel [eds.]. 1969. Engineering aspects of thermal pollution. Vanderbilt Univ. Press.

Patrick, R. 1969. Some effects of temperature on freshwater algae. In P. A. Krenkel and F. L. Parker [eds.], Biological aspects of thermal pollution. Vanderbilt Univ. Press.

Proffit, M. A. 1969. Effect of heated discharges upon aquatic resources of White River at Petersburg, Indiana. Indiana Water Resour. Center Rep. No. 3.

Raney, E. C., and B. W. Menzel. 1967. A bibliography: Heated effluents and effects on aquatic life with emphasis on fishes. Philadelphia Elec. Co. Ichthyol. Assoc. Bull. I.

Ross, F. F. 1959. The operation of thermal stations in relation to streams. J. Proc. Inst. Sew. Purif. 16:2-11.

Ross, F. F. 1970. Warm water discharges into rivers and the sea. Proc. Annu. Conf. Inst. Water Pollut. Contr.

Savage, P. D. 1970. Some aspects of the phytoplankton of Southampton water. (1965) Brit. Phycol. Bull. 2:51-516.

Sprules, W. M. 1947. An ecological investigation of stream insects in Algonquin Park, Ontario. Publ. Ontario Fish. Res. Lab. No. 56.

Steel, E. 1961. op. cit., Mann (1965).

Strangenburg, M., and M. Z. Pawlaczyk. 1961. Nank Pol Wr. Wroclaw No. 10. Inzyn. Saint. Wat. Poll. Abst. 1:67-106.

de Sylva, D. P. 1969. Theoretical considerations on the effects of heated effluents on marine fishes, p. 229-293. *In* P. A. Krenkel and F. P. Parker [eds.], Biological aspects of thermal pollution.

Swale, E. M. F. 1964. A study of the phytoplankton of a calcareous river. J. Ecol. 52:433-446.

Tinker, J. 1970. Daily Telegraph Colour Supplement, Sept. 1970.

Trembley, F. J. 1965. Effects of cooling-water discharge from steam-electric power plants on stream biota. *In* Biological Problems in Water Pollution. 3rd Seminar, August 3-17th, 1962. U.S. Dept. Health Educ. & Welfare. Public Health Service Publ. 999-WP-25.

Winterbourne, M. J. 1968. The faunas of thermal water in New Zealand. Tuatara (New Zealand) 16:111.

Wurzt, C. B. 1969. The effects of heated discharges on freshwater benthos, p. 199-228. *In* P. A. Krenkel and F. P. Parker [eds.], Biological aspects of thermal pollution. Vanderbilt Univ. Press.

—————, and T. Dolan. 1960. A biological method used in the evaluation of the effects of thermal discharges in the Schuylkill River. Proc. 15th Ind. Waste Conf., Purdue, 1960, p. 461-472.

EFFECTS OF PESTICIDES AND INDUSTRIAL WASTES
ON SURFACE WATER USE

William A. Brungs

Too frequently, considerations of the effects of pesticides and industrial wastes on river ecology are limited to aquatic life. Modern man cannot be separated from his dependence on fresh surface waters, and his needs as a species must be taken into account. I realize that the title of this symposium might imply a more narrow framework for discussion, but such a perspective might be less productive, in terms of environmental protection, than the one I propose. Within a broader context man's dependence on his water supplies becomes even more critical than if we limit his dependence to those benefits involving aquatic life. The more man realizes his dependence upon his aquatic environment, the more responsible he will become with regard to its protection.

With this point in mind, let us consider what are generally accepted as the principal uses of the aquatic environment. Uses involving aquatic life are, of course, to many of us, purists as well as practically oriented people, the most appealing and significant uses to be protected. Aquatic life can provide not only an aesthetic value, but also a direct-use value for sport and commercial fishing. However, man's dependence on fresh surface waters should also be related to such other uses as municipal water supplies, irrigation, or other agricultural needs, and recreational uses such as water-contact sports. One additional use of the aquatic environment which deserves consideration is that for industrial purposes. Rarely is a significant emphasis placed on such a use. At present industry is required to invest large sums of money for equipment, manpower, and maintenance to insure water supplies of acceptable quality for a variety of needs. How much better it will be when more of this effort and money can be spent to prevent contamination of upstream water supplies rather than to improve an unacceptable water supply.

At least two principal approaches can be followed in any presentation of the effects of pesticides and industrial wastes on the aquatic environment. Probably the most acceptable and dramatic technique would involve a thorough discussion of many facts and data on materials that are better understood and about which specific conclusions and recommendations can be made. A few of the more glamorous might be DDT, mercury, detergents or phosphates, and so on. I'm sure such a discussion would be well received and enjoyable for most of the audience. Unfortunately, such an approach, I feel, would be less meaningful at the present time. We already know quite a bit about these much discussed problems, and research and field investigations on them are at various stages of completion. The public is concerned after much prodding from the now-concerned news media. I suppose the latter must be convinced of a potential or real problem before the public can be adequately informed and therefore alarmed. As a result of these conditions, I'm sure nothing I could relate to you here about these well-publicized contaminants would benefit the public or yourselves.

To be more specific, let me mention some examples that will indicate another approach to the problem and that will place the remainder of my presentation in its proper perspective. A slight, but ecologically acceptable increase in water temperature could result in the reduction by a downstream sewage treatment plant effluent of the available dissolved oxygen concentration from an acceptable range to an unacceptable one because of an increased oxidation rate in a given volume of diluted effluent. At many locations a superficially acceptable environment may contain a productive potential sport fishery that is limited by unpleasant taste or odor. Or another example, a situation resulting in frequent misinterpretation, is an increasing abundance of fish, which does not necessarily imply an improved or acceptable aquatic ecosystem. Closer inspection may reveal that this increase is a result of proliferation of undesirable or forage species accompanied by the decline of typically predatory fish species of interest to sport fishermen. Admittedly, considerations such as these may not be tomorrow's headlines, but such assorted subtleties nevertheless result in an unacceptable environment by today's standards, aesthetic or legal.

With this approach in mind, we can consider two general classes of effects on water uses. Probably the principal causes of these dramatic effects are carelessness or accidents. Railroad tank cars derail and dump cyanide into a ditch leading to a stream. A massive fish kill results, and domestic water supplies are unsafe for a few days. There are innumerable examples of plant personnel turning the wrong valve and causing similar ecological crises. Oil tankers traveling in fog when they should be at anchor collide and foul beaches. In time this careless or accidental behavior by man will reduce in frequency as the people involved develop a care and concern for the environment, whatever their motivation may be, that will be commensurate with their care and concern for profit.

Fortunately, these ecological rapes are healed in time by a sometimes forgiving nature. But a more demanding consideration is that class of water-use effects which involves continuous, chronic, sublethal degradation. This degradation is typically unnoticed, except by fishermen, ecologists, and taxonomists, since the effect is not dramatic or well publicized. No gory pictures, such as those of Great Lakes beaches covered with dead fish or large, colorful blotches of dyes, oils, and debris, are available for scrutiny. Advertising gimmicks are not helpful to describe a change in biological populations due to artificial selection for undesirable species. They say Lake Erie is dead, and although this is a symposium on river ecology, Lake Erie provides an adequate example to support my contention. The now well-known changes in this environment have not occurred overnight. Many decades of study have demonstrated well this gradual degradation. Studies of nutrient levels and the gradual decline of dissolved oxygen at the bottom-water interface during periods of thermal stratification and studies on the drastic changes in bottom organisms related to siltation and enrichment showed us many years ago the eventual fate of this lake. It is now time to consider a gradual decline in the desirability of aquatic environments as justification to reverse the pattern.

It is not sufficient to prevent accidents and carelessness which lead to fish kills when chronic, sublethal changes provide a similar fate over a longer time span. Reductions in the rate of reproduction by aquatic species or subtle, but biologically significant changes in food-chain patterns are adequate and justifiable reasons for controlling industrial and municipal discharges causing gradual degradation. In most aquatic environments of this country, this will require improvements in water quality, not just prevention of additional contamination.

Research into the effects of pesticides and industrial wastes has progressed at such a rate that any presentation is outdated before it can be read. Evidence for this is found in the number of references cited here that are characterized as "in press." Another factor that indicates continuing change is the recent revision of the 20-year-old bioassay method in the new edition of Standard Methods for the Examination of Water and Wastewater (American Public Health Association, 1971). The so-called "green book" (Federal Water Pollution Control Administration, 1968) is undergoing revision by the National Academy of Sciences under contract with the Environmental Protection Agency and will be available in early 1972. In April 1970 the Congress assigned a deadline of April 1972 for a report to be made available to the states that will present recommendations to be used for setting water-quality standards for pesticides. This partial list could be expanded, but is included to provide an idea of the present status of conditions in the field of water-pollution control.

Any current discussion of river ecology must consider, when available, the latest information and trends, and it is impossible to be up to date on all aspects. The topics I shall emphasize are selected not because of national priority, but because of my degree of experience in certain areas. The pre-1960 literature is

355

founded on short-term studies and used mortality as an end point. Such evidence is appropriate if we wish only to prevent fish kills. Fortunately, this is no longer true; the public desires a balanced aquatic ecosystem. To protect this ecosystem from gradual degradation, we must provide criteria that will protect the entire life cycle of the desirable species as well as the food chains on which these species depend. A significant reduction in available food or reproductive success will result in a condition similar to that after a fish kill. Criteria must therefore be based on chronic or life-cycle studies that may also permit extrapolation to untested species and toxicants. Extrapolation is necessary when chronic tests with fish require usually one year of continuous exposure to the toxicant. Food-chain organisms require less time. The number of species and the variety of toxicants, excluding combinations of toxicants, preclude testing of all possible pairings.

At this time it is appropriate to outline the research program at the Environmental Protection Agency's National Water Quality Laboratory at Duluth, Minnesota. This program includes the efforts of two field stations—the Western Fish Toxicology Laboratory at Corvallis, Oregon, and the Newtown Fish Toxicology Laboratory at Cincinnati, Ohio. The mission of these three laboratories is to determine permissible limits for any use and the impairment that can be expected if these limits are exceeded in fresh water.

For municipal, industrial, and agricultural uses, the water is, or can be, removed from the watercourse and treated to achieve the needed quality. For water-based recreation and production of aquatic life, the water must meet minimum requirements in the watercourse. For this reason, the permissible limits are generally more restrictive for recreation and aquatic life use than for the other uses, and the greater research effort is needed in these two areas. The research needed to determine even one criterion requires considerable expenditures of money and time. But the results may save a body of water for a future use or may enable a municipality or industrial plant to select an alternate and cheaper method of treatment or even an alternate and cheaper method of treatment or even an alternate site. If either happens once, the savings could equal the cost of the research for several years.

Although the Duluth laboratory's research ranges across the whole spectrum of requirements for water uses, its initial investigative concern has been to determine the water quality necessary for production of sport and commercial fish and for recreational uses. Research into requirements for recreation focuses on health and aesthetic factors. Some impurities do not adversely affect water quality, but do mar the beauty and uses of lakes and streams with discolorations and offensive smells. Algal growths, which not only discolor water a vivid green, but might also cause skin irritations and unpleasant odor, must be identified and controlled. The subject of "safe" levels for swimming has been the target of controversy among water-quality experts. Established bacterial counts considered unsafe for body-contact water sports are arbitrary and need refinement. Research on chemical pollutants that may affect the eyes, ears, and skin is quite limited.

Aquatic life research involves chronic or long-term testing of all or nearly all of the life cycle of the test animals, both fish and fish-food organisms. We feel that to obtain some measure of confidence in estimated safe concentrations, all life stages need to be studied. This includes, among other things, an evaluation of toxicant effects on parental growth and survival, reproduction, hatchability, and growth and survival of offspring.

It was realized early that we would have to limit the number of fish species to obtain adequate information on the more important ones. Assistance was obtained from outside authorities to establish our list of important fish species. Their commercial or sport-fishing importance, their use as forage species, and their distribution were some of the factors used for evaluation. The following list includes species with a sufficiently wide geographic distribution that they are exposed potentially to a wide variety of toxicants:

Common Name	Scientific Name
Threadfin shad	*Dorosoma petenense*
Brook trout	*Salvelinus fontinalis*
Rainbow trout	*Salmo gairdneri*
Northern Pike	*Esox lucius*
Emerald shiner	*Notropis atherinoides*
Fathead minnow	*Pimephales promelas*
White sucker	*Catostomus commersoni*
Channel catfish	*Ictalurus punctatus*
White bass	*Morone chrysops*
Bluegill	*Lepomis macrochirus*
Largemouth bass	*Micropterus salmoides*
Yellow perch	*Perca flavescens*

A second list contains species with more limited distributions that might be exposed potentially only to selected toxicants:

357

Common Name	Scientific Name
Coho salmon	*Oncorhynchus kisutch*
Lake trout	*Salvelinus namaycush*
Lake herring	*Coregonus artedii*
Mountain whitefish	*Prosopium williamsoni*
Rainbow smelt	*Osmerus mordax*
Smallmouth bass	*Micropterus dolomieui*
Walleye	*Stizostedion vitreum vitreum*

Also, some work is being done to use a goldfish as a "white rat." Representatives of important groups of food-chain organisms, such as scuds, amphipods, mayflies, and cladocerans, are also tested for their entire aquatic life cycle.

The toxicants and environmental factors studied to date include several heavy metals, pesticides, detergents or components of detergents, sewage plant effluent, reduced pH and dissolved oxygen, and elevated temperature. This is certainly a limited representation of the adverse situations possible in the environment. Work with these factors provides some immediate answers, however, and furnishes the potential for extrapolation to others.

For many years the basic toxicity test was limited, at best, to 96 hours with the end point being mortality. Some investigators recommended that by multiplying the 96-hr TL_{50} (the calculated concentration that would result in 50% survival after 96 hours) by an application factor of 1/10, a "safe" concentration could be estimated. Several years ago, when we conducted our first chronic tests for the complete life cycle, we compared the maximum acceptable toxicant concentration (MATC) to the 96-hr TL_{50} for the same species and toxicant. The resultant calculated application factor was much less than 1/10, and we decided that the latter was an inaccurate general estimate. The simple formula we use tentatively now is:

$$\text{Application factor} = \frac{\text{MATC}}{\text{96-hr } TL_{50}}$$

A selection of several application factors will indicate the wide range observed by experimentation:

358

Toxicant	Application Factor
Linear alkylate sulfonate (LAS)	1/4 (Pickering and Thatcher, 1970)
Copper	about 1/8 (Mount and Stephan, 1969)
Malathion	1/45 (Mount and Stephan, 1967)
2,4-D	1/20 (Mount and Stephan, 1967)
Zinc	1/150 (Brungs, 1969)
Cadmium	1/200 (Pickering, in press)
Sevin	1/40 (Carlson, in press)

Several immediate questions developed after this information was known. Numerous environmental variables (e.g., temperature, hardness, pH, relative species sensitivity) affect acute toxicity (96-hr TL_{50}). If it affected the chronic toxicity (MATC as used here) to a greater or lesser extent, the resultant application factors for a single toxicant would vary. Several tests have been conducted to evaluate this potential problem, and it is beginning to appear that a single application factor may be adequate for a single toxicant. More tests, as well as work with mixtures of toxicants, will be necessary to obtain a satisfactory level of confidence in this approach. If this approach is useful and an application factor for a specific toxicant is known, a discharger could conduct a simple acute toxicity test with the receiving stream water and locally important species and calculate a reasonable estimate of the "safe" concentration that would allow for the natural factors affecting toxicity.

We fully realize that a one-year test that investigates all of the various life stages is not a simple study to be routinely conducted by industries and municipalities. Many possible short-cut tests are being compared to the chronic test. these include, but are not limited to, physiological and metabolic changes, changes in blood chemistry and enzyme systems, and activity of the fish and oxygen consumption. So far, none looks very promising as an early indicator of the MATC.

Another early question was raised concerning the validity of applying safe concentrations as determined under controlled laboratory conditions to natural environments. In the laboratory fish are fed *ad libitum*. They are treated prophylactically as needed, and there are no predators, no competition for spawning areas, and no exposure to extremes of natural water quality. We are now conducting a study in a natural stream to determine the usefulness of laboratory information for protecting aquatic life in natural environments.

Toxicological data provide at present very little concerning the effects of industrial effluents on aquatic organisms. We have much evidence on lethal effects of single, usually simple, reagent-grade chemicals. Our own emphasis is similar, but does include chronic effects as I have just described. Before we can begin to understand the toxicology of mixed effluents containing innumerable complex chemicals, we must be able to design and interpret appropriate testing procedures. Sufficient expertise is probably available now to develop the procedures for testing industrial wastes directly. An in-house committee at the National Water Quality Laboratory has been established for this purpose.

The water-quality standards program will never include all possible toxic materials specifically. There are too many for nonindustrial research to provide all the essential criteria necessary to specify standards. Much of the responsibility will of necessity be that of industry. This involves in-plant toxicity tests with receiving stream water (biomonitoring). At present two general approaches appear to be possible: to carry out chronic life-cycle tests such as I have described previously, or, if this is not feasible or desirable, to accept an application factor based on similar discharges, allow for some safety factor, and conduct acute tests.

There has been so much policy discussion concerning an effluent-standard program that it would be inadvisable at this time to describe it in detail. Whatever I could mention would very likely be changed. However, the Ohio River Valley Water Sanitation Commission (ORSANCO) is developing an effluent-standards system pertaining to sewage and industrial wastes discharged into the Ohio River. The commission specifies limiting concentrations for the most common chemicals in the discharges, and for other materials the concentrations in the river shall not exceed 1/20 of the 96-hr TL_{50} for aquatic life.

Much greater emphasis is placed on the elimination of tainting of fish flesh that would limit the usefulness of the fishery resource, even though the resource itself is protected. ORSANCO includes the comment in their proposed recommendations that there will be no "tainting of fish flesh."

Whenever pesticides, industrial wastes, or municipal effluents enter a river system, a Pandora's box full of problems can be, and frequently is, opened. Where multiple discharges exist, many chemical reactions may occur in the receiving water that intensify or reduce the toxic effect of the original materials. Not only is specific knowledge concerning the effluent characteristics required, but also a knowledge of the potential chemical and physical changes is necessary to estimate better the effects of multiple effluents on the environment. In addition, we must comprehend the potential of combined stresses on aquatic life that result in an effect that cannot be explained on the basis of a single contaminant. A simple example is a change in the toxicity of a heavy metal when the pH of the river is changed. Another more common situation is that of an environment barely acceptable with regard to dissolved oxygen which becomes unacceptable if the temperature is continuously increased, resulting in an increase in oxygen demand by the aquatic life.

A problem involving pH and metal toxicity came to light recently in West Virginia. Attempts were made to reopen and expand deep coal mines on one of the principal trout streams. This stream is borderline in quality because the pH is low and because iron enters the entire watershed from abandoned mines. The sport fishery was maintained by a put-and-take operation supported almost exclusively by a federal fish hatchery dependent on that same stream water. The reopening of mining operations would introduce sufficient acid into the stream to make it unfit for stocking fish. The justification for attempting to limit any further reduction in pH was not that a lethal condition due to pH would result, but rather any slight reduction—in this case 0.3 pH units—would cause the existing iron concentration in the stream to become toxic under certain predictable and expected flow conditions. Fortunately, this situation involved only two marginally toxic conditions. Many industrialized areas compound the factors to be evaluated.

The effects of municipal sewage-plant effluents on the aquatic environment have not received the attention a problem of this magnitude requires. The principal problem is a result of the oxygen demand of this waste. A summarization of fish kills in the United States (U.S. Department of the Interior, 1968) indicates that more than 50% of the fish killed by pollution from 1960 to 1968 died as the result of oxygen-demanding, nontoxic wastes. The sublethal effects of reduced oxygen cannot be estimated. A significant percentage of these fish kills occur as the result of industrial and agricultural discharge of wastes from paper mills, animal feedlots, food processing plants, and slaughterhouses.

The amount of research on the effect of reduced oxygen on fresh-water fishes is probably greater than for any other toxic condition, including temperature. Doudoroff and Shumway (1970) critically reviewed this literature, and with one exception no data are available on the chronic effect of low dissolved oxygen concentrations on fecundity and related biological processes. Brungs (1971) found that fathead minnows exposed continuously for 11 months to constant low oxygen concentrations spawned satisfactorily at 3 mg/l, but fry survival was limited greatly at 4 mg/l, and fry growth was reduced at all concentrations below nearly saturated control conditions. More chronic data are available on aquatic insects presumably because of their smaller size, availability, and relatively short life cycle. Cold-water mayflies and stoneflies cannot tolerate dissolved oxygen levels much below 5 mg/l (Gaufin, 1971). Other insects such as mosquitoes, which are not dependent on the dissolved oxygen content of the water, can live where the oxygen is near zero at all times. Nebeker (1971) recently completed chronic studies on the effect of low dissolved oxygen on nine species of aquatic insects, and the estimated safe concentrations for survival and emergence ranged from less than 0.6 mg/l for a midge to slightly less than saturation for emergence of a burrowing mayfly.

A more recent development (Fetterolf, 1970) indicates present and potential problems concerning the effect of chlorine used for disinfection of sewage effluent. To ensure adequate disinfection of effluent for public health purposes, a chlorine residual is necessary and is not uncommonly excessive, but does result in damage to aquatic life. For example, when live cages with fathead minnows were placed as far as 3 miles below a sewage treatment plant using chlorination and contributing 14% of the river flow, all the test fish were killed. Only 25% survived 4 miles below the point of discharge. One result of these findings is the establishment of a project at the National Water Quality Laboratory, Environmental Protection Agency, to determine the acute and chronic toxicity of chlorinated sewage plant effluent, as well as raw effluent, dechlorinated effluent using sulfur dioxide, and ozonated effluent for fish and several macroinvertebrates. A comparable study in two sewage treatment plants under actual operational conditions is under development.

An interesting condition with reference to the chlorine content of the water prevails at many laboratories attempting to conduct toxicity tests with aquatic life. Many of these locations are not equipped with large volumes of clean, natural water for experimentation. Many are limited to the use of municipal water supplies which are routinely chlorinated. The practice until now has been to dechlorinate by carbon adsorption. Unfortunately, some chloramines are always present due to ammonia present in these water supplies. Carbon adsorption alone will usually not reduce chloramine concentrations to a level acceptable for chronic toxicity testing. Arthur and Eaton (1971) observed adverse effects on amphipod reproduction at chloramine concentrations as low as 3.5 µg/l. Fathead minnow spawning was reduced at concentrations greater than 16.5 µg/l. These authors cite preliminary data for *Daphnia magna* which indicate that chloramines kill all individuals within 3 to 5 days at 1 µg/l. Data are not yet available to relate laboratory chloramine toxicity to observed toxicity below chlorinated effluent discharges from sewage plants.

The use of anti-fouling and slime-control chemicals, such as chlorine among others, to clean power-plant cooling devices is under attack at the present time. Recommendations for Lake Michigan call for mechanical cleaning devices as a substitute for chemicals. This trend will probably extend to other parts of the country.

As mentioned earlier, congressional action requires that the states be provided information with which to develop water-quality standards for pesticides with the Environmental Protection Agency. As a consequence, probably more research that will attempt to establish water-quality criteria is just beginning than has been completed in the past. Although we have an abundance of pesticide toxicity data, very little long-term toxicity information is available. Many earlier studies involved the effect of application techniques, emulsifiers, or doses and were oriented principally to agricultural-use procedures. Mount and Stephan (1967) studied malathion and the butoxyethanol ester of 2,4-D, and Carlson (in press) has determined the chronic effects of Sevin.

Ettinger and Mount (1967) conclude, as many others have, that a wild fish available for capture should also be a wild fish which is safe to eat. This problem is probably more related to pesticides than to any other group of toxicants. Mercury is a recent problem involving tissue-residue levels. With regard to aquatic life criteria for materials that cause off-flavor or taste or result in unsafe residue levels, the multiple-use criterion would be more restrictive. Fish may be un-affected by levels of phenols, mercury, and pesticides that would result in the fish being unacceptable for human consumption. The criterion would then become more restrictive.

Some work has been done to assist in the solution of questionable fish kills. Mount, Vigor, and Schafer (1966), Brungs and Mount (1967), and Mount and Boyle (1969) reported critical concentrations of endrin and parathion in fish blood and indicated that when a particular concentration of a pesticide is exceeded in the blood of fish, these fish will probably die. Information such as this assisted in assigning the cause of the Mississippi River fish kills in the early 1960's (Mount and Putnicki, 1966).

In conclusion, to emphasize again the importance of recognizing the causes of gradual and unspectacular declines in the quality of aquatic life, I will cite an example of 2,4-D effects that could easily be overlooked in any evaluation of environmental quality. A study conducted at the Bureau of Commercial Fisheries pesticide laboratory at Gulf Breeze, Florida, indicated that fish exposed to 2,4-D for 1 to 5 months grew and survived as well as the control animals. How-ever, the exposure apparently lowered the general body resistance to a micro-sporidian parasite and a massive invasion of the central nervous system of the fish resulted. Very few control fish were infected.

Dramatic effects such as fish kills are the more easily corrected ones, but the solution to this problem may provide a false sense of security about a job sup-posedly well done. Solution of the chronic, sublethal problems requires much more than the elimination of carelessness or improperly designed plants. It requires a willingness on the part of industry and the paying public to expend time and money to obtain a satisfactory environment. The number of different kinds and mixtures of discharged wastes does not permit us to develop all the needed criteria. The responsibility for determining criteria for uncommon effluents and mixed effluents will probably belong to the originator of these wastes. If this becomes the rule, then the techniques presently being developed during our research could provide the basis for testing these mixed or uncommon effluents, or both.

It is very obvious by now that any incomplete effort to protect and improve our aquatic environment will result in little benefit. Once a significant start has been made, there is no turning back. Removing or detoxifying most effluents when one remains that can degrade the environment is useless. The effort and the responsibility are so great that they cannot be assumed by one agency, one industry, or one state. Only by preventing gradual degradation by sublethal, chronic causes can we provide the aquatic environment desired by an ever-increasing number of concerned users.

Literature Cited

American Public Health Association. 1971. Standard methods for the examination of water and wastewater. 13th ed. New York, 874 p.

Arthur, J. W., and J. G. Eaton. 1971. Toxicity of chloramines to the amphipod, *Gammarus pseudolimnaeus* Bousfield, and the fathead minnow *Pimephales promelas* Rafinesque. J. Fish. Res. Board Can. (in press).

Brungs, W. A. 1969. Chronic toxicity of zinc to the fathead minnow, *Pimephales promelas* Rafinesque. Trans. Amer. Fish. Soc. 98:272-279.

—————. 1971. Chronic effects of low dissolved oxygen concentrations on the fathead minnow (*Pimephales promelas* Rafinesque). J. Fish. Res. Board Can. (in press).

—————, and D. I. Mount. 1967. Lethal endrin concentration in the blood of gizzard shad. J. Fish. Res. Board Can. 24:429-432.

Carlson, A. R. 1971. Long-term exposure effects of Sevin on survival, growth and reproduction of the fathead minnow *Pimephales promelas* Rafinesque (in press).

Doudoroff, Peter, and D. L. Shumway. 1970. Dissolved oxygen requirements of freshwater fishes. Food and Agriculture Organization of the United Nations, Fish. Tech. Paper No. 86, 291 p.

Ettinger, M. B., and D. I. Mount. 1967. A wild fish should be safe to eat. Environ. Sci. Tech. 1:203-205.

Federal Water Pollution Control Administration. 1968. Water quality criteria report of the National Technical Advisory Committee, 234 p.

Fetterolf, C. 1970. Summary of staff investigations into the toxicity of chlorine to fish. Presented to the Water Resources Commission, Muskegon, Michigan.

Gaufin, A. R. 1971. Studies on the tolerance of aquatic insects to low oxygen concentrations (in press).

Mount, D. I., and H. W. Boyle. 1969. Parathion—Use of blood concentration to diagnose mortality of fish. Environ. Sci. Tech. 3:1183-1185.

——————, and G. J. Putnicki. 1966. Summary report of the 1963 Mississippi fish kill. Trans. N. Amer. Wildl. Natur. Resour. Conf. 31:177-184.

——————, and C. E. Stephan. 1967. A method for establishing acceptable toxicant limits for fish—malathion and the butoxyethanol ester of 2,4-D. Trans. Amer. Fish. Soc. 96:185-193.

——————. 1969. Chronic toxicity of copper to the fathead minnow (*Pimephales promelas*) in soft water. J. Fish Res. Board Can. 26:2449-2457.

Mount, D. I., L. W. Vigor, and M. L. Schafer. 1966. Endrin: Use of concentration in blood to diagnose acute toxicity to fish. Science 152:1388-1390.

Nebeker, A. V. 1971. Effect of low oxygen on survival and emergence of aquatic insects. Can. Entomol. (in press).

Pickering, Q. H. 1971. Chronic toxicity of cadmium to the fathead minnow, *Pimephales promelas* Rafinesque (in press).

——————, and T. O. Thatcher. 1970. The chronic toxicity of linear alkyate sulfonate (LAS) to *Pimephales promelas* Rafinesque. J. Water Pollut. Contr. Fed. 42(2) Part 1:243-254.

U.S. Department of the Interior, Federal Water Pollution Control Administration. 1968. Pollution-caused fish kills. 9th annual report, 17 p.

365

RADIONUCLIDES IN RIVER SYSTEMS

D. J. Nelson, S. V. Kaye and R. S. Booth

Man's historic dependence on rivers as a source for water and as a disposal site for unwanted materials has continued into the atomic age. Site selection criteria used by the Manhattan Engineering District during the development of atomic technology includes the presence of large volumes of high-quality cooling water, hydroelectric power, and means for dilution of effluents. A site on the Clinch River in eastern Tennessee was chosen for pilot plant operations, and a major production facility was located on the Columbia River in eastern Washington.

The potential hazard of radioactive wastes was recognized from the beginning because of previous experience with other forms of ionizing radiation such as x-rays and radium used for medical purposes. Consequently the general philosophy with respect to handling radioactive wastes was to "concentrate and contain" or "dilute and disperse." When wastes were dispersed to the environment, recommendations of the International Commission on Radiological Protection were followed to assure that humans ingesting water would not be exposed to greater than recommended doses of radiation. These early guides were based on fragmentary information, and there has been a continual updating of standards as more data have become available. Presently, a significant body of data with respect to the biological effects of ionizing radiation is used to set standards for drinking water. The nuclear industry is somewhat unique in the annals of man in that protection standards with respect to radioactive waste effluents were developed before technology became widespread.

We are moving rapidly into an era when nuclear energy will be used increasingly to generate electrical power. Because of transmission costs, it is more economical to locate generating stations near load centers. Reactors will be located almost exclusively on rivers or coastlines. Suitablilty of a particular site for a boiling water or pressurized water nuclear power station (the two major types built in the United States) depends on availability of cooling water.

*Research sponsored by the U.S. Atomic Energy Commission under contract with the Union Carbide Corporation.

Favorable sites along our rivers may become scarce before electrical power demands of the next decade are met. Excessive thermal loading of rivers from heated discharges may impose restrictions on the growth of a nuclear economy in the United States, but alternative heat exchanger systems such as cooling towers, closed systems using special reservoirs, and different coolants (liquid metals and molten salts) could change this prognostication.

The purpose of this paper is to discuss sources of radionuclides released to rivers, fate and transport of the released radioactivity, and potential radiation doses to man utilizing the Clinch River, Tennessee, for drinking water, fisheries, and recreation.

Sources of Radionuclide Releases

Radionuclide releases to the environment originate from one of the following potential sources:

1. Nuclear power stations
2. Fuel reprocessing plants
3. AEC-sponsored production and research laboratories
4. Nuclear weapons testing
5. Other potential sources including medical institutions, university and industrial research laboratories and reactors, and peaceful nuclear explosives

Nuclear Power Stations. The waste management system at a nuclear power station is designed for maximum safety, even in the unlikely event of a malfunction or accident of some type. The system is designed to concentrate and contain radionuclides under all operating conditions. Routine releases of radioactivity into the river system are under the complete control of reactor operations personnel, whose job is to maintain the concentration of radioactivity in effluents as far below the permissible limits as practicable.

Before river water can be used in the primary cooling loop of a nuclear reactor, it must be purified by demineralization to extract natural impurities and chemical pollutants which might become activated by neutron bombardment. Additional activation products may be expected to appear in the water from activation of corrosion products originating from the coolant system, leakage of fission products through the fuel element cladding, diffusion of tritium through stainless steel cladding, and activation of the water itself. At some nuclear power stations chemical additives in the cooling water or tramp uranium contamination on fuel elements add to the level of activation and fission products (Blomeke and Harrington, 1969). Radioactivity from these various sources must be removed from the coolant water before it is used again or discharged to the environment. Water decontamination is accomplished by demineralization (ion exchange), filtration, or evaporation and is dependent on reactor type. Since none of these processes is 100% effective in removing radioactivity, some reactor produced radioisotopes are passed on to the river system through the coolant discharge canal. Radioactivity discharged to this canal comes from various holding tanks. The release rates are predicated from the assay of the tank contents (μCi of each radionuclide/ml) and the flow rate and mixing characteristics of the river system to insure that the legal discharge limits will not be exceeded.

The U.S. Public Health Service has completed an extensive environmental surveillance program at the Dresden I Nuclear Power Station and its local environment. Kahn, et al. (1970) reported that the purpose of this program was " . . . to identify and quantify the radionuclides in effluents and in pathways from the point of discharge in the environment to man in order to provide the technical basis for improving surveillance programs." Besides gaseous releases through a 300-ft-high stack, liquid wastes are released into a cooling-water discharge canal which empties into the Illinois River below the confluence of the Kankakee and Des Plaines rivers. On the basis of the Public Health Service's measurements over a nine-month period, the radiation dose delivered to the surrounding population through consumption of food (including fish) and water was not measurable.

Only a slight amount of radioactivity is added to the discharge canal water from operating the reactor (Table 1). The difference between discharge concentration and intake concentration is presumably due to the Dresden I Nuclear Power Station. A continuous flow ion-exchange apparatus was used to sample the ionic species in the water. Since clay minerals are also noted for their ionic exchange capacity, silt was sampled separately in relation to water volume and analyzed for radioactivity. Therefore, the total concentration of radioactivity in the Dresden intake and discharge canals is the sum of the ionic (ion-exchange) and silt fractions reported in Table 1 on a per-liter basis. The role of clay minerals in sorption of radioactive ions is widely known, and the selective sorption of radiocesium from radioactive waste solutions has been studied in some detail by Tamura and Jacobs (1960 and 1961). Under environmental conditions, for instance, cesium cations are more readily sorbed to clay minerals than are strontium cations (Tamura, et al., 1961). Partitioning of cesium and strontium between the ionic and silt fractions tabulated in Table 1 follows this expected pattern (cesium higher in silt; strontium higher in ionic sample).

Table 1. Concentrations of radionuclides in coolant canal water of Dresden I Nuclear Power Station, August 20, 1968. (Data taken from Kahn, et al., 1970)

| Radionuclide | Concentration, $\mu\mu$Ci/liter water | | | |
| | Intake | | Discharge | |
	Ionic[a]	Silt[b]	Ionic[a]	Silt
^{58}Co	< 0.1	0.0	0.2	0.1
^{60}Co	< 0.1	0.02	0.2	0.7
^{89}Sr	< 0.2	0.0	<0.2	0.0
^{90}Sr	0.3	0.005	0.5	0.04
^{131}I	< 0.2	0.0	<0.2	0.0
^{134}Cs	0.1	0.0	<0.1	0.1
^{137}Cs	0.1	0.07	0.2	0.3
^{140}Ba	< 0.1	0.0	0.2	0.1
^{3}H	200[c]		200[c]	

[a]Samples collected with continuous flow ion-exchange device

[b]Suspended silt from coolant-canal water

[c]Water analyzed

Fuel Reprocessing Plants. The second potential source of radioactivity in rivers results from a process intended to conserve nuclear fuel and permit economical operation of nuclear power plants. Fuel elements containing uranium, thorium, plutonium and fission products are periodically reprocessed at a fuel reprocessing plant, which may be located at a great distance from the reactor. A series of chemical procedures removes the fission products which must be stored and disposed of, but the fuel is refabricated into fuel elements and eventually shipped back to the reactor for loading into its critical assembly. Nuclear fuel must be reprocessed after approximately 5 to 10% of the fuel has been fissioned (King, 1964).

Although several fuel reprocessing plants are currently in the planning and/or construction stages, only the Nuclear Fuel Services, Inc., facility in western New York State (town of Ashford, Cattaraugus County) is in operation. This plant started to reprocess fuel elements in April 1966, and is licensed to discharge gaseous wastes through a 200-ft-high stack and liquid wastes to Cattaraugus

Creek. The operational license granted by the AEC stipulates that yearly averaged concentrations shall not exceed the concentrations listed in the Code of Federal Regulations (Title 10, Part 20) for releases to unrestricted areas and that, at any time, the actual concentrations in Cattaraugus Creek shall not exceed twice these concentrations (U.S. AEC Technical Specifications, 1967). The pathway of liquid discharge is from holding ponds at the reprocessing plant and from a solid waste burial ground to Franks Creek, which flows into Erdman Brook to Buttermilk Creek, and then to Cattaraugus Creek, which empties into Lake Erie approximately 28 miles downstream.

The environmental surveillance program for the Nuclear Fuel Services Plant has continually shown that the maximum allowable concentrations in Cattaraugus Creek are not exceeded. However, Kelleher (1969) reported ^{90}Sr concentrations of one-tenth to one-third the allowable limit were not uncommon for short periods. Concentrations of radionuclides in the three feeder streams (all on the 3,300-acre, fenced Nuclear Fuel Services Reservation) were reported to be much greater than the 10 CFR 20 limits for release to unrestricted areas during the first two years of operation. Fish from the larger Cattaraugus Creek have access to these smaller streams as do game animals.

Initial surveillance planning for this facility did not anticipate some of the potential discharge rates for liquid effluents, and the major monitoring program was designed to keep close surveillance on gaseous discharges through the stack. After monitoring of creek water and biota uncovered a potential problem, operations at the plant were improved to reduce radioactivity releases, and a more extensive surveillance program was undertaken. Disposal of liquid wastes by the hydraulic fracturing and injection method (de Laguna, 1969) has been studied for use on the Nuclear Fuel Services Reservation, and if employed on a routine basis, this method of disposal would result in a substantial reduction of radioactivity released to Cattaraugus Creek.

It appears that operation of fuel reprocessing plants will pose more potentially hazardous situations from liquid releases than will nuclear power reactors. We feel this situation might develop because fuel reprocessing plants will have large inventories of radioactivity, e.g., fuel elements from many reactors on site at one time, and much of the fuel, fission products, and activation products will not be protected by the fuel element cladding. After the United States swings more toward a nuclear power economy, at any one time over 95% of the contained radioactivity associated with electrical power production will be located at fuel reprocessing plants.

AEC-Sponsored Production and Research Laboratories. Some of the major AEC-sponsored laboratories have been in operation for almost 30 years. During this time, considerable volumes of low-level wastes have been safely released to rivers at three locations: Oak Ridge National Laboratory, Oak Ridge, Tennessee; Savannah River Laboratory, Aiken, South Carolina; and Pacific Northwest Laboratory, Richland, Washington. The radionuclides in the waste releases from each installation differ considerably and reflect the type of operations carried out onsite. Culler (1959) estimated that of all the low-level activity discharged to the environment in the United States during the 15-year period 1944-1959, the Pacific Northwest Laboratory (formerly Hanford Atomic Works) released 96.4% and Oak Ridge National Laboratory released 3.4%. Because of changes in policy and procedures during the past decade releases at these laboratories have been reduced, but the relative percentages probably have not changed appreciably.

Nuclear Weapons Testing. Since the President, with the consent of the Senate, ratified the Limited Test Ban Treaty for the United States in October 1963, there has been no above-ground testing of nuclear weapons by adherers to this treaty. France and Communist China, however, are nonadherents and have conducted numerous atmospheric tests since 1963. Nuclear weapons detonated above ground, or underground detonations that vent to the atmosphere, deposit radioactivity in the environment as "fallout." When this vented activity reaches the stratosphere, the fallout may remain aloft for two or three years before it is deposited globally (Machta, 1965). Only a small fraction of global fallout falls directly into rivers, but redistribution through runoff and erosion contributes sizable inputs of radioactivity in some cases.

Morgan and Stanbury (1961) reported on contamination of three Thames tributaries with fission products from fallout by comparing catchment basin input (fallout) to the output to tributaries (concentration in river water). Corresponding rain and river water samples showed that natural decontamination factors were smallest for ^{90}Sr and greatest for ^{144}Ce. The results of this study suggested that less than 1% of the fallout radioactivity leaves the rivers, but in periods of heavy rains and high flow rates, this amount may be increased considerably.

Other Potential Sources of Release. Medical institutions, research laboratories, research reactors, and peaceful nuclear explosives are considered under this category. The total quantity of radioactivity released to rivers from these potential sources is minute compared with actual or potential releases considered previously.

Most nonsealed sources used in industry and medical institutions are for tracer purposes and generally do not contain large quantities of radionuclides. *In vivo* applications of tracers require radionuclides with short radioactive half-lives so that the radiation dose is minimized. Only extremely small quantities of radionuclides are released routinely via drains to sewers, domestic waste treatment plants, and rivers. Since the total volume of wastes originating at medical institutions is small, these wastes are usually stored and disposed of at an approved repository. Industrial liquid wastes are more varied and may include some releases directly to rivers.

Research reactors of the types typically built by universities do not release liquid effluents. Some liquid wastes do originate at these installations, but they are transported for storage or disposal at an AEC-approved repository.

Peaceful nuclear explosives have been proposed for several large engineering applications involving rivers; river diversion, dam formation, harbor excavation, and canal excavation are the most frequently mentioned projects. None of these applications has been carried out in the United States, but such projects are very appealing because of substantial cost savings for the larger project (Nordyke, 1969). If any of these projects were conducted, the detonations would add radioactivity to surface waters from vented as well as nonvented radioactivity that might be flushed out by flowing water. Such potential releases would have to be considered from the health and safety standpoint in each project's feasibility study.

Environmental Fate of Radionuclides

Radionuclide Transport in Streams. The transport of radioactivity by streams is primarily by hydrodynamic processes which include dispersion by turbulent diffusion and gradient diffusion, convective flow of dissolved radionuclides, transport of suspended loads, and movement of bed sediments. However, movement is also dependent upon physical and chemical characteristics of the radionuclide and of the stream system such as the chemical character of the bottom sediments and aqueous system, the surface concentrations of radionuclides on bed sediment and on biomass, and the amount and type of suspended matter (White and Gloyna, 1970, Yousef, et al., 1970).

Mass-balance analyses covering a two-year period and 160 river miles showed that the majority of ^{90}Sr, ^{137}Cs, ^{60}Co, and ^{106}Ru released from Oak Ridge National Laboratory was carried by the Clinch River in the water phase with negligible quantities stored in the biota and small quantities in the sediments, except perhaps for ^{137}Cs where the results were subject to error (Parker, et al.,

373

1966). The river system investigated had an average flow rate of 4,560 ft^3/sec at the major source of radionuclide input, low organic content (28 ppm volatile solids), low nutrient content (1.5 mg/l NO_3; 0.1 mg/l PO_4), slight basicity (pH 7.7) and moderate hardness. However, it was concluded that a catastrophic large-scale release of radioactive materials from the sediments of a river system of this type would not occur, even as a result of major changes in flow, in pH, or in oxidation-reduction potential.

The fraction of released radioactivity retained by bed sediments and biota is most important to hazard evaluations and is dependent upon physical, chemical, and environmental conditions. Radioactivity in the water phase is divided between that in solution and in suspension. The amount of radioactivity associated with suspended solids varied from 8% for ^{85}Sr to 54% for ^{65}Zn in one measurement (Yousef, et al., 1970) and between 10% and 90% in another using I, SO_4, Sr, Cu and Fe (Sayre, et al., 1963) with the most important characteristic affecting the extent of sorption being particle size. Particle size appears to be of general importance for the transport of metallic ions in rivers (deGroot, et al., 1971). When the suspended solids settle, radioactivity is deposited on the bottom sediments or plant surfaces where it can be stored, transported by sediment movement, or resuspended into the aqueous phase. Thus, settling velocities and sediment transport, which can vary considerably at nearby points in the same stream, are important factors in determining the fraction of released radioactivity retained by the aquatic system and therefore available for uptake in food chains leading to man.

Interception of radionuclides in the Clinch River bottom sediments after release from ORNL during nearly 20 years of laboratory operation was 21% for ^{137}Cs, 9% for ^{60}Co, 0.2% for ^{106}Ru and 0.2% for ^{90}Sr. Ninety percent of the radioactivity was retained in the 15 miles of the river immediately downstream from the release point. In this stream, velocity and turbulence decrease, and consequently sediment transport capacity decreases in the downstream direction. On the Columbia River, the amounts of radionuclides from the Hanford reactors contained in bed sediments were generally much higher downstream than upstream and was proportional to the amount of bed sediment downstream that was fine grained (Nelson and Hausheld, 1966).

An interesting aspect of radionuclide sorption to sediments is that it is only partially or slowly reversible, the desorbing rate being much slower than the buildup rate. On the other hand, pure ion-exchange, plant sorption, and dead zone sorption usually follow a reversible exchange model (Yousef, et a., 1970).

From the above considerations it is evident that predicting the source terms (radioactivity in biomass, bed sediment, and in solution) for a food chain pathway to man after radionuclide deposition in a stream must include temporal and spatial change from hydrological effects, sorption and desorption on biomass and bed sediments, and bed sediment transport. One predictive model of this type is presented by White and Gloyna (1970). The basic equation of the model is

$$\frac{\partial C}{\partial t} = D_L \frac{\partial^2 C}{\partial x^2} - \bar{u} \frac{\partial C}{\partial x} + \sum_{j=1}^{m} S_j \tag{1}$$

where

C = radionuclide concentration at time t and point x downstream from the source

D_L = coefficient of dispersion in the downstream direction

$\bar{u}$ = fluid velocity in the downstream direction

j = the index number for each sorption phase, and

S_j = uptake or release from the jth of m sorption phases

Dispersion is accounted for by the first term on the right-hand side of Eq. (1), convection by the second term, and sorption and desorption by bottom sediments and aquatic plants by the third term. The parameter D_L is often found to depend upon other parameters of the stream. Yousef, et al. (1970), expressed D_L as proportional to $\bar{u}$ raised to a power, but one reference is indicated where D_L depends only upon the stream's vertical velocity gradient and another where D_L is a function of the average shear velocity. Difficulties in finding a single expression for D_L are perhaps caused by the dependence of this parameter upon turbulent diffusion which, in turn, is a complex function of the stream's three flow velocities and its physical properties.

The sorption terms are of the form

$$S_j = K_j [W_j - g_j (C)] \tag{2}$$

where

K_j = the mass transfer coefficient for phase j

$g_j(C)$ = the transfer function relating the concentration of activity in water to the equilibrium level in phase j_1 expressed as $g(C) = KC^n$ where K and n are constants and

$$W_j = \text{the specific activity in the } j^{th} \text{ of m sorption phases}$$

The time rate of change of W_j is expressed as

$$\frac{dW_j}{dt} = -S_j \qquad (3)$$

Thus, if $W_j < g_j(C)$ (the current concentration in the bottom sediments or plants being less than its equilibrium concentration) then sorption occurs; and if $W_j > g_j(C)$, desorption occurs.

Numerical solutions to Eq. (1) for an impulse source at $x = 0$, with appropriate boundary conditions and parameter values, exhibit the general character of a backward-skewed and expanding normal distribution moving downstream with a velocity $\bar{u}$. Comparisons of numerical solutions of Eq. (1) with measurements in a model river (flume) indicated general agreement, with dispersion well predicted and velocity only slightly overestimated.

Another model of radioactive transport in streams (RADAC) treats radioactive waste and stream properties as stochastic variables of the problem parameters (Thomas, 1965). The river is divided into a series of reservoirs (areas with similar hydrologic and physical characteristics) with the outflow of one reservoir being the input of the next one downstream. The numerical values of problem variables at time $t + \Delta t$, the flow rate in the reservoirs, for example, are determined from their values at time t and from a transition matrix. The transition matrix for radionuclide concentrations contains elements which depend upon hydrologic parameters. The numerical results presented by Thomas (1965) are, however, for radionuclide flow independent of hydrology, and no comparisons with measurements are presented. The authors believe that this stochastic modeling technique can be used to formulate general laws of transport, mixing, and dilution of radioactive wastes not available from simulation or optimization analysis.

Simulation techniques have also been used to predict the movement of radioactivity in a stream. In one model, the concentrations of pollutants at downstream use-points were determined by simulating streamflow, dilution, mixing, nuclear decay, uptake and release from biomass, water withdrawal, and water treatment (Thomas, 1965). This model begins with a vector diagram of the flow of water and pollutants in the river system, which includes stream interconnections, flow rates, and gains and losses. The condition of the system at $t + \Delta t$ is determined from its condition at t by continuity equations, conservation laws, and empirical functions which relate water flows and pollutant flows. Thomas presented numerical results for a study of the Clinch-Tennessee River system where the relative importance of various input parameters was investigated.

Radionuclide Concentration by Aquatic Organisms. Rivers may be considered flow-through systems with respect to radionuclides, since only small quantities are stored in sediments or biota. However, the small quantity that accumulates in aquatic organisms is of potential concern to humans who utilize the biota for food. Thus, knowledge of environmental pathways is essential to evaluate possible hazards to humans utilizing fishery resources. There are data with respect to the interaction of a large number of radionuclides with aquatic organisms. However, as the data in Table 1 indicate, there are a limited number of radionuclides released from power reactors. At least 100 radionuclides have been identified in effluents released to the Columbia River from the Hanford Works at Richland, Washington. Most of these are short-lived activation products from the cooling water which is exposed to the neutron flux in the reactor.

Organisms accumulate some radionuclides such as ^{90}Sr, ^{140}Ba, ^{24}Na and ^{45}Ca directly from the water. Others such as ^{137}Cs, ^{32}P, ^{51}Cr, ^{64}Cu, ^{76}As, and the rare earths are primarily accumulated by ingestion of contaminated food. Initial entry into the food chain with the latter group of radionuclides occurs largely by adsorption on algae or other particulate matter. Subsequent movement through the food chain occurs through assimilation. The proportion of an ingested radionuclide that is assimilated may vary significantly depending on the adventitious materials associated with the food. When ingested ^{137}Cs is on algae, 70% may be assimilated; but when there are inorganic clay sediments associated with the ingested food, assimilation is less than 10% (Kolehmainen and Nelson, 1969). This appears to be a major factor contributing to the relative accumulation of ^{137}Cs by fish which are carnivorous (Nelson, in press). For instance, largely carnivorous bluegills contained less ^{137}Cs than did carnivorous gizzard shad because of the inclusion of adventitious clay sediments in bluegill food which strongly adsorbed ingested ^{137}Cs. Specific food chains appear to be important since Gallegos (1969) showed rainbow trout from a montane lake in Colorado assimilated 20% to 40% of the ^{137}Cs associated with feldspars and quartz in their digestive tracts.

In aquatic systems the relative level of radionuclides in organisms can be related to the concentration of radioactivity in water through the use of concentration factors. The concentration factor is generally expressed as the ratio of activity per g of wet tissue to activity per ml of water. These factors can also be determined on the basis of stable chemical measurements on organisms and

377

water. Radiation protection recommendations issued by the ICRP and NCRP give maximum permissible concentrations for each radionuclide in water and guidance is given also for maximum permissible daily intakes of radionuclides from all sources. Thus, knowledge of concentration factors for the several radionuclides may be used to assess potential intakes of radionuclides due to the consumption of fish or other edible aquatic organisms.

Concentration factors for the radionuclides associated with waste releases from light water reactors vary depending upon the radionuclide and the organism (or tissue) being considered. Since fish are potentially a significant pathway for transfer of radionuclides to humans, a great deal of study has been associated with radionuclides in fish. Only recently have we shown that fish concentrate cobalt from 25-50 times that in water (Nelson, in press). Concentration factors for ^{90}Sr or ^{89}Sr range from one to about 100. Strontium does not ordinarily accumulate in muscular tissue of fish, and the low values come from samples where care was taken to exclude bone or scales from the samples analyzed. However, in fish such as suckers and carp which have many small bones, bony tissue included in the sample frequently results in the higher concentration factor. Fish bone alone has a concentration factor of 3,000 to 4,000 for Sr. Very few data are available for ^{131}I, but a common concentration factor is from 25 to 50. Cesium isotopes include ^{134}Cs and ^{137}Cs with the latter more important because more is produced and it has a longer radioactive half-life. Reported concentration factors for ^{137}Cs range from 125 to about 10,000. The higher number was obtained from experimental ponds where a single spike was followed for several months. A true steady state was never attained and the high concentration factor was the result of the transient conditions. Generally, ^{137}Cs concentrations are higher in water having less dissolved mineral elements. The most common value from peak concentration factors appears to be about 1,000. The concentration factor for ^{3}H has caused considerable concern in the last two years but it now appears clear that the concentration factor for ^{3}H is unity. There are no data for ^{140}Ba concentration factors except for the early report by Prosser, et al. (1945), where concentration factors of 10-150 were reported.

Concentration factors for radionuclides found in effluents from nuclear power reactors are summarized in Table 2. These data show that the concept that concentrations of radionuclides increase with increasing trophic level in aquatic food chains is generally not tenable. This has been an attractive assumption and only recently have we had sufficient data on specific food chains to show that this was not the case. The data in Table 2 suggest that the highest concentration factors typically occur at the lower trophic levels. With ^{60}Co, concentration factors of 2,500-6,200 occur in algae, about 300 in invertebrates, and 25-50 in fish. Similarly, ^{137}Cs may be concentrated as much as 25,000 times by

Table 2. Concentration factors of selected radionuclides in aquatic organisms. These radionuclides are important in reactor effluents and a range of concentration factors is given reflecting environmental and biological variables. All data are on a fresh weight basis. (Modified from Reichle, et al., 1970)

	^{90}Sr	^{137}Cs	^{60}Co	^{131}I	^{140}Ba	^3H
Water	1.0	1.0	1.0	1.0	1.0	1.0
Algae	10-3000	50-25000	2500-6000	200		
Vascular plants	100- 350	50- 1000		60	75-400	
Invertebrates						
Detritus feeder	10-4000	60-11000		20-1000		
Herbivore	1.0	600	325			
Carnivore		800				
Fish (flesh)						
Omnivore	1- 100	160- 1200	50	25- 50	150	1.0
Carnivore	1- 100	120- 1400	25- 30			1.0

algae but only approximately 1,000 times by fish. It is apparent that surface adsorption phenomena are important for some organisms, especially those having a large surface volume ratio (Cushing, 1970). Subsequent movement through the food chain depends upon how readily the radionuclide is assimilated and retained by various body tissues.

The dispersion of radionuclides through the movements of aquatic organisms has been the subject of speculation since Peredelski and Bogatyrev (1959) suggested emergent insects would remove radioactivity from reservoirs and contaminate the surrounding landscape. Calculations based on Clinch River data showed that general contamination of the landscape by biotic vectors is unlikely. Chironomids are the most abundant insect living in the bottom sediments of the Clinch River. Using the ^{90}Sr concentration in emergent adults and their estimated productivity we were able to show there would be approximately 45 times as much ^{90}Sr entering the river directly from fallout as was removed from the river by emerging insects.

We also estimated the removal of ^{90}Sr in mollusk shells which are harvested commercially. The annual harvest of shells ranged from 5,000 to 10,000 tons. Using the maximum harvest and an average concentration factor of 4,000 for ^{90}Sr, an estimate for ^{90}Sr removal of 64 mCi/yr was obtained. The total mollusk population contained about 79 mCi/yr or approximately 0.36% of the ^{90}Sr

released to the river annually. We can safely assume that aquatic organisms will not generally effect a significant mass movement of radioactivity from aquatic environments nor will they at any one time contain a great proportion of the radioactivity in the water. This conclusion is in keeping with one geochemical principle that the biota contain at any one time a small proportion of the mineral elements in the environment.

The Specific Activity Concept. Radionuclides of an element are not unique chemical species of that element and, as a result, organisms metabolize radioactive atoms and stable atoms of the same element in a similar manner. The ratio of radioactive atoms to total atoms of the same element is defined as the specific activity. Specific activity relationships may also be expressed as curies (Ci) per gram of element. The distinction here being between units of radiation and mass since the mass quantities of radioactive atoms are ordinarily small and are not measured by quantitative chemical analyses. Further, curie quantities are an important factor required to determine dose to man or other organisms exposed to radioactive materials.

The specific activity concept is important because it is one method which permits an *a priori* prediction of radionuclide uptake. In its simplest form, the concept assumes that the distribution of the radioactive chemical species between an organism and water will be the same as for the stable chemical form. Thus, on the basis of stable chemical analyses of organisms (or their tissues) and water, the steady state distribution of radionuclides can be estimated for chronic releases of radionuclides. These relationships may be applied directly when the radioactive half-life (T_r) of the radionuclide involved is much longer than the biological processes (e.g., growth and biological half-time) under consideration. For instance, good agreement between predicted and observed specific activities was found for ^{90}Sr $(T_r = 28$ yr) and ^{137}Cs $(T_r = 30$ yr) in white crappie (*Pomoxis annularis*) from the Clinch River, Tennessee (Nelson, 1967, 1969).

There are several assumptions and factors that should be considered in applying the specific activity concept. Perhaps the most important requirement is that the stable and radioactive atoms of the element be in the same chemical form and available to organisms in the same proportion they occur in water. The simple ratio ignores such parameters as biological half-time and radioactive half-life, growth, and time; this leads to an overestimate of the potential radionuclide accumulation in aquatic organisms. The effect of these parameters on estimates of radionuclide accumulation was discussed in detail by Kaye and Nelson (1968).

380

Mineral Metabolism by Organisms. A further understanding of the relationships between the metabolism of radio- and stable nuclides may be obtained by considering how the chemical composition of organisms is affected by the chemical composition of their environment. As a framework for reference we can consider two types of responses by organisms to waters of different chemical quality (Figure 1). In this diagrammatic presentation the environmental concentration of an element increases on the abscissa and the resultant concentrations in organisms are expressed as µg/g tissue or as concentration factors.

Indeterminant concentrations of elements in organisms are represented by a constant concentration factor and a tissue concentration that is directly proportional to the environmental concentration. This proportionality will not continue

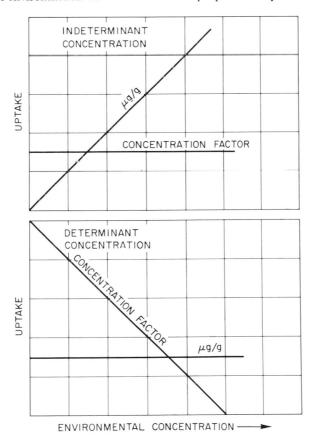

Figure 1. A diagrammatic presentation of the biogeochemical responses of organisms to their chemical environment. Indeterminant and determinant concentrations of some chemical elements in organisms have been identified

for unreasonable environmental concentrations of an element. Reed and Nelson (1969) found that the quickly exchanged fraction of Sr in bluegills was proportional to environmental concentrations, even at concentrations greater than those normally found in nature. With indeterminant concentrations of elements, the organisms do not maintain homeostatic control over the element content of their bodies. Toxic elements would be considered in this category. However, some elements such as Zn or Co are essential at trace quantities but toxic at higher concentrations. Generally, it appears that many trace elements have indeterminant concentrations.

Chemical elements with determinant concentrations in organisms are metabolically regulated, and their concentrations are homeostatically controlled. Elements included here are probably the macroconstituents of living tissue. A good example where data are available is K in fish (Kolehmainen and Nelson, 1969). Concentrations of other elements such as C, H, and N in soft tissues and Ca in bones do not vary significantly and may be considered determinant.

Many of the elements of concern from the hazard standpoint in the nuclear industry are trace elements. These elements occur in trace quantities in environment as well as in organisms. From the standpoint of radiological protection the dose (rems) delivered by the quantity of radioactivity (Ci) present is the essential point. Specific activities expressed as atom ratios for ^{137}Cs : Cs in white crappie from the Clinch River were 1:10^8 (Nelson, 1969). Similarly, atom ratios for ^{90}Sr:Sr were 1:10^{10} in white crappie flesh (Nelson, 1967). The concentrations of Cs and Sr were 2.5 x 10^{-5} and 0.07 µg/ml, respectively. Trace quantities of stable Sr and Cs were present in the water but the radionuclide content of the water and fish was chemically insignificant. Suggestions have been made that radionuclide uptake could be reduced by altering the environment chemically. Chemical amendments to water to reduce biological uptake of radionuclides would not be effective with elements having indeterminant concentrations in the biota. Further, the atom ratios clearly show that the dilution effect of stable element amendments would have little effect with elements having determinant concentrations. These data point out the salient fact that control of radionuclides at the source is essential. Current AEC regulations with respect to reactor effluents are emphasizing source control in the "concentrate and contain" philosophy.

Historic Experience with Radionuclides in the Clinch River. At Oak Ridge, liquid effluents are released via White Oak Creek to the Clinch River. Present releases are minimal, but in past years considerable quantities of ^{90}Sr, ^{137}Cs, ^{106}Ru, and ^{60}Co were released at *less than* (MPC)$_w$ concentrations to White Oak Creek. Further dilution was achieved as this water flowed into the Clinch River.

Since this practice had continued from 1944 to 1963, a detailed interdisciplinary study was conducted to evaluate the dose to populations downstream from operations at the Oak Ridge National Laboratory (Struxness, et al., 1967).

Based on knowledge of water utilization downstream, Cowser and Snyder (1966) evaluated the following potential exposure situations: (1) consumption of contaminated water and fish; (2) consumption of agricultural products irrigated with river water; (3) external exposure from contaminated water and bottom sediments; (4) external exposure to accumulated sludge and deposits in water purification systems utilizing river water. Consumption of contaminated water and fish was assessed as the critical exposure pathway in this study (Table 3).

Table 3. Estimated cumulative dose (rem) received by critical organs of males from use of Clinch River and Tennessee River[a] (Cowser and Snyder, 1966)

Critical Pathway	Clinch River		Tennessee River	
	Skeleton	Total Body	Skeleton	Total Body
Drinking water	1.4	0.11	0.38	0.030
Recreation	0.081	0.019	0.003	0.003
Fish	1.8	0.14	0.070	0.0057
Total	3.2	0.27	0.45	0.039
Maximum permissible dose[b]	60	10	20	1.0

[a] Aggregate exposure for the period 1944 to 1963.

[b] As recommended by ICRP, the annual dose rates for continuous occupational exposure are reduced to 1/10 and applied to the Clinch River and are reduced to 1/30 for bone as critical organ and to 1/100 for total body as critical organ and applied to the Tennessee River.

External exposure from sludge at water treatment plants was a small fraction of total dose, as was internal dose from consumption of produce grown in soil irrigated with river water. Crop irrigation was not common, but Cowser and Snyder (1966) speculated that if irrigation becomes a common practice in the future, the dose commitment from consumption of crops irrigated with contaminated water could become the critical exposure pathway. Environmental studies of radionuclide movement in the semiarid environs of production reactors at the Pacific Northwest Laboratory also showed that irrigation of crops with Columbia

River water did not constitute a critical exposure pathway because irrigation was not used extensively (Foster and Soldat, 1966).

The unique experience on the Clinch and Tennessee rivers for a period of almost 20 years shows that nuclear facilities can utilize rivers for the disposal of small quantities of radioactive waste effluents without excessive exposure to downstream populations. Current releases to the Clinch River are about a factor of 10 less than those which occurred in the years covered by the study. Many of the radionuclides in the Clinch River are the same as those released by the current generation of power reactors.

Research experience with radioactive waste releases to the Clinch River provides considerable knowledge with respect to the behavior or radionuclides in surface waters. The Clinch River can be considered similar to many rivers which will receive low-level waste releases in the future. Rivers act as input-output systems with respect to mass balance of most radionuclides. Retention of radioactivity by the sediments is generally a small proportion of the quantity released and biotic reservoirs are insignificant. Nevertheless, release rates of radionuclides should be reduced by control at the source because of potential critical pathways to man through consumption of fish or water.

Literature Cited

Blomeke, J. O., and F. E. Harrington. 1969. Management of radioactive wastes at nuclear power stations. ORNL-4070, 100 p.

Code of Federal Regulations, Title 10, Part 20, Appendix B, Standards for Protection Against Radiation.

Cowser, K. E., and W. S. Snyder. 1966. Safety analysis of radionuclide release to the Clinch River. ORNL-3721, Supp. 3 to Status Report No. 5 on Clinch River Study. 115 p.

Culler, F. L., Jr. 1959. Statement of F. L. Culler, Jr. *In* Industrial Radioactive Waste Disposal, Hearings before the Special Subcommittee on Radiation of the Joint Committee on Atomic Energy, Congress of the United States (86th Congress) I:32-57, U.S. Gov. Print. Office, Washington, D.C.

Cushing, C. E. 1970. Periphyton: autoradiography of zinc-65 adsorption. Science 168:576-577.

de Groot, A. J., J. J. M. de Goeij and C. Zegers. 1971. Contents and behavior of mercury as compared with other heavy metals in sediments from the rivers Rhine and Ems. Geologie en Mijnbouw (*In* special issue on: Research on Sedimentology and Sedimentary Geology in the Netherlands, in press).

de Laguna, W. 1969. Disposal by hydraulic fracturing. *In* Health Physics Div. Ann. Progr. Rept. for Period Ending July 31, 1969, p. 9-11. ORNL-4446. 342 p.

Foster, R. F., and J. K. Soldat. 1966. Evaluation of the exposure resulting from the disposal of radioactive wastes into the Columbia River. *In* Proc. Symp. on Disposal of Radioactive Wastes into Seas, Oceans and Surface Waters, IAEA, Vienna, 1966, p. 683-696. 898 p.

Gallegos, A. F. 1969. Radiocesium kinetics in the components of a montane lake ecosystem. Ph. D. Thesis. Department of Radiology and Radiation Biology, Colorado State University.

Kahn, B., et al. 1970. Radiological surveillance studies at a boiling water nuclear power reactor. U.S. Dept. of Health, Education, and Welfare Report BRH/DER 70/1. 116 p.

Kaye, S. V., and D. J. Nelson. 1968. Analysis of specific activity concept as related to environmental concentration of radionuclides. Nuclear Safety 9(10):53-58.

Kelleher, W. J. 1969. Environmental surveillance around a nuclear fuel reprocessing installation, 1965-1967. Radiolog. Health Data and Reports 10:329-339.

King, C. D. G. 1964. Nuclear Power Systems. The Macmillan Company, New York. 480 p.

Kolehmainen, S. E., and D. J. Nelson. 1969. The balances of ^{137}Cs, stable cesium, and the feeding rates of bluegill (*Lepomis macrochirus* Raf.) in White Oak Lake. Ph.D. Thesis. Zoology Department, University of Tennessee. ORNL-4445. 115 p. (1969).

Machta, L. 1965. Status of global radioactive-fallout predictions, p. 369-391. *In* Proc. of the Second Conf. on Radioactive Fallout from Nuclear Weapons Tests, Germantown, Md. (A. W. Klement, Jr., ed.), USAEC CONF-765. 953 p.

Morgan, A., and D. G. Stanbury. 1961. The contamination of rivers with fission products from fallout. Health Phys. 5:101-107.

Nelson, D. J. Concentration factors and assimilation of ^{137}Cs and ^{60}Co in bluegills, goldfish, and gizzard shad. *In* Radionuclides in Ecosystems, Proc. Third National Symp. Radioecology, Oak Ridge, Tennessee (May 10-12, 1971). U.S. AEC CONF-71050, NTIS, Springfield, Va. (in press).

——————. 1967. The prediction of ^{90}Sr uptake in fish using data on specific activities and biological half lives, p. 843-851. *In* Proc. Int. Symp. "Radioecological Concentration Processes," Stockholm, Sweden (April 25-29, 1966).

——————. 1969. Cesium, cesium-137, and potassium concentrations in white crappie and other Clinch River Fish, p. 240-248. *In* Proc. Second National Symp. Radioecology, Ann Arbor, Michigan (May 15-17, 1967).

Nelson, J., and W. L. Hausheld. 1966. Accumulation of radionuclides in bed sediments of the Columbia River between the Hanford Reactors and McNary Dam. *In* Proc. Symp. on Disposal of Radioactive Wastes into Seas, Oceans, and Surface Waters, Vienna, Austria (May 16-20, 1966).

Nordyke, M. D. 1970. Peaceful uses of nuclear explosions, p. 49-107. *In* Proc. Panel on The Peaceful Uses of Nuclear Explosions, Vienna, Austria, IAEA, 1970. 454 p.

Parker, F. L., et al. 1966. Dilution, dispersion and mass transport of radionuclides in the Clinch and Tennessee Rivers, p. 33-55. *In* Proc. Symp. on Disposal of Radioactive Wastes into Seas, Oceans, and Surface Waters, IAEA, Vienna, Austria, 1966. 898 p.

Peredelski, A. A., and I. O. Bogatyrev. 1959. Radioactive contamination of land spaces by insects flying out from contaminated basins. Izvestia Akad. Nauk. U.S.S.R., Seriya Biologiches Kaya 2:186-192.

Prosser, C. L., W. Porvinsek, J. Arnold, G. Svihla, and P. C. Tompkins. 1945. Accumulation and distribution of radioactive strontium, barium-lanthanun, fission mixture and sodium in goldfish. ORNL Document CH-3233-C.

Reed, J. R., and D. J. Nelson. 1969. Radiostrontium uptake in blood and flesh in bluegills, p. 226. *In* Proc. Symp. on Radioecology, Ann Arbor, Michigan (May 15-17, 1967).

Reichle, D. E., P. B. Dunaway, and D. J. Nelson. 1970. Turnover and concentration of radionuclides in food chains. Nuclear Safety 11(1):43-55.

Sayre, W. W., H. P. Guy, and A. R. Chamberlain. 1963. Uptake and transport of radionuclides by stream sediments. Geol. Surv. Profession Paper 433-A (prepared in cooperation with USAEC).

Struxness, E. G., P. H. Carrigan, Jr., M. A. Churchill, K. E. Cowser, R. J. Morton, D. J. Nelson, and F. L. Parker. 1967. Comprehensive report of the Clinch River Study. ORNL-4035. 115 p.

Tamura, T., and D. G. Jacobs. 1960. Structural implications of cesium sorption. Health Physics 2:391-398.

—————. 1961. Improving cesium selectivity of bentonites by heat treatment. Health Physics 5:149-154.

Tamura, T., et al. 1961. Fundamental studies of sorption by minerals. *In* Health Physics Div. An. Progress Rept. for Period Ending July 31, 1961, p. 27-42. Oak Ridge National Laboratory, ORNL-3189.

Thomas, H. A., Jr. 1965. Operations research in disposal of liquid radioactive wastes in streams. Harvard University. NYO 10447.

U.S. Atomic Energy Commission. 1967. Technical Specifications, Change No 4, Revision, Specification 4.1.1, AEC Docket No. 50-201 (June 8, 1967).

White, A., and E. F. Gloyna. 1970. Radioactivity transport in water—mathematical simulation. Tech. Rept. No. 19 to the U.S. Atomic Energy Commission.

Yousef, Y. A., A. Kudo, and E. F. Gloyna. 1970. Radioactivity transport in water — Summary Report. Tech. Rept. No. 20 to the U.S. Atomic Energy Commission.

NUTRIENTS

Walter M. Sanders, III

From the very beginning of our evolutionary history, life and man have been inevitably tied to the rivers of the world. Archeological findings place the habitats of Mesolithic man in the valleys close to flowing streams. Modern anthropologists speculate that the changing climate brought on by the receding ice caps, the disappearing rain forests, and the diminishing rivers guided the evolution of pre-man into thinking and speaking beings, mutually dependent upon each other for survival.

Diggings at Jericho, one of the oldest continually inhabited cities, have unearthed a water cistern with supply channels in the original wall constructed about 7000 B.C. By 5500 B.C. man was building complex irrigation systems between the Tigris and Euphrates rivers. Egyptian hieroglyphics relate the accomplishment of the first Pharaoh, "Menes," 4400 B.C., in developing dams and canals to drain the swamps along the Nile between Memphis and Cairo. In China, the Tukiungyien multipurpose water project used rocks and bamboo structures to divert the flow of the Min River into many canals which irrigated 1.5 million acres of the Upper Chinese plains below Tibet.

These early ancestors learned not only to use the water but also the rich nutrient-laden sediments brought down from the mountain watershed by the spring rains and monsoon floods. These organic sediments increased the fertility of the land; however, continual irrigation with the mineral-rich waters produced the first recorded pollution. Continual irrigation increased the concentrations of mineral salts in the groundwater causing large areas of once fertile land in Mesopotamia and India to become toxic to domesticated agricultural crops. Thus, 1000 years before Christ, man experienced both the beneficial and detrimental effects of nutrients in aquatic ecosystems.

Three thousand years later man is just beginning to realize that the key to water quality management is the development and maintenance of a desirable balance between the flow of energy and materials in water and the numbers and kinds of biological organisms. In this context, nutrients occupy a major role, and understanding their sources, transport, requirements, storage, cycling and losses is essential to judicious ecosystem management and water pollution control.

The current state of knowledge concerning nutrients in aquatic ecosystems is woefully inadequate and at a deplorable level. Many scientists are even unable to agree on which materials should be considered as nutrients. Biologists interested in specific classes of organisms often view nutrients in terms of either the major requirements of those organisms or the major sources of the materials. Thus, scientists interested in plant growth usually refer to nutrients as inorganic materials. To some bacteriologists, nutrients are organic compounds and inorganic materials are called basal salts or trace elements. To many fishery biologists, nutrients may represent conglomerate organic and inorganic materials such as detritus and living organisms. Thus, to satisfy the diverse interests of this symposium, nutrients are defined as "elements and their organic derivatives necessary to stimulate and sustain the growth of flora and fauna in river ecosystems." The elements generally accepted as nutrients are listed in Table 1, divided into two groups according to the relative proportions required for plant growth.

Table 1. Elements generally cited as nutrients

Macronutrients		Micronutrients	
Carbon	Calcium	Iron	Sodium
Hydrogen	Magnesium	Manganese	Molybdenum
Oxygen	Sulfur	Copper	Chlorine
Nitrogen	Phosphorous	Zinc	Vanadium
Potassium		Boron	Cobalt
		Silicon	

After reviewing some of the recent literature concerning the impact of man on nutrient sources and concentrations in river ecosystems, this paper discusses current concepts related to nutrient cycling and management in aquatic ecosystems. It also discusses carbon cycling data indicating problems associated with single factor limitation concepts.

River Ecosystem Characterization

Odum (1959) describes an ecosystem as a functional unit including both organisms and the abiotic environment, each influencing the other and both necessary for the flow of energy and materials to sustain life. In this context, boundaries of the abiotic compartment of a river ecosystem may be selected or established according to the magnitude and purpose of the nutrient investigation (Figure 1). The abiotic compartment should be described in terms of the following characteristics:

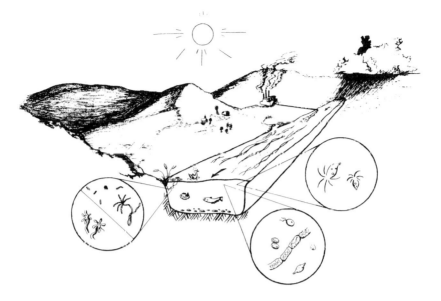

Figure 1. River Ecosystem

1. Physiography, i.e., a convenient length or reach of a river or an entire basin

2. Major sources and sinks of energy and materials such as nutrients. These are:

   — the watershed adjacent to the system
   — the atmosphere
   — the riverbed
   — the advective water (downstream flow into and out of the system)

The biological compartment of the system may also be divided into three functional subsystems:

1. The producers, consisting of autotrophic organisms using solar energy and inorganic nutrients to produce organic matter

2. The macroconsumers, including heterotrophic organisms such as the primary or plant-eating (herbivore), and the secondary, or animal-eating (carnivore), consumers

3. The decomposers, consisting of aquatic bacteria, fungi, and other microscopic heterotrophic organisms responsible for the decomposition of organic matter

The river ecosystem is distinct from other aquatic systems because the mass transport of nutrient materials is greatly affected by the degree of turbulent mixing of the water. Unlike the lake or impoundment systems, the nutrient pool is in rapid motion through the ecosystem, while many of the producers and consumers remain attached to or live in the sides and bottom of the river channel. Thus, many of the interdependent biological and chemical reactions are subject to and affect different nutrient pools as the organisms progress through their life cycles. Concomitantly, the biological and chemical reactions confined to the moving water mass in a river system may exhibit downstream cycling similar to time sequences reported for lakes and impoundments. In lakes and impoundments aquatic organisms influence their environments in many ways; however, in most river systems the primary stress seems to be exerted on the organisms by the abiotic environment and man's related activities.

Nutrient Sources and Sinks

Atmospheric Sources. Nutrients may enter a river ecosystem from the atmosphere as:

— materials dissolved in rain and snow
— particulates falling on the water surface
— aerosols and gases

Airborne nutrients originate from a variety of natural processes of the land masses and in the sea, and from man-made air pollution. The concentration and relative significance of these atmospheric inputs vary with the geographical location of the system, the time of occurrence, and the physical state or nature of the nutrient.

Weibel et al., (1966) described the organic constituents in rainwater collected in Cincinnati, Ohio, in 1963. Chemical oxygen demand (COD) ranged from 4.6 to 59 mg/l with an average of 16 mg/l. Average suspended solids, volatile suspended solids, and organic chlorine concentrations of 13.0 and 0.28×10^{-3} Mg/l, respectively, were reported. The organic chlorine fraction contained DDT, DDE, and BHC.

Fisher et al., (1968) reported seasonal variations in sulfate, nitrate and hydrogen ion concentration of precipitation collected in the Hubbard Brook watershed of New Hampshire. Rain accounted for most of the 30-50 kilograms of SO_4 per hectare carried by the streams annually and levels were highest during summer months. Data presented by Brezonik (1971) for North Central Florida showed that concentrations of nitrogen and phosphorous in rainwater decrease with

392

increasing duration of rainfall. Analysis of historical records for that area fails to show an increasing trend in rainfall nitrogen from increased fertilizer use or from other cultural sources.

Table 2 provides a range of nutrient concentrations in rainwater reported within the past 20 years.

Table 2. Ranges of reported values for nutrients in rainwater. (References are given in "Literature Cited - B")

Nutrient	Range (mg/l)	Nutrient	Range (mg/l)
TOC	--	Cu	- -
N	0.01 - 1.85	Zn	- -
K	0.03 - 0.60	B	0.01
Ca	0.1 - 10.0	Na	0.29 - 1.85
Mg	0.06 - 0.35	Mo	- -
S	0.8 - 2.0	Cl	0.5-3 - 3,920
P	0.033 - 0.058	Vn	- -
Fe	0.02 - 0.10	Co	- -
Mn	--	SiO_2	0.4

Particulate nutrients from atmospheric sources range in size from bacteria to leaves and vary widely in their relative significance. Matheson (1951) reported that 39% of the total atmospheric nitrogen input (5.8 lb/acre/year) collected daily for 18 months at Hamilton, Ontario, occurred on days without precipitation. Particulate data collected by the National Air Sampling Network, DHEW (1962), show atmospheric concentrations decreasing from a mean high of 176 micrograms air per cubic meter over large metropolitan cities to a mean low of 21 micrograms air per cubic meter over uninhabited mountain regions. According to Heaney and Sullivan (1971), atmospheric particulate fallout over a section of Chicago decreased from 19,300 kg/km^2/month in 1953-54 to 12,900 kg/km^2/month during 1966, thus indicating a net reduction in particulate atmospheric pollution.

Generally, the most significant atmospheric source for river ecosystems is the transfer of gaseous O_2, H_2, N_2, CO_2, CH_4, etc., at the air-water surface. Using pairs of gases in a laboratory reactor, Tsivoglou et al., (1968) demonstrated that the gas transfer coefficient, k_2, is inversely proportional to the molecular diameter of the gas molecule. Field measurements using the Kr tracer techniques (Tsivoglou, personal communication) give values of k_2 for oxygen ranging from 250 per day for waterfalls to 0.1 per day for sections of the Chattahoochee River near Atlanta, Georgia. His data also indicate that the ratio of k_2 carbon dioxide to k_2 oxygen is 0.89. Values for gas exchange are affected by chemical pollutants, the turbulent mass transport, and water temperature.

Nutrient losses from river systems to the atmosphere may be significant for ecosystems where high rates of production of O_2 by photosynthesis or CO_2 by heterotrophic decomposition supersaturate the water. Kerr, Paris, and Brockway (1970) report that bubbles reaching the surface of Flat Creek, a small sewage-enriched stream tributary to Lake Lanier, Georgia, contained 8-9% CO_2 and 6% O_2 despite high algal production and no measurable dissolved CO_2 in the water. Many authors report O_2 values greater than 150% saturation during periods of peak photosynthetic activity indicating net transfers of O_2 from the water.

Watershed Sources. The concentration and form of nutrients originating from watersheds are functions of the drainage basin geochemistry, hydrology, weather patterns, and cultural land use. These are discussed in terms of forest, agricultural, and urban sources.

1. Forest sources. Researchers at the Hubbard Brook Experimental Forest, New Hampshire; Stroud Water Research Center, Pennsylvania; the Rocky Mountain Forest and Range Experiment Station, Arizona, the Coweeta Hydrologic Laboratory, North Carolina; and the Walker Branch Watershed, Tennessee, are providing excellent experimental data on base-line conditions for nutrient balances and cycling within forested and clear-cut watersheds.

Johnson et al., (1969) found that 58% of the precipitation received by the Hubbard Brook watershed leaves as streamflow and that most of this water infiltrates and percolates through the soil before reaching the receiving streams. Bormann, Likens, and Eaton (1969) report that the annual losses from undisturbed Hubbard Brook Forest watersheds averaged 25 kg/ha of particulate matter and 139 kg/ha of dissolved substances. Of this, particulate and dissolved organic matter accounted for average annual losses of 13 kg/ha and 40 Kg/ha, respectively. They conclude that chemicals may be exported from various ecological compartments in the forest as particulate or dissolved organic matter from the organic compartment, dissolved inorganic substances from the available nutrient compartment, and inorganic particulate matter from the soil and rock-mineral compartment. They state that the export of individual elements depends upon the geochemistry of the element, its biological utilization potential, and its differential accumulation with the system.

394

At the Stroud Water Research Center, Pennsylvania, Vannote (1969) found that the leaf input into three stream ecosystems ranged from 200 to 800 $gm/m^2/$ year (dry weight) and was highest during autumn leaf fall; high gradient streams with steep slopes had both the highest input and retention. Attiwill (1968) reported inorganic nutrients per 100 gm (dry weight) of leaf litter (*Eucalyptus obliqua*) of 15-50 mg phosphorus, 118-333 mg magnesium, 115-729 mg calcium, 47-146 mg potassium, and 16-42 mg sodium.

In addition to the inorganic nutrient inputs, falling leaves also add large quantities of organic matter and serve as substrate and shelter for the growth of many organisms indigenous to small stream ecosystems. Vannote's (1969) 24-gram leaf packs functioned as effective water filters, accumulating 8-10 gm ash-free dry weight of organic detrital particles less than 500 microns in diameter during a seasonal study; the energy content of a mixed deciduous leaf pack was reported to be 4655 cal/gm dry weight. Fisher (1970) reported atmospheric organic materials (mostly leaves) represented 44.2% of the total organic load entering a Hubbard Brook stream.

Swank and Helvey (1970) observed increased streamflows following the clear-cutting of a 16-hectare forest basin at Coweeta in 1939. After 23 years of regrowth, flows were still slightly higher than the precut base line. Hibbert (1969) noted that streamflows were inversely related to grass production on a Coweeta watershed and that, in general, grasses use less water than trees and brush. Thus, both logging operations and the conversion of forests to grass are accompanied by increased streamflow.

At Hubbard Brook, Likens et al., (1970) showed that clear-cutting a forested area increased the concentration of all major ions in the receiving stream except ammonia, sulfate and bicarbonate; sulfate and bicarbonate ions decreased. The resulting export of 75-97 metric tons dissolved solids per square kilometer per year, exclusive of organic matter, was 6-8 times greater than that reported for an undisturbed area. While stream turbidity did not increase significantly, particulate matter increased fourfold in mass and 26% in inorganic content.

Clear-cutting forested areas effectively eliminates leaf-fall inputs, changing not only the chemical load but also the organic habitats.

2. Agricultural sources. Agricultural activities affect both the sources and transport of nutrients from the watershed and are as complex as the terrestrial ecosystem itself.

The chemical characteristics of waters relate closely to the soil characteristics of the drainage basins and the soils themselves reflect geological, climatological, and topographical conditions. The wastes entering waters from agricultural areas are highly dependent upon local circumstances such as irrigation and fertilization practices, crop types, and animal populations. Bailey (1968) describes the important roles of soil type, particle size, pH, reduction potential, temperature, organic content, flow-through time and evaporation in nutrient retention on the soil. Total phosphorus concentrations in natural soils range from 100 to 2000 mg/l, averaging about 1000 mg/l. The inorganic phosphorus fraction varies from 25 to 97% of the total and normally ranges from 50 to 75% (Armstrong and Rohlich, 1969). Most phosphorus-bearing compounds react vigorously with the soil, thus preventing losses into groundwater until available adsorption sites have been filled (Bailey, 1969). Dissolved inorganic phosphorus concentrations in soil solutions given by Biggar and Corey (1969) seldom exceed 0.2 mg/l and normally range from 0.01 to 0.1 mg/l. They also state that phosphorus concentrations in soil solutions of subsoil layers are frequently less than 0.01 mg/l.

Biggar and Corey (1969), reviewing the literature on agricultural drainage, state that 95% or more of the soil nitrogen may occur in organic matter. Armstrong and Rohlich (1969) report that as much as 50% of this organic nitrogen may be present in the amino form. Organic nitrogen is converted to inorganic forms through a series of microbial metabolic processes resulting in the formation of ammonium (NH_4^+), nitrite (NO_2^-), and then nitrate NO_3^-) compounds. Ammonium ions are held on the cation exchange sites in soils so the concentration of ammonium in soil solutions should not be very high. Nitrate ions are highly soluble and thus are more subject to leaching and transport in groundwater. More than 99% of the soluble nitrogen was in the nitrate form in lysimeter studies in New York and California (Bailey, 1968).

Fertilizer usage in the United States increased approximately 3 million tons from 1967 to 1970 (Farm Chemicals, 1970) with primary nutrients reaching a level of 16.1 million tons and secondary and micronutrients 1.8 million tons. Nitrogen applications have increased thirtyfold in the past 23 years with a current application rate of 7 million tons per year for the entire United States (Martin, 1969). Local application rates vary according to the location and crop type; however, rates as high as 300 lb/N/year have been reported. During 1968 the consumption of P_2O_5 in fertilizers in the United States reached 4.6 million tons (Fertilizer, 1968) with application rates as high as 40 lb/P/year.

Vollenweider (1968) calculated the maximum average loss of nutrients to agricultural waters in the United States to be 2.5 gm N/m^2/year and 0.075 gm P/m^2/year.

Brezonic (1971), Bailey (1968), Smith (1967), Biggar and Corey (1969), and the FWPCA report, "Agricultural Practices and Water Quality" (1968) all serve as excellent source material for the mechanisms and quantities of nutrients originating from agricultural lands.

The practice of confining animals on feedlots results in tremendously concentrated sources of nutrients. Animal wastes in the United States are estimated to be as much as 1.6 billion tons per year (Wadleigh and Britt, 1969) with 50% of this amount originating from feedlots. One feedlot near Greeley, Colorado, holds approximately 90,000 cattle on 320 acres. Cattle are capable of producing 156 kg N and 17 kg P per 1000 kilograms body weight per year, with 1000 head of beef producing the equivalent organic load from 6000 people (Loehr, 1969).

The AWWA Task Group (1967) estimated that 60% of the nitrogen and 42% of the phosphorus entering the nation's watercourses came from agricultural lands, with nonagricultural lands contributing 8% of the nitrogen and 29% of the phosphorus.

Miner and Willrich (1969) report trace element composition of various types of animal manure; however, as indicated in Table 3, information is incomplete for minor nutrients found in agricultural drainage waters.

Table 3. Ranges of reported values for nutrients in agricultural wastewaters. (References are given in "Literature Cited - B")

Nutrient	Range (mg/l)	Nutrient	Range (mg/l)
TOC	- -	Cu	*7.6 - 40.8
N	5.0 - 25	Zn	*43.0 - 247
K	4.0 - 11	B	4.0 - 15.0
Ca	160 - 390	Na	620 - 2050
Mg	70 - 230	Mo	*0.84 - 4.18
S	1500 - 3900	Cl	310 - 640
P	0.13 - 0.33	Vn	- -
Fe	0.035 - 0.9	Co	*0.25 - 4.70
Mn	*75 - 549	SiO_2	- -

*Data obtained from mixed farmyard manures

3. Urban sources. On a total volume basis, municipal and industrial discharges represent the second most significant sources of nutrients for river ecosystems. Urban runoff and nonsewered wastes, such as septic tank discharges, may be significant in some areas.

Vollenweider (1968) provides a detailed discussion and literature review regarding phosphorus and nitrogen contents of various types of wastes. He states that recent sewage analyses indicate that the quantity of phosphorous is two to three times greater than the human physiological requirement and may represent detergent and water-softening reagent sources. He also states that concentrations of 1.5-3 gm P/capita/day and 12 gm N/capita/day were selected as average load values for domestic sewage in the United States.

The AWWA Task Group (1967) states that domestic sewage contributes 1.6 billion lb N/year and 0.5 billion lb P/year to the nation's surface waters.

Wollenweider (1968) also provides a table showing concentrations of BOD_5: nitrogen, P_2O_5, and phosphorus for a number of industrial sources. Highest reported values were 1,900 mg N/1 for distillery wastes and 274 mg P/l and 39,000 mg BOD/l for beet sugar factory wastes.

Weibel et al., (1966) found average values of 111 mg COD/l, 1 mg inorganic N/l and 1.1 mg total hydrolyzable P/l in urban stormwater runoff from 27 acres in Cincinnati, Ohio. The type of sewage system, i.e., separate or combined storm sewers, affects the nutrient inputs from urban runoff since sewage overflow contributions from combined sewers may be as high as 10%. Sylvester (1961) reports 208 µg/l total P and 2.01 µg/l Kjeldahl-N in highway runoff measured 30 minutes after the start of a heavy rain. He points out that sodium chloride used for de-icing roads during winter months may represent significant sources of this compound in some areas. Table 4 shows the nutrient content of urban waste waters.

Riverbed Sources. Man's activities, through pollution or poor land management, can directly affect nutrient sources, sinks, and transport relative to the riverbed. Such activities may also have both a direct and indirect effect on the biological populations involved in nutrient cycling.

Sveum (1970) reports an excellent example of sediment effects in the Cuyahoga River where the phosphorus concentrations in sediments three miles upstream from the Cleveland Harbor averaged 4 mg/gm dry weight. He speculates that the phosphorus in the river water was precipitated by the iron-containing wastes from local steel mills. Sediment COD of 270 mg/gm and nitrogen of 5 mg/gm were reported for the same area.

398

Table 4. Ranges of reported values for nutrients in urban wastewaters. (References are given in "Literature Cited - B")

Nutrient	Range (mg/l)	Nutrient	Range (mg/l)
TOC	--	Cu	0.04 - 0.54
N	7.0 - 18	Zn	0.07 - 1.09
K	7.0 - 15	B	0.4
Ca	15 40	Na	40 - 70
Mg	15 - 40	Mo	--
S	15 - 30	Cl	20 - 50
P	2 - 150	Vn	--
Fe	0.01 - 0.20	Co	--
Mn	--	SiO$_2$	--

Sediment deposits and heavy attached biological growth resulting from organic nutrients can cover and isolate the original riverbed materials, thus changing the geochemical reactions in the system. Sanders (1964) discussed the development of attached bacterial slimes on submerged surfaces and reported interactions between growth of slimes and the organic nutrient concentration of the water, temperature and water velocity. These biological slimes are important sinks and sources of nutrients in many river systems.

Advective River Sources. When the boundaries of river ecosystem do not include the headwaters or estuary, nutrient loads carried downstream into and out of the system may represent the major sources and sinks. Thus, the nutrient quality of the influent water represents a summation of all upstream ecological processes. Nutrient values reported in the available literature for rivers, lakes and springs are summarized in Tables 5, 6, and 7.

Concepts in Nutrient Cycling and Management

Nutrients originating from man's activities and those transported into a river ecosystem from all sources become part of the "nutrient pools." They may be classified as available or unavailable relative to their chemical state and the biochemical requirements of the organisms. Nutrient cycling then is defined as "the flow of essential materials between the nutrient pools, the biological compartments, and the abiotic environment." In this context, the rate and magnitude

Table 5. Ranges of reported values for nutrients in rivers of the United States. (References are given in "Literature Cited - B")

Nutrient	Range (mg/l)		Nutrient	Range (mg/l)	
TOC	1.0	- 12.0	Cu	0.001 - 0	.280
N	1.0	- 54.0	Zn	0.002 -	1.182
K	0.4	- 26.0	B	0.001 -	1.800
Ca	1.0	- 54.0	Na	1.9 -	79.0
Mg	0	- 45.0	Mo	0.001 -	1.100
S	6.0	- 12.0	Cl	1.0	- 118
P	0.002	- 5.040	Vn	0.001 -	0.300
Fe	0.001	- 0.952	Co	0.001 -	0.048
Mn	0.0003	- 3.230	SiO_2	3.0	- 18.2

Table 6. Ranges of reported values for nutrients in lake waters of the United States. (References are given in "Literature Cited - B")

Nutrient	Range (mg/l)		Nutrient	Range (mg/l)	
TOC	2.0	- 160.0	Cu	0.003 -	0.056
N	0.001	- 0.8	Zn	0.001 -	0.295
K	0.5	- 1.7	B	0.008 -	0.031
Ca	1.3	- 42.9	Na	0.002 -	12.2
Mg	5.7	- 8.0	Mo	0.001 -	0.011
S	9.3	- 27.5	Cl	6.6	- 118.0
P	0.001	- 0.39	Vn	0.001 -	0.010
Fe	0.006	- 0.084	Co	0.001 -	0.003
Mn	0.0008	- 0.004	SiO_2	0.46 -	2.5

Table 7. Ranges of reported values for nutrients in spring water. (References are given in "Literature Cited - B")

Nutrient	Range (mg/l)	Nutrient	Range (mg/l)
TOC	- -	Cu	- -
N	0.15 - 0.19	Zn	0.03 - 0.80
K	0.4 - 0.6	B	0.01
Ca	1.4 - 68.0	Na	1.2 - 2.9
Mg	9.6	Mo	- -
S	34.0	Cl	7.8
P	0.006	Vn	- -
Fe	0.02	Cu	- -
Mn	0.005 - 0.026	SiO_2	- -

of the exchange of nutrients between the "available pools" and the producers, consumers, and decomposers become extremely important. Odum (1963) states:

As with energy, it is evident that the rate of movement or cycling may be more important in determining biological productivity than the amount present in any one place at any one time.

In natural, "desirable" aquatic ecosystems the flow of energy and material is low and a high diversity of organisms exists in the biological compartments. However, as specific nutrients or pollutants are added, those organisms directly or indirectly stimulated will increase in number, whereas other species unable to compete or adapt may disappear completely. In cases where the materials are toxic, all biological life may be suppressed. Pollutants stress the system, disrupting the normal flow of energy and materials, and generally produce less desirable systems. Thus, it is necessary to define the structure, functions, and dynamics of desirable aquatic ecosystems and then manage and control environmental pollution to maintain them in the optimum state.

The current, most popular, concept relating to ecosystem management, originating about 130 years ago, states that biological productivity may be controlled by limiting one of the essential nutrients. While studying the effects of inorganic fertilizers on plant growth, Liebig (1940) expressed his now famous "law" of the minimum, which states that the rate of growth is dependent on the nutrient present in the minimum quantity in terms of need and availability. Liebig's law was modified by Taylor (1934) to include factors other than nutrients; Monod (1942), the French microbiologist working with bacteria, expanded the concept by developing a mathematical expression relating the specific growth rate of an organism to the concentration of an essential nutrient in a liquid medium.

$$k = Km \left(\frac{c}{K+c} \right)$$

k   =   specific growth rate

Km =   growth rate constant

c   =   concentration of limiting nutrient

K   =   saturation concentration for limiting nutrient

Monod's model was also modified by Herbert, Elsworth, and Telling (1956) and Rich (1963) who applied these concepts to continuous-flow microbial systems and added a concentration term for the basal metabolic requirement of the organism.

The basic concept and its many corollaries are useful when applied to simple systems where the organisms and environmental factors may be carefully analyzed and controlled. However, in a complex aquatic ecosystem application of the limiting concept to a single nutrient or factor is made difficult, if not impossible, by several practical considerations. Too little is known concerning organism tolerances, genetic potentials, selectivity, adaptability, and factor interactions. In addition, limitations are imposed by restrictive analytical capabilities.

The results from laboratory and field nutrient-cycling experiments, described below, demonstrate that:

1. The mechanisms involved in nutrient regulations are extremely difficult to isolate and quantify in mixed autotrophic-heterotrophic communities

2. Results from short-term nutrient enrichment experiments may provide erroneous information concerning biological stimulations or limitations when extrapolated to natural ecosystems

3. Apparent growth regulation of organisms in one compartment may be the result of secondary or tertiary reactions occurring within other compartments of the ecosystem. Direct cause-and-effect relationships can therefore be projected for a population only when the mechanisms have been properly identified

402

Kerr et al., (1970) reported data (Figure 2) showing the carbon requirement of the blue-green alga, *Anacystis nidulans*, over an 11-hour cell cycle. During this period the metabolic requirements for inorganic carbon varied from a minimum at the time of division to a maximum approximately seven hours into the cell cycle.

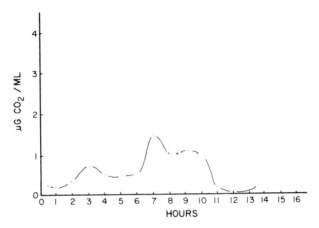

Figure 2. CO_2 uptake by the blue-green alga Anacystis nidulans, during one cell cycle

The total CO_2 requirement for growth and cell division was calculated to be 1×10^{-5} µg CO_2/cell by integration of one area under the demand curve. Laboratory screening indicated that *A. nidulans* does not have the capability to utilize internally stored carbon nor externally supplied organic carbon compounds for growth measured by cell division. According to Chang and Tolbert (1970), who followed the metabolic path of $^{14}CO_2$ in *Ankistrodesmus braunii* (Figure 3), as much as 70% of the inorganic carbon fixed by the cells was excreted as organic carbon by-products during a cell cycle. Since the rate of excretion depended upon the physiological state of the cells, it varied throughout the cell cycle.

On the other hand, the phosphorus taken in by algal cells is used primarily in cell reproductive processes and little, if any, is excreted during the active life of the cells. Most algae also have the capability of storing excess supplies of phosphorus, when available, for internal use during periods when external supplies are minimal or below actual metabolic requirements. Research by Kerr et al., (1970) on the uptake and storage of phosphorus by *A. nidulans* (using ^{32}P as an isotope tracer) (Figure 4) indicates that, whereas 0.3×10^{-8} µg phosphorus per cell is the minimum cellular concentration permitting cell division, these cells can take up and store a maximum of 3×10^{-8} µg phosphorus per cell prior to division. Those cells that had stored the maximum amount of phosphorus were able to continue through three cell divisions when placed in a growth medium devoid of phosphorus.

403

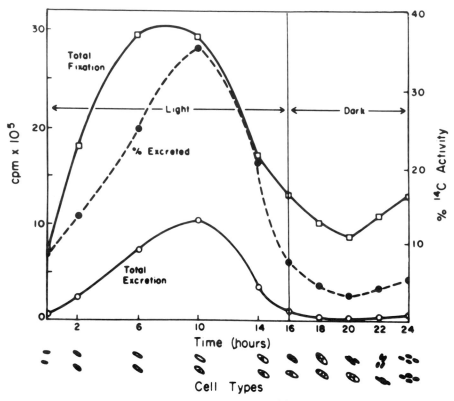

Figure 3. Changes in the activity of photosynthetic $^{14}CO_2$ fixation and ^{14}C excretion during the life cycle of A. braunii. □ : ^{14}C activity fixed in 5 min/ml 1% cell suspension; O: ^{14}C activity excreted in $\overline{5\ min}$/ml 1% cell suspension; ●: percentage ^{14}C excreted of the total ^{14}C fixed. (Chang and Tolbert, Plant Physiol., Vol. 46, 1970, p. 380)

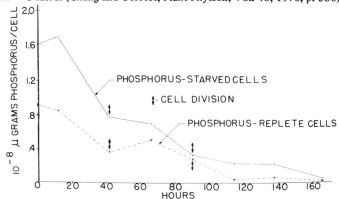

Figure 4. Utilization of stored phosphorus during growth of
Anacystis nidulans in Tris medium

Using the carbon and phosphorus values for *A. nidulans* measured by Kerr et al., the ratio of inorganic carbon required to sustain a cell through division to the minimum cellular phosphorus concentration permitting cell division is 10,000 to 1. The ratio of the carbon requirement to the maximum stored cellular concentration of phosphorus is 1,000 to 1. These values greatly exceed the normally accepted 100 to 1 ratio of cellular carbon to phosphorus.

It is true that the carbon and phosphorus requirements for algae will vary among species and even among locally selected populations within the same species. However, there is an urgent need to better identify metabolic mechanisms and reevaluate the requirements of the organisms which must be controlled in order to maintain desirable aquatic ecosystems.

Laboratory nutrient competition experiments were performed by Kerr et al., (1970) in which natural populations of heterotrophic and autotrophic organisms were cultured together in sterilized pond water (containing less than 5 µg/l of nitrogen and phosphorus). They found that additions of inorganic nitrogen, phosphorus, and potassium stimulated algal growth within 30 hours (Table 8). The experiments also showed, however, that a similar stimulation was achieved by adding only glucose (an organic carbon source considered ineffective in promoting algal growth) to the system. Detailed analysis of the bacterial growth, inorganic carbon production, and individual community stimulation studies clearly demonstrated that increased algal growth resulted from a symbiosis with the bacteria in the system. Additions of both organic and inorganic nutrients stimulated the growth of the bacterial community, increasing the rate of conversion of organic carbon to inorganic carbon which, in turn, resulted in an overall increase in the number of algae. It is hypothesized that different bacterial populations within the heterotrophic community were regulated by each added nutrient. When both materials were added together, the resulting increase in population size was greater than the sum of the two populations obtained when the materials were added separately. This indicates that for any given ecosystem composed of several populations within each biological compartment, more than one regulating nutrient may be operating at the same time, each affecting different organisms but all producing similar total system responses.

To amplify these laboratory studies, Kerr et al., (1971) reported on a small pond dosed with the same proportions of reagent-grade inorganic nitrogen, phosphorus, and potassium (20-20-5 fertilizer) used in the previous laboratory experiments. The measured total bacterial and algal populations as a function of time are shown in Figure 5. The nutrients were added on the 17th day and the bacteria in the water responded immediately. However, maximum algal growth did not occur until four days after the initial bacterial stimulation. The authors

Table 8. Growth of indigenous bacterial and algal populations in an infertile pond water (5 µg/l each of N and P) following additions of organic carbon and/or inorganic fertilizer

Medium	Total Bacterial Count/ml		mg CO_2 + HCO_3^-		Total Algal Count/ml	
	0 Hour	30 Hours	0 Hour	30 Hours	0 Hour	30 Hours
Pond water	27.5×10^5	8.5×10^4	28.0	24.0	2.0×10^5	1.6×10^5
4 g 20-20-5/1	27.5×10^5	*	26.3	35.9	2.0×10^5	6×10^5
2.5 g glucose/1	27.5×10^5	40×10^5	25.3	45.4	2.0×10^5	5×10^5
4 mg 20-20-5/1 2.5 g glucose/1	27.5×10^5	*	17.3	22.6	2.0×10^5	12.5×10^5

*Too numerous to count

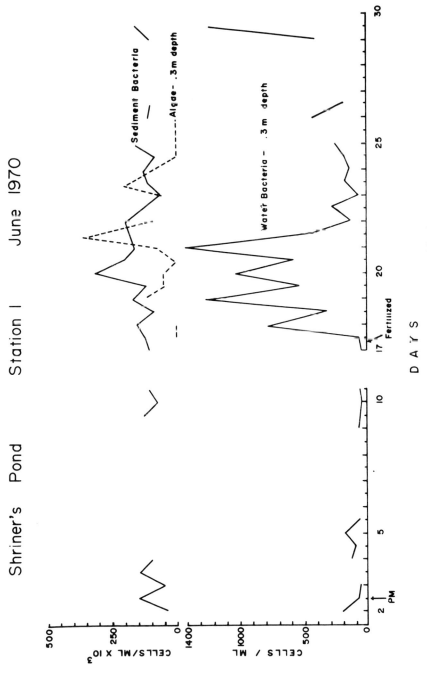

Figure 5. Bacterial and algal growth after fertilization

407

reported that the increased bacterial growth was accompanied by increases in the CO_2 concentration in the pond water. The corollary nutrient regulation reported by Brockway et al., (1971) indicated that when inorganic carbon as gaseous CO_2 was introduced continuously into a plastic enclosure (10 ft x 3 ft x water depth) located in the same pond, the algae responded immediately and the number of cells in the enclosure increased to a maximum within three days (Figure 6). The bacteria, however, showed little response until six days into the experiment. This experiment demonstrates further the symbiotic relationship between biological communities and the secondary or indirect nutrient effects which must be considered for ecosystem management.

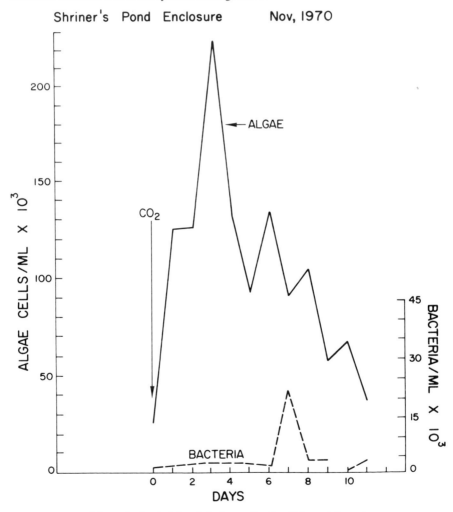

Figure 6. Bacterial and algal growth after CO_2 enrichment

Conclusions

Effective water quality management requires the consideration of streams and rivers on a total ecosystem basis, including all of the significant biological compartments, the abiotic environment and their dynamic interrelationships. Within the various compartments the wide diversity of organisms differ in their nutrient requirements, their tolerances for various environments, their genetic potential for adaptation and selection and their symbiotic relationships. It may therefore be expected that additions of several nutrients to the same system will result in similar overall biological responses even though individual species of organisms will respond in different ways. Thus, to achieve and maintain desirable aquatic ecosystems all nutrients including organic energy sources must be maintained at an optimum level.

Acknowledgments

I wish to acknowledge the assistance of Mr. Bruce Ferguson, Microbiologist, National Pollutants Fate Research Program, in compiling the ranges of nutrient values for different waters, and Mr. David Lewis, Microbiologist, National Pollutants Fate Research Program, for his artistic and drafting services.

Literature Cited - A

Armstrong, D. E., and G. A. Rohlich. 1969. Effects of agricultural pollution on eutrophication, p. 314-330. *In* Agricultural practices and water quality, proc. of a conf. concerning the role of agriculture in clean water. Fed. Water Pollution Contr. Admin., U.S. Dept. Int. Water Pollution Contr. Res. Series 13040EYX.

Attiwill, P. M. 1968. The loss of elements from decomposing litter. Ecology 49(1):142-145.

Bailey, G. W. 1968. Role of Soils and Sediment in Water Pollution Control. Part I. Reactions of Nitrogenous and Phosphatic Compounds with Soils and Geologic Strata. Fed. Water Pollution Contr. Admin. U.S. Dept. Int., Southeast Water Lab., Athens, Georgia. 90 p.

Biggar, J. W., and R. B. Corey. 1969. Agricultural drainage and eutrophication, p. 404-445. *In* Eutrophication: causes, consequences, correctives, National Acad. Sci., Washington, D.C.

Bormann, F. H., G. E. Likens, and J. S. Eaton. 1969. Biotic regulation of particulate and solution losses from a forest ecosystem. Biosci. 19(7):600-610.

Brezonik, P. L. 1971. Nitrogen: sources and transformations in natural waters. Presented at the 161st National American Chem. Soc. Meeting, Los Angeles, Calif., March 28-April 2, 1971.

Brockway, D. L., D. F. Paris, P. C. Kerr, and J. T. Barnett, Jr. 1971. Biological responses following nutrient additions to a small pond. Presented at the 26th Ann. Purdue Industr. Waste Conf., Purdue University, Lafayette, Ind., May 4-6, 1971.

Chang, W., and N. E. Tolbert. 1970. Excretion of glycolate, mesotartrate and isocitrate lactone by synchronized cultures of *Ankistrodesmus braunii.* Plant Physiol. 45:377-385.

Federal Water Pollution Control Administration. 1969. Agricultural Practices and Water Quality, Proceedings of a Conference Concerning the Role of Agriculture in Clean Water. FWPCA, U.S. Dept. Int. Water Pollution Control Res. Series 13040EYX. 407 p.

Farm Chemicals Handbook. 1970. Gordon L. Berg (Ed.). Meister Publishing Co., Willoughby, Ohio. 496 p.

Fisher, D. W., A. W. Gambell, G. E. Likens, and F. H. Bormann. 1968. Atmospheric contributions to water quality of streams in the Hubbard Brook Experimental Forest, New Hampshire. Water Resources Res. 4(5):1115-1126.

Fisher, S. G. 1970. Annual energy budget of a small forest stream ecosystem: Bear Brook, West Thornton, New Hampshire. PhD thesis, Dartmouth College, Hanover, New Hampshire. 97 p.

Heaney, J. P., and R. H. Sullivan. 1971. Source Control of Urban Water Pollution. J. Water Pollution Control Fed. 43(4):571-579.

Herbert, D., R. Elsworth, and R. C. Telling. 1956. The continuous culture of bacteria. J. Gen. Microbiol. 14:601-622.

Hibbert, A. R. 1969. Water yield changes after converting a forested catchment to grass. Water Resources Res. 5(3):634-640.

Johnson, N. M., G. E. Likens, F. H. Bormann, D. W. Fisher, and R. S. Pierce. 1969. A working model for the variation in stream water chemistry at the Hubbard Brook Experimental Forest, New Hampshire. Water Resources Res. 5(6):1353-1363.

Kerr, P. C., D. L. Brockway, D. F. Paris, and W. M. Sanders III. The carbon cycle in aquatic ecosystems. Presented at the 161st National Amer. Chem. Soc. Meeting, Los Angeles, Calif., March 28-April 2, 1971.

Kerr, P. C., D. L. Brockway, D. F. Paris, and J. T. Barnett, Jr. The interrelation of carbon and phosphorus in regulating heterotrophic and autotrophic populations in an aquatic ecosystem, Shriner's Pond. Presented at a symposium of the American Soc. of Limnol. and Oceanogr., Kellogg Biological Sta. (Michigan State Univ.), February 10-13, 1971.

Kerr, P. C., D. F. Paris, and D. L. Brockway. 1970. The interrelationships of carbon and phosphorus in regulating heterotrophic and autotrophic populations in aquatic ecosystems. Fed. Water Qual. Admin., U.S. Dept. Int. Water Pollution Contr. Res. Series 16050 FGS 07/70. Supt. of Documents, U.S. Govt. Printing Office, Washington, D.C. 53 p.

Liebig, Justus. 1840. Chemistry in its application to agriculture and physiology. Taylor and Walton, London (2nd Ed., 1842). 400 p.

Likens, G. E., F. H. Bormann, N. M. Johnson, D. W. Fisher, and R. S. Pierce. 1970. Effects of forest cutting and herbicide treatment on nutrient budgets in the Hubbard Brook watershed ecosystem. Ecol. Monographs 40:23-47.

Loehr, R. C. 1969. Animal wastes—a national problem. Amer. Soc. Civil Engineers, J. Sanitary Engrg. Div. 95(SA2):189-221.

Martin, W. P., W. E. Fenster, and L. D. Hanson. 1969. Fertilizer management for pollution control. p. 142-158. *In* Agricultural practices and water quality, proceedings of a conference concerning the role of agriculture in clean water. Fed. Water Pollution Control Admin., U.S. Dept. Int. Water Pollution Contr. Res. Series 13040EYX.

Matheson, D. H. 1951. Inorganic nitrogen in precipitation and atmospheric sediments. Canad. J. Technol. 29:406-412.

411

Miner, J. R., and T. L. Willrich. 1969. Livestock operations and field-spread manure as sources of pollutants. p. 231-240. *In* Agricultural practices and water quality, proceedings of a conference concerning the role of agriculture in clean water. Fed. Water Pollution Contr. Admin., U.S. Dept. Int., Water Pollution Contr. Res. Series 13040EYX.

Monod, J. 1942. Recherches sur la croissance des cultures bacteriennes. 1958. edition, 110 p. Hermann and Cie, Paris.

Odum, E. P., in collaboration with H. T. Odum. 1959. Fundamentals of Ecology, 2nd Ed. W. B. Saunders Co., Philadelphia. 546 p.

Odum, E. P. 1963. Ecology. Holt, Rinehard & Winston, Inc., N.Y. 152 p.

Rich, L. G. 1963. Unit Processes of Sanitary Engineering. John Wiley and Sons, New York, N. Y. 190 p.

Sanders, W. M., III. 1964. Oxygen utilization by slime organisms in continuous culture. Proc. 19th Indus. Waste Conf., Part Two, Engrg. Ext. Series No. 117, 986-1010, Purdue Univ., Lafayette, Ind.

Smith, G. E. 1967. Fertilizer nutrients as contaminants in water supplies, p. 173-186. *In* Agriculture and the quality of our environment. A Symposium presented at the 133rd Meeting of the Amer. Assoc. Advancement Sci., December 1966. AAAS Publication 85. Nyle C. Brady (Ed.). 460 p.

Sveum, D. L. 1970. The quantity and quality of sediments deposited in Cleveland harbor at Cleveland, Ohio. Paper No. 8. Proc. seminar on sediment transport in rivers and reservoirs, April 7-8, 1970. The Hydrologic Engrg. Center, Corps of Engineers, Davis, Calif.

Swank, W. T., and J. D. Helvey. 1970. Reduction of streamflow increases following regrowth of clearcut hardwood forests, p. 346-360. *In* IASH-UNESCO symposium on the results of research on representative and experimental basins, Wellington, N. Z., December 1970.

Sylvester, R. O. 1961. Nutrient content of drainage water from forested, urban and agricultural areas. *In* Algae and metropolitan wastes, U.S. Publ. Health Svc., SEC TR. W61-3:80.

Task Group 2610P. 1967. Sources of nitrogen and phosphorus in water supplies. J. Amer. Waterworks Assoc. 59:344-366.

Taylor, W. P. 1934. Significance of extreme or intermittent conditions in distribution of species and management of natural resources, with a restatement of Liebig's Law of the Minimum. Ecol. 15:274-379.

Tsivoglou, E. C., J. B. Cohen, S. D. Shearer, and P. J. Godsil. 1968. Tracer measurement of stream reaeration. II. Field studies. J. Water Pollution Contr. Fed. 40(2), Part I, 285-305.

U.S. Dept. Health, Educ. and Welfare. 1962. Air Pollution Measurements of the National Air Sampling Network. Pub. Health Svc. Publ. No. 978. 217 p.

Vannote, R. L. 1969. Detrital consumers in natural systems. Tech. Report No. 7, W. K. Kellogg Biolog. Sta. and Inst. of Water Res., Boston, Mass.

Vollenweider, R. A. 1968. Scientific fundamentals of the eutrophication of lakes and flowing waters, with particular reference to nitrogen and phosphorus as factors in eutrophication. Organisation for Economic Cooperation and Development. DAS/CSI/68.27, Paris.

Wadleigh, C. H., and C. S. Britt. 1969. Issues in food production and clean water, p. xix-xxvii. In Agricultural practices and water quality, proceedings of a conference concerning the role of agriculture in clean water. Fed. Water Pollution Contr. Admin., U.S. Dept. Int. Water Pollution Contr. Res. Series 13040EYX.

Weibel, S. R., R. B. Weidner, A. G. Christianson, and R. J. Anderson. 1966. Characterization, treatment, and disposal of urban stormwater, p. 329-343. In Third international conference on water pollution research, proceedings. (Munich, Germany.) Water Pollution Contr. Fed., Washington, D. C.

Weibel, S. R., R. D. Weidner, J. M. Cohen, and A. G. Christianson. 1966. Pesticides and other contaminants in rainfall and runoff. J. Amer. Waterworks Assoc 58(8):1075-1084.

Literature Cited - B

Agriculture and the Quality of our Environment. 1967. Nyle C. Brady (Ed.). Amer. Assoc. Advancement Sci. Publ. 85. The Plimpton Press, Norwood, Mass. 460 p.

Brezonik, P. L., W. H. Morgan, E. E. Shannon, and H. D. Putnam, 1969. Eutrophication factors in north central Florida lakes. Bull. Ser. No. 134. Water Resources Res. Cen. Publ. No. 5. Florida Engrg. and Industr. Exp. Sta. Univ. of Florida, Gainesville.

Federal Water Pollution Control Administration. 1968. Lake Erie surveillance data summary 1967-1968. FWPCA, U.S. Dept. Int., Great Lakes Region, Cleveland Program Office. 65 p.

―――――――. 1968. Eutrophication of surface waters - Lake Tahoe, bioassay of nutrient sources. FWPCA, U.S. Dept. Int. First Progr. Rept., Grant No. WPD-48. Lake Tahoe Area Council, South Lake Tahoe, Calif.

―――――――. 1969. Agricultural practices and water quality, proceedings of a conference concerning the role of agriculture in clean water. FWPCA, U.S. Dept. Int. Water Pollution Contr. Res. Series 13040EYX.

―――――――. 1969. Feasibility of joint treatment in a lake watershed. FWPCA, U.S. Dept. Int. Water Pollution Contr. Res. Series DAST-17. 113 p.

Harriss, R. C. 1967. Silica and chloride in interstitial waters of river and lake sediments. Limnol. Oceanogr. 12:8.

International Lake Erie Water Pollution Board and the International Lake Ontario-St. Lawrence River Water Pollution Board. 1969. Report to the International Joint Commission on the pollution of Lake Erie, Lake Ontario and the international section of the St. Lawrence River. Vol. 2, 316 p.; Vol. 3, 329 p.

Johnson, N. M., G. E. Likens, F. H. Bormann, and R. S. Pierce. 1968. Rate of chemical weathering of silicate minerals in New Hampshire. Geochimica et Cosmochimica Acta 32:531-545.

Kopp, J. F., and R. C. Kroner. Released 1969. Trace metals in waters of the United States. A five-year summary of trace metals in rivers and lakes of the United States (Oct. 1, 1962 - Sept. 30, 1967). Fed. Water Pollution Contr. Admin., U.S. Dept. Int., Div. Pollution Surveillance, Cincinnati, Ohio. 32 p.

Lake Tahoe Area Council. 1963. Comprehensive study on protection of water resources of Lake Tahoe basin. Prepared by P. H. McGauhey, R. Eliassen, G. Rohlich, H. F. Ludwig, and E. A. Pearson, Consulting Engineers, Engineering Science, Inc. 157 p.

Martin, J. B., B. N. Bradford, and H. G. Kennedy. 1970. Factors affecting growth of *Nujas* in Pickwick Reservoir. Nat. Fertilizer Develop. Center, Tennessee Valley Auth., Muscle Shoals, Ala. Bulletin Y-11. 47 p.

Reid, G. K. 1961. Ecology of inland waters and estuaries. Reinhold Publishing Corp., New York. 375 p.

Tamblyn, T. A., P. L. McCarty, and P. P. St. Amant. 1969. Bacterial denitrification of agricultural tile drainage. *In* Collected papers regarding nitrates in agricultural wastewater. Fed. Water Qual. Admin., U.S. Dept. Int. Water Pollution Contr. Res. Series 13030 ELY. 185 p.

A COMMENTARY ON "EFFECTS OF RIVER USES"

T. P. VandeSande, Session Chairman

After a long history of river use for the benefit of a small portion of the population, river quality has now deteriorated to an extent that the public can no longer ignore. The public believes that with present technology the disruption of the river ecology can be minimized and that the society is responsible for the cost of this corrective effort—a cost that is fiscally within reason. Man is beginning to show a concern for all the uses of a river and to develop the knowledge necessary for an evaluation of the effects of these uses.

You, as members of the American Fisheries Society, are a special group with sufficient biological education and experience to make ecological forecasts of the effects of river use. You have a responsibility to protect the natural values of public waters and you also have a responsibility to the public. You must make the public aware of that which has occurred in the past, of the changes presently taking place, and of the alternatives available for the future.

SESSION IV

RATIONALIZATION OF MULTIPLE USE
Chairman: David J. Allee

RATIONALIZATION OF MULTIPLE USE OF RIVERS

John Cairns, Jr.

On Friday, October 30, 1970, the Virginia Tech Daily Bulletin carried a notice of an Ecofestival, billed as a get-together with nature, to be held on Saturday, October 31, 1970, in the amphitheater near the Duck Pond—in case of bad weather the meeting would be held in Burruss Auditorium. Evidently the arrangers of the Ecofestival saw nothing incongruous in their willingness to meet nature only under suitable conditions and, when these conditions became unsuitable, to retreat without hesitation into the shelter of one of the products of the industrial society against which some of them at least were protesting. (It rained all day and the meeting was held in Burruss Auditorium!) Apparently even the most dedicated environmentalists are not always willing to take nature as it is, but frequently wish to modify it to a more amenable state. There is, of course, a considerable difference between modifying it to provide shelter from rain and discharging a toxic material into a stream. Nevertheless, it is well to realize that most of us are so accustomed to modifying nature toward certain ends that we are not even aware that we take this for granted. Therefore the basic question is not whether or not to modify the natural environment, but to what degree it should be modified and by whom.

Most people seem to feel that the environment should be modified to optimize the benefits to the human race. A small but vocal minority are now saying that man should not be the measure of all ecological problems, but that he should take into consideration the well-being of many other species as well. The focal point does not seem to be whether or not the welfare of mankind should be the primary consideration, but rather whether some temporary sacrifices of human well-being to preserve certain other species is in the long-term best interests of mankind. Thus the members of the Sierra Club feel that it is in the best interest of mankind to preserve life spans of redwood trees and they oppose those persons who would utilize the stands of redwoods for picnic tables, fence posts, and the like. If we accept that the well-being of mankind should include both short-term and long-term goals and that sometimes these will be in conflict we are at least in a position to begin considering environmental use.

Unfortunately, the decision-makers are now in a position somewhat comparable to that of a spaceship captain who has taken off on an interplanetary voyage with too many passengers and little control over their use of the spaceship. If we could easily solve existing environmental problems without affecting any people or industries, this conference would not be necessary. Our conference goal is to suggest ways to make maximum beneficial use of the environment while preserving as much as possible in satisfactory condition for continuing use and at the same time restoring as rapidly as possible those areas that are so badly damaged that they are either useless or dangerous.

A common view of multiple use is a group of industries, farmers, sportsmen, outdoor recreation enthusiasts, municipalities, and federal governmental agencies, all cooperatively using a single water supply without encroaching upon the rights of others. Unhappily, this utopian situation does not now exist anywhere and is not likely to exist. Even if technology and management practices for preserving water quality were exemplary throughout all the user groups, and capital for adequate treatment were readily available whenever necessary, it would still be virtually impossible to use water for one purpose without interfering with other beneficial uses. Any use of the water modifies either its quality, quantity, or esthetic characteristics; and frequently modifies all three of these. It is highly unlikely that these modifications would enhance other uses, and it is quite likely that they would detract from other uses. For example, a steam electric-power plant will discharge water at a temperature somewhere around 21 to 14 degrees above ambient, and even the chemical quality will be altered slightly (e.g., the dissolved oxygen concentration generally decreases as temperature increases). The quantity of the water may be slightly affected by evaporative loss, particularly if some of the heat is removed with evaporative cooling towers before the coolant water is returned to the stream. The increased temperature of the water may have little or no effect upon the aquatic community or it may substantially alter it by excluding heat-sensitive fishes, through effects on the metabolic activities, and even survival, of aquatic organisms. Because the power company likes to have a dependable supply of water, a dam may be built to insure regular flow, further altering the characteristics of the water in that drainage basin. Although dams are commonly used for flood control, there is often a conflict between retaining the water pool at maximum full level to insure a dependable supply of coolant water, and the maintenance of a reserve capacity in the dam to counteract sudden surges of water due to floods. Even in a simple case such as this, one must order the priorities for that region and resolve the conflicts in such a way that overall management of the entire system becomes possible. We are now ordering priorities through the adversary system, which usually polarizes the entire region on each issue, rather than focusing attention on an overall management program.

422

Relation to Human Population Growth

At the December 1970 annual meeting of the American Association for the Advancement of Science, Professor Barry Commoner stated his belief that population was not a major factor in the environmental problems of the United States, and therefore we should address ourselves to controlling pollution and changing the life style that leads to pollution (*Time*, Jan. 11, 1971, p. 56). Professors Paul R. Ehrlich of Stanford University and Garrett Hardin of the University of California in Santa Barbara took the opposing view that the population problem is a major determinant in the environmental crisis, but they agreed with Professor Commoner that our life style needs to be changed. For example, the use of energy per person in the United States, which has recently risen at a spectacular rate, along with the less dramatic, but nevertheless significant increase in our population, have produced an astonishing demand for electrical energy. Large-scale "brown-outs" and occasional power failures for large regions, both of which have already occurred, may become commonplace in the future if these trends continue unabated.

Even if each woman in the United States has only two children, this will not immediately produce zero population growth. If the breeding part of the population becomes committed to two children per family (something not likely, according to recent surveys), the population in the United States will probably stabilize at around 300 million. Therefore, even the most optimistic planners should develop plans based on a population in the United States ranging between 300 and 350 million. If this figure is too high, the planners will merely have provided an environment in which the population does not exist on the edge of either disaster or major inconvenience.

A substantially larger population than the 205 million now inhabiting the United States will undoubtedly result in increased environmental stress even if our life style undergoes drastic changes. We are, after all, in an industrial society with most of our population living in urban or suburban areas, dependent on a vast transportation system to deliver fuel, goods, building materials, and raw materials and to transport people. Three hundred million people living on the land mass now occupied by 205 million cannot abandon mechanized farming, nor can they eliminate the demands for energy and raw materials, although these could be substantially reduced. More people and more industry on the same land mass means more multiple use.

Relation to Maintenance of Industrial Society

Pollution is in many ways like sin. Everyone is against it in other people but sees the same actions as virtues in oneself. For example, owners of some of the older industrial plants will admit, when pressed, that they are polluting the environment. They rationalize this sin by pointing out that they employ large numbers of people who would be out of work if the plant were to shut down, which would happen if it were forced to spend the money necessary for the improved pollution abatement. Their motives could be humanistic, or merely an attempt to squeeze the last dollar of profit from an obsolescent plant. Since a case similar to the one just described is probably known in almost every county in the United States, it is quite evident that we cannot always depend on good local stewardship to save the environment—at least not where there is a "muck-is-money" attitude. The only way we can satisfactorily cope with environmental problems, whether they be technological or political, is to deal with them on a regional basis in such a way that all of the complex interactions and consequences of environmental pollution are evident to the people of that region. Clearly showing the costs paid by society would probably result in closing an obsolete plant, rather than permitting it to operate as a "social benefit." If we are to maintain a viable industrial society with no frontiers to exploit, these "hidden" costs must be incorporated realistically into the basic costs of industrial production. Rational multiple-use programs must be based on realistic cost accounting which includes environmental costs as well as the ones traditionally used. These attempts will be sloppy at first but will improve rapidly—the "pocketbook nerve" is the most sensitive nerve humans have.

Management for Multiple Use

All industries discharging waste are using the environment as an extension of their waste-disposal system. If this were not true, each industry could recycle waste water since its quality would be equal to that of the water entering the plant. Once this simple fact is recognized and accepted by the general public and industrial management groups, a new perspective on our relationship with environmental problems is possible. Conservationists may not like industrial and municipal use of the environment as a receiving system for wastes, but this situation is the rule rather than the exception in all industrialized societies and many developing societies as well. However, each group must compromise if we are to have a gradual transition from our present environmental situation to a more amenable one.

If we are willing to regard the environment as an extension of the plant waste-disposal system, then it seems quite reasonable to extend the quality-control practices which exist within the plant to the area outside the plant. The essence of quality control inside the plant is the continuous or regular collection of data regarding the process quality at different points in the operation and alteration of the management practices as necessary to maintain the desired quality. The same principle can be applied in the natural environment, although the techniques would be somewhat different. For example, in the figure below there are industrial, agricultural, and wooded areas, as well as one small city. There are two reservoirs marked X and Y in the upstream areas and several small tributaries, one running through an agricultural area containing two fruit orchards and a hayfield, and another running adjacent to a hayfield which is next to Industry 3. Two types of biological or environmental monitoring systems (Cairns, et al., 1970) could be used: (1) An in-plant system uses a mixture of the plant waste and water from the receiving river to see if this will prove harmful to one or more representative aquatic organisms. The in-plant monitoring systems are designed to tell something about the biological quality of the wastes before they actually enter the receiving river or stream. (2) The in-stream monitoring units provide data from the receiving system itself because our ultimate objective is to protect it. In-stream monitoring units should utilize ecological data that are derived from the complex interlocking cause-effect pathways characteristic of natural systems. At present these are based on diversity indices or modifications of these (one infers that a high diversity of species indicates a large number of cause-effect pathways), although ideally they should be based on the function of the system as well as on the structure. The justification for maintaining the diversity characteristic of a particular locale has been discussed elsewhere (Cairns, 1967).

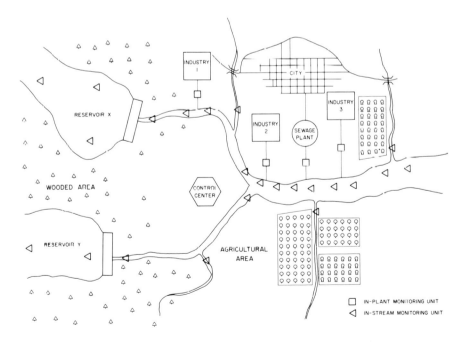

Mixed industrial and agricultural area

Data obtained from a series of these in-plant and in-stream monitoring units could be fed into a central data control center to obtain correlations between changes in the inputs into the system and changes that have occurred in the system itself. Eventually, prediction models could be made that would be useful to planners; even in the initial stages the system would be better than those now existing.

There are a number of immediate benefits even if only isolated industries use in-plant and in-stream monitoring systems. For example, if an industry with an operative in-plant monitoring unit had a signal which indicated the appearance of some deleterious substance in the waste stream, it would have several options; (1) shunt the waste to a holding pond, (2) recycle the waste for further treatment, or (3) reduce operations (and therefore waste volume) sufficiently to eliminate the unfavorable signal. These options could be used either singly or in combination and would, even without the large information system shown in the figure, provide additional safety to the receiving system by insuring that the wastes from that industry were not lethal to aquatic organisms. However, if the entire drainage basin were being operated as a system, there would be additional options when an unfavorable signal appeared. Release from the upstream reservoirs to dilute the waste further so that it would not reach harmful concentrations in the river. This operation could be simulated in the in-plant waste system with a series of in-plant monitoring units at different dilution ratios, some of which would have a signal because the concentration of waste going through that particular unit was high, and others of which would not have the signal because the concentration of waste was low enough. These latter concentrations would be matched in the receiving stream by increasing the release from upstream reservoirs. Another option would be to decrease discharges from other industries in the system since the deleterious effects of wastes upon receiving systems are more likely to act collectively than individually.

Both in-plant and in-stream monitoring units would be of considerable economic value to individual industries. Modifications of waste-treatment design and plant operations could also be made much more efficiently since a continuous flow of information about the biological quality of the waste would enable management to identify those periods of operation, as well as those particular types of operations, most likely to cause trouble.

The in-stream system can serve additional management purposes, even for a single industrial plant. If the monitoring system is installed before the plant begins operation, it will show the present condition of the receiving system—it is extremely important to establish the baseline conditions, whatever they may be. Also, the establishment of one or more reference or control stations upstream

426

from the effects of the plant discharge would enable a fairly reliable estimate to be made of the effects of upstream discharges, operations, and other existing deleterious effects. Thus, if a major spill occurred upstream from the industry, this would be detected by an upstream monitoring unit. A further advantage is that regular information about the quality of the biota in the receiving stream would enable the plant to make prompt corrections in its waste treatment system if, in fact, it turned out to be inadequate. The company could build up a series of waste control management practices based on information feedback from both in-stream and in-plant units which would develop correlations between waste quality and biological changes. This should substantially increase the efficiency of the waste treatment, as the industry could identify those particularly deleterious practices and periods and make certain that the waste-treatment dollar was spent most effectively.

As was the case with the in-plant units, still further advantages would ensue from the in-stream monitoring units if the entire drainage basin were operated as a single quality-control unit. For example, if all of the industries, municipalities, and other groups affecting the stream treated the maintenance of environmental quality as a collective problem, it might be desirable for Industry 3 to pretreat its process water by partially subsidizing the waste-treatment facilities of Industry 1. If Industry 2 or 3 were discharging zinc, and the receiving capacity of the stream for zinc was significantly lowered by the effect of poorly treated sewage from the sewerage plant lowering the dissolved oxygen concentration of the stream, the chances are that the industries involved would be less indifferent to the performance of the local sewage-treatment plant than most are now.

Viewed as a problem in systems optimization, spaceship earth is not very well run. Everyone from fishermen to industrial managers expects to make "free use" of the environment in a time when quality environments are in scarce supply. In addition, environmental demands are greater than the resources available to meet them. Therefore we must reduce the demands by developing specific conditions under which use may occur and prohibiting any use which does not meet these conditions. In the not-too-distant future a charge for all kinds of environmental use seems inevitable.

Social attitudes will be very important in determining how these changes are implemented. The "fish hog" has the same basic attitude as the industrial polluter. Both will get as much as they can of the environmental resources without regard for the rights of others. Because social changes occur so slowly and technological changes occur with frightening rapidity, we cannot count on the educational process to supply the motivation necessary to correct these problems. We must, therefore, divide the environmental "pie" as equitably as possible, recognizing that controls are necessary in order to insure a fair distribution. A carefully programmed multiple use of our scarce environmental resources will insure a more equitable distribution of these resources than is currently possible.

It will not be easy to change from the frontier philosophy of life upon which our country was founded to the spaceship philosophy which it must have to permit a quality life in the future, since it is especially difficult for people to accept restraints upon uses of the environment which have been free traditionally. The question is not whether regulation and multiple use are necessary, but rather what form these should take.

Forces Antagonistic to Multiple-Use Concepts

Other than the natural inclination of humans to resist the restrictions upon their generally accepted "rights," there are a number of other forces which are antagonistic to well-programmed multiple use. A few of these follow:

(1) Conflicts between government agencies. These result from a number of causes, but the primary force seems to be resistance to change and blind following of already established programs. For example, the U.S. Department of Agriculture pays farmers to take cotton land out of production and recommends and tolerates the use of sprays to protect the cotton from insect pests on the lands that are in use. At the same time Fisheries and Wildlife, U.S. Department of the Interior, is worried about the effects of pesticides upon fish, and the Food and Drug Administration has condemned fish with concentrations of pesticides considered unsafe for human consumption. Each of these organizations would feel threatened if their programs were drastically altered in the interest of an overall environmental plan and would probably resist vigorously. Yet this is exactly what must happen in order to achieve effective, rational multiple-use programs.

(2) The poverty/environment conflict. Recently, in the Carolinas the German industry BSAF attempted unsuccessfully to get permission to build a rather large chemical plant. They finally abandoned the plan voluntarily to withdraw from what undoubtedly was going to be a long and costly battle. The unique environment of the area is worth preserving, and it seemed highly probable that the operation of the plant would have had some deleterious influence on this environment. On the other hand, despite the affluence of many residents of the area, there are sizable numbers of people with shockingly low incomes who might have benefited from the additional jobs that would have been created by this industry. People with financial security naturally have a different attitude toward further use of the environment from that of people at poverty level. One group is already enjoying a good standard of living and wants to enjoy a quality envir-

428

onment as well; the other group has inadequate housing, food, and education and is willing to sacrifice environmental quality in order to improve these. The attitude of each group is understandable and must somehow be given serious consideration if large-scale, regional, environmental-quality-control programs are to be accepted and understood by the general public.

(3) Discipline-oriented training. Most of our educational training becomes more specialized as it becomes more advanced. To achieve success in a highly specialized field, one is usually forced to direct all his attention to this field, ignoring most others. Consequently, a technology has been developed in which the specialists are unable to make predictions about the ecological consequences of the technological innovations they have produced. In most cases they are even unaware of the need to make these predictions. Without doubt, the training of specialists will continue to be the primary thrust of universities and other institutions involved in education and training. Nevertheless, these institutions have a responsibility to produce generalists capable of a broad overview encompassing a number of disciplines. Ultimately, these generalists should be capable of seeing complex natural and industrial ecological systems in their entirety, rather than as a series of individual components.

(4) Change in public attitude. This is going to be the most serious problem. Because of Earth Day, television programs on the environment, and the like, the general public in the United States is better informed about environmental problems than it has been in the past. However, there is no strong evidence that this information has initiated a substantial change in life style. In Los Angeles it is quite clear that the major cause of the smog, which is affecting the health of its citizens and disturbing the environment, is automobile exhaust. The problem could probably be reduced to manageable levels if everyone simply drove half as much. It might be reduced to tolerable levels if a public transportation system were established, if people used smaller cars, drove less, and walked or rode bicycles whenever possible. These changes, while substantial, would not be an insufferable sacrifice for the citizens of Los Angeles and would produce an immediate beneficial effect. Nevertheless, there is much talk and legislation and little positive action, because apparently the citizens of Los Angeles, California, are not willing to make these changes on their own initiative. I seriously doubt that they are very different from the citizens of the rest of the country. Therefore the strong controls which will make rational multiple use possible must come from governmental action, rather than from voluntary changes in life style. People may be willing to tolerate these changes if they know that everyone must

comply and that they are not making an individual sacrifice. The right to pollute without restriction is still regarded as more important than the right to be free of pollution caused by the actions of others. Until a change in social attitude occurs, caused by one or more environmental catastrophies, effective legislation for multiple use will probably not be possible.

Literature Cited

Cairns, J., Jr. 1967. The use of quality control techniques in the management of aquatic ecosystems. Water Resour. Bull. 3(4):47-53.

——————, K. L. Dickson, R. E. Sparks, and W. T. Waller. 1970. A preliminary report on rapid biological information systems for water pollution control. J. Water Pollut. Contr. Fed. 42:685-703.

MULTIPLE USE OF RIVER SYSTEMS:
AN ECONOMIC FRAMEWORK

James Crutchfield

The field of water resource management and development offers an uncommonly large number of examples of "Gresham's Law of the Intellect"—bad talk will quickly drive out good. In particular, the field has come to be dominated by nonoperational, emotion-laden words which are all good to some people, all bad to others, and of little real use to the serious scientist, engineer, or economist concerned with maximizing the benefits of a resource system to the society of which he is a part.

"Conservation" certainly must be considered one of these catch phrases. A second one, equally bland and equally nonfunctional, is "multiple use." To the extent that it implies something socially beneficial about usage of a river system to include the largest numbers of participants, it is simply incoherent (except perhaps, as a rational way of maximizing the security of the administrator by minimizing the number of excluded complainers). If we stop short of this extreme interpretation we find it necessary to give content and dimension to the term. How many users? How much water to each? How should water allocations respond to changes in demand and technology?

The second section of this paper spells out an economic framework which goes far to make "multiple use" an operational concept, despite the formidable theoretical and empirical difficulties discussed in the third section. The final section raises a series of troublesome questions as to the adequacy of economic efficiency as a guiding criterion for river development and indicates possible avenues for their resolution.

Before proceeding to these points, however, it is important to specify three substantive constraints on the formal systematics of multiple-purpose development of a river basin system.[1] First, the objective function to be maximized cannot be defined except by reference to the geographic area concerned. If the national welfare is taken as the guiding criterion, a set of meaningful efficiency

---

[1]For brevity, the term "river" is used below as shorthand for the more comprehensive "river basin system."

conditions can be defined, and other objectives—income distribution, reduction of unemployment, or environmental impacts, for example—become at least manageable in an empirical sense. If, on the other hand, the geographic area in question is narrowed down (to the state level, for example), it becomes literally impossible to distinguish efficiency effects from mere transfers of income and employment from other areas. Widely different appraisals of the desirability of any system for allocating river water (and for future development) will result if local parochial interests dominate as compared to the answers that would be forthcoming if the question were viewed in the context of national benefits and costs.

Second, river development projects are severely constrained by the complex and slow-moving process through which water rights are established and transferred. Since multiple-purpose development of a river basin on a systematic basis never reaches the planning stage until many of the valuable locations and claims to water have already been staked out, it is inevitably consigned to a search for second-best solutions. If water rights could be traded freely on an established market, this constraint would be far less serious. But since this is rarely the case, either under law or administrative procedure, multiple-purpose utilization of a river must normally take as given certain uses at certain levels of water consumption (with their associated effects on water quality) and—hopefully—optimize with respect to subsequent adjustments that can be made to those constraints.

Finally, it must be acknowledged that a key element in the design of optimal allocation systems for river water—the ability to measure incremental values of water in its various uses—is seriously restricted. Reasonable dollar values can be placed on some uses (e.g., net returns to irrigation, transportation, waste assimilation, and industrial cooling), but for others (e.g., outdoor recreation, wildlife, and scenic beauty) valuation is still in the early research stage.

A Simple Static Approach to Optimal Water Allocation

Let us ignore these complications for the moment and establish a set of conditions defining optimal multiple-purpose use of a river under static assumptions; i.e., with no changes over time in demand, price, or physical production functions for the various economic services that can be derived from the river. An important corollary to these assumptions is that relative values of outputs are assumed unchanging over time. As a first approximation "optimal" is assumed to be synonymous with economically efficient. Thus, the question posed is: "What allocations of water, in what quantities, will maximize the net economic benefits obtainable from the river?"

At the risk of belaboring the obvious, it must be noted that the output of a typical river system is made up—at least potentially—of several useful services, some of which are competitive, some neutral, and a few complementary. The output mix from a large, developed western river, for example, would include: irrigation, power generation, navigation, municipal and industrial water supply, support for commercial fish runs, habitat for sport fish, wildfowl and wild animals, waste assimilation, and (with a negative sign) flood damage. Most of these are related in complex ways. With only trivial exceptions, preservation of water quality and flow conditions favorable for fish production is likely to be complementary to requirements for wildfowl and wild animal habitat. Navigation requirements are largely neutral, though they may provide some additional benefits to the users by regularizing flows, and may result in some degree of quality deterioration. Diversion for irrigation inevitably involves some consumptive use of water, together with some quality degradation. Municipal and industrial users also return backflows that are smaller in quantity and lower in quality than water taken in. Hydropower generation and flood control may be complementary, but both may conflict sharply with "preservation demands" for water.

The ultimate ideal of water resource allocation would be represented by a complete model of usage in which competing allocations of water will yield the same return at the margin, net of associated costs. In other words, ideal allocation of a given annual quantity and time pattern of water in a river would be realized when it is no longer possible, by transferring water from one use to another (or by shifting its use in time) to achieve larger net economic benefits. A maximum of this sort requires only that marginal benefits from all uses be constant or declining (a condition that would normally prevail) and assumes that all of the outputs of the system can be measured in a common monetary unit. It is worth emphasizing that it is not necessary to deal with the wholly intractable problem of measuring *total* value from the river. Maximization under the assumptions specified requires measurement—and equalization—of benefits from marginal quantities in each use.

It should be clear that marginal net benefits accruing from uses that are complementary or neutral are additive, whereas those associated with water usage that diminishes the output of other elements in the system require netting. In simple, formal terminology, optimal allocation requires that the marginal net benefits of any consumptive use be equated to the sum of the marginal net benefits of nonconsumptive uses.

It need hardly be added that this condition is rarely approximated at present: in part for lack of knowledge, in part because of public policies that dictate a zero price for water *in situ* for some very important uses. In spite of these limitations, engineering-economic analysis can still provide boundary values for marginal allocations of water. If, for example, it must be taken as given that Columbia River water will be used for irrigation and municipal and industrial purposes up to the point where marginal net benefits in those usages approach zero, it is still possible to indicate the marginal value of water in the remaining unconstrained uses and allocate accordingly. The resulting marginal value figures will be too high, since our initial assumptions guarantee that too much water will go to the "open-end" users, but they can provide a basis for useful second-best decisions.

Similarly, the knotty problem of evaluating water in "amenity uses" not priced in the marketplace can be given a Gordian knot treatment that may produce a rather good second-best solution. Referring again to the Columbia, specification of flow and quality standards at levels that safeguard anadromous fish runs will provide adequate protection for a number of other unmeasurable uses that might otherwise be sacrificed in an effort to maximize the quantifiable net benefits.

The next step is obviously to expand the allocation process to include investment over time; i.e., to allow for the possibility of altering the pattern of outputs available from the river by investment in facilities that alter the quantity, quality, and/or timing of water available. This is the domain of benefit-cost analysis, and its usefulness and problems need not be recapitulated here. Suffice it to say that it is at least theoretically possible to develop estimates of the present value of alternative streams of benefits and costs in a multiple-purpose system. By techniques now reasonably well established and tested, a cluster of technically feasible alternatives can be identified from which economic analysis can select the most efficient level, composition, sequence, and timing of projects. It is also possible to indicate the efficiency costs of project sets that will meet other objectives. The political decision-maker may reasonably expect economic analysis to provide a major reduction in the number of alternatives that merit consideration.

Problems! — Problems!

This completes the easy part of the discussion. When we turn to the actual task of measuring net benefits from alternative allocations of a given stream flow, not to mention the further complication of modification of the system through subsequent investment, a frightening array of difficulties appears. Many of the socially valuable outputs of a water system simply cannot be valued directly in monetary terms, even in principle. Scenic beauty, preservation of areas of unique historical and scientific interest, and the reservation of wilderness areas fall in this category. In other cases, it is theoretically possible to assign economic values, but the market system, because of pervasive external effects, will not provide such information directly. The provision of such public goods as flood control and improved water quality fall in this category. Finally, there are outputs, such as outdoor recreation, that could be valued by using prices or fees, but which have traditionally been supplied at zero or nominal cost to users, with noneconomic rationing devices of various types to allocate limited supplies of the service.

These difficulties are no less formidable because they carry the label "empirical" rather than "analytical." The stochastic nature of river flows, coupled with the need to assess the effects of different allocations on complex biological systems, makes for extraordinarily difficult problems of estimating physical output alternatives. And it is sad but true that putting dollar signs on bad numbers rarely improves them.

There are also analytical problems of formidable proportions, which can only be mentioned here. The most important is the treatment of time. Economists are still wrestling with the problem of defining the appropriate discount rate to be used in calculating present values of benefits and costs with different time-profiles. The lack of general agreement on this crucial number is particularly galling since in most cases it is of critical concern in proper allocation of existing water and the assessment of projects.

A second conceptual problem involves the implicit assumption that values of different outputs from a river will change at identical rates. It seems likely, for example, that the demand for the "amenity" uses of river waters is highly responsive to changes in real per capita incomes, in mobility, and leisure time as well as to increases in numbers of people. It has been estimated that the demand for outdoor recreation has been increasing at a rate between six and eight percent per annum over the last three decades, and the rate of increase is even higher for several types of recreational activity that place the highest demands on water quality and on the relatively unspoiled character of land-water systems used for highly valued recreational purposes. On the supply side, these are precisely the kinds of uses of resource systems for which substitutes are most difficult to come by.

On the other hand, a modern market-oriented economy has a remarkable ability to fend off the effects of increasing prices for certain types of inputs by substituting other raw materials, by changing processes, or by changing final outputs to make greater use of more abundant resources. This is as true of industrial and agricultural uses of water as it is of other factors of production. On balance, then, we might expect that the marginal value of water in these uses would remain quite stable over time, while the marginal contribution of water to the value of recreational and "preservationist" outputs of the system would increase over time.

On reasonable assumptions as to appropriate discount rate and relative rates of change in demand, it can be demonstrated that the effect of this assumed but quite reasonable behavior of relative prices over time can lead to drastic errors in water allocation if one follows conventional valuation procedures. The principal reason is, of course, the fact that the present value of future income streams for which demand is growing more rapidly is much more heavily weighted by values twenty years and more in the future. In short, the use of conventional economic analysis in terms of static assumptions about relative values of a complex pattern may well have caused substantial underinvestment in facilities and programs required to protect water quality and to provide for a growing public preference for recreational and amenity uses as compared to agricultural and industrial uses of given water systems.

This approach to multiple use has important implications for water policy. First, it is obviously vital that "river basin planning" be viewed as a continuous process, subject to periodic review and revision, rather than as an exercise to produce a document cast in concrete as solid as the structures it recommends. Second, the construction of both major projects and ancillary facilities should be undertaken with clear recognition of the irreversibility of many decisions involving allocation of water among competing users. Regardless of the possibility of physical reversal, the economics of undoing an erroneous decision involving major structural changes in a river system are forbidding, to say the least. Since we must live with our longer lived mistakes, it would seem the essence of wisdom to provide for maximum flexibility in modifying water usage over time in a given river.

These conclusions are reinforced by the limited capacity of regional economics to forecast with any great degree of precision changes in the level and composition of regional economic activity, and related changes in demographic factors. The ability to provide, reasonably far in advance, estimates as to where people will locate, what activities they will perform, and what the relative weight of their demands upon available water supplies will be is not even at the art stage, much less a science. Serious consideration must be given to the desirability of sacrificing a considerable amount in static efficiency for greater ability to alter the timing, sequence, and pattern of water projects as the future moves into a shorter and more accurate planning horizon.

In this connection, it is interesting to point out the difference in attitude of public and private planners toward estimating future demand and technological changes. The managers of most major corporations are under no illusions as to their ability to make accurate forecasts 50 to 100 years hence. They know they will always be producing the wrong products in the wrong places, to some degree, with the wrong techniques. The problem is to balance the marginal expected gains from optimal design, given the expectations of year t, against the present value of the losses imposed by the fact that the planner is fairly well locked into that design for t + 50 years.

Crude as it is, this preliminary formulation can clear away much of the rank underbrush surrounding the economics of multiple use. For example, the common tendency to designate, in state law as well as in operating procedures of water agencies, certain uses of a given river system as superior to others is manifest nonsense. While it may be perfectly reasonable to argue that household use of a minimal amount of water is crucial to life itself, the amount of water that can be drunk, used for cooking and washing, or flushed down the toilets of one household is severely limited. Once these critical requirements have been met, *additional* units of water will obviously carry progressively lower values. Eventually the marginal value of water for municipal and industrial use must fall below that of even such frivolous things as outdoor recreation. Similarly, the objective (never specified openly but frequently pursued vigorously) of maximizing the number of users in a given resource system would produce optimal allocation among usages only by an incredible combination of miscalculation and coincidence.

At the other extreme, the tendency (illustrated with shocking clarity in some portions of the report of the Public Land Law Review Commission) to designate as "dominant" one usage of a given land-water system, to the exclusion of others, is equally inconsistent with the marginal principle. Regardless of the extent to which one set of outputs may "dominate" total output from the system, the critical comparison is the contribution of the *marginal* unit of water to output in any alternative use. Even if the term "dominant" is construed less rigorously, it is no more operational as a guide to allocation among several users than the instruction to maximize, within reasonable limits, the number of people who can be accommodated within a given resource system.

Finally, the task of allocating water to multiple users has been made much more difficult by our inability or unwillingness to use pricing rules that have served us well in the private sector. Instead we have chosen to discriminate systematically (in favor of agriculture, for example) and in random fashion (through local competition for payrolls and plants by blatant giveaways of municipal water). And since water is sold almost everywhere at prices lower than its net economic value, a variety of nonprice rationing devices have developed to equate demand and quantities available.

437

JAMES CRUTCHFIELD

Objectives — One or Many?

As indicated above, even if economic efficiency of a proposed investment is the sole criterion, there are severe and (in some cases insurmountable) obstacles to accurate economic measurement of some outputs and costs of a water system that are of legitimate concern to society. Moreover, an increasing number of competent resource economists have given general, if somewhat grudging, recognition to the fact that economic efficiency is not the only basis for ordering of public investments. Other social objectives (for example, income distribution, alteration in regional growth rates, reduction in structural unemployment, and environmental protection), though not always measurable in purely economic terms, may reflect real social values that must be considered in assessing alternative public investments. It is vitally important to the nation, and to every component part of it, that major water projects be efficient in both engineering and economic senses. It is also essential that redistribution of income and other secondary effects be identified, measured, and evaluated with respect to their impact on the public welfare. Even efficient projects (in the narrow economic sense) may involve such grossly inequitable distribution of benefits and costs as to raise serious doubts as to their desirability. Similarly, if alternative projects or project systems will produce significantly different effects on the environment, it is essential that these be made clear to the people of the nation and of the individual states concerned. It may not be possible to measure all of the social costs and benefits of water supply systems in terms of a single monetary unit; but it is possible and mandatory that we weigh economic benefits against incremental deterioration of the environment, income redistribution, unemployment and other related effects to make sure that the mix has, in fact, been chosen after thorough public scrutiny of the alternatives.

We repeat, however, that recognition of nonefficiency considerations that may affect social welfare does not diminish in any sense the primary importance of evaluating the economic efficiency of competing patterns of water usage and projects designed to alter production possibilities over time. It is trite, but bears repetition, that projects with larger net benefits provide more resources from which nonefficiency objectives can be met. In addition, the essential political decisions with respect to the weighing of economic efficiency versus other social objectives cannot be made intelligently unless the trade-offs, in terms of national income foregone to achieve such objectives, are clearly spelled out. Economists are on solid ground in stressing the primary importance, nationally and regionally, of impartial, nonpolitical, and *complete* analyses of economic costs and benefits associated with the allocation of river waters among users cast in the general framework of impact on national economic values.

Nevertheless, the critics of a pure efficiency criterion for water utilization have made a number of perfectly valid points. Public decisions about the usage of major river systems *do* shift wealth from one region to another, from one occupational group to another, and from people at one income level to others at higher or lower levels. Society *does* care about gainers and losers; one need only look to the endless variety of special legislative actions designed to make such transfers of income possible. And no one concerned with national resources today can be blind to the upsurge of public concern about the impact of public and private decisions on environmental quality. At the same time, it is clear that none of these other objectives are costly. How much national economic output should be sacrificed to relieve the plight of Appalachia?

It is one thing to be aware of these nonefficiency objectives and quite another to weave them into a coherent framework for decision-making—about water use or any other public activity. The recent effort by the Water Resources Council to integrate four sets of objectives into a quantifiable objective function left a faintly sour taste in many mouths. Part of this reaction reflected a lingering suspicion that the effort was designed to restore feasibility to many federal projects shot down by the Office of Management and Budget insistence on more realistic (i.e., higher) discount rates on water programs. But much of the uneasiness stemmed from serious concern over both conceptual and measurement issues, and the fear that policies based on a first pass at unresolved problems might become enshrined in agency dogma—particularly if they enhance the prospects for job security.

The emphasis on regional economic growth as an objective is particularly vulnerable. I know of no analytic link that would permit assessment of the *national* interest in allocating water in a way that accelerates growth in Region A. And there is as yet no real possibility of measuring the reduction in income and employment in Regions B, C, and D that must follow, since much of Region A's gains are clearly transfers. It is hard indeed to place much faith in a set of accounts with such shaky logical foundations and with a built-in bias toward overstatement of social gains.

The concern over income distribution and environmental effects of different allocations of water is more solidly based, but efforts to find a single dimension in which marginal trade-offs among efficiency, income redistribution, and environmental quality can be compared directly, have an unexciting track record to date. The issues are far too complex for thorough treatment in this paper. Suffice it to say that economic analysis has much to contribute: in estimating the economic cost of achieving any given set of distribution, employment, environment or other nonefficiency goals; and in generating resistance to the tendency

439

to invent economic values to justify such goals on the ground that even bad dollar estimates are better than none at all. Perhaps most important, it is incumbent on the economist to make certain that all alternative paths to each objective are explored. After all, there are many ways to redistribute income, reduce chronic unemployment or improve the environment other than to reallocate water from river systems.

It is inevitable, and probably useful, that professionals in *all* disciplines concerned with wise use of the nation's water should concentrate on the weaknesses and limitations of present practice. Yet a closing note of optimism is not out of order. A great deal of solid progress has been made in delineating the outlines of optimal river basin development, and much has been learned about the critical gaps in both theory and information. Effective communication among specialists is no longer viewed as a genteel intellectual exercise. It is a powerful, and indeed essential, economizing device in ordering concepts and data to use our rivers to greatest advantage; and we have moved a long way down that road over the past decade.

POLITICAL ASPECTS OF MULTIPLE USE

John D. Dingell*

Other speakers have dealt with a tremendously varied number of aspects of environmental problems relating to our river systems.

By the Conference program, it now falls to my lot to discuss the impast of politics on river systems. I, too, would like to stress the river as an environmental system. It is with regard to what have come to be called environmental values that our priorities are out of balance.

In the best sense of the word, politics is the science and art of government. It is through this science and art that we set our priorities and seek to implement them in complex systems like rivers. It is here that rationalization of multiple use must be achieved. There is no other choice.

It is unfortunate, but true, that environmental and ecological considerations have weighed but little in our political decisions until the recent past. And these recently changed priorities have not yet resulted in enough change in our actions.

There is some evidence, however, that we are beginning to learn that politics and the environment are indivisible—that our political decisions are in fact environmental decisions. Indeed, management of the environment must be government's responsibility by the very nature of the way the problems come about and the kinds of solutions called for.

It is impossible to discuss the politics of river systems or the solution to the problems of preserving and protecting our river systems and their environment apart and separate from other activities of man or from other environmental matters. Rivers are but one of many natural systems that transmit the effects of man's activities. Even where the natural systems are not interrelated man's activities are. Thus, if we protect one system the problems may be shifted to another.

---

*These remarks reflect an important input of ideas from David J. Allee, who, when I was kept in Washington by an important vote, presented this portion of the Symposium, and assisted in the completion of the final draft.

Because of our failure in the past to understand these interrelationships, the nation finds itself in a difficult situation with regard to environmental matters.

Our air is so contaminated that it poses a very present danger to stone obelisks in New York, ladies' nylon hose in Detroit, vast forest areas in the vicinity of Los Angeles and to the health of a substantial portion of our citizenry. Airborne toxic wastes seem to have found their way to every wild area left.

Our land has been despoiled to a shocking degree. We have seen unplanned residential, commercial and industrial developments proliferate across our countryside. We have ripped apart vast areas to build highways to speed people and commodities from one place to another. At the same time we have allowed our railroad system to deteriorate to an alarming degree.

Our rivers, streams and lakes have been used as dumping grounds for municipal and industrial wastes. In some areas our waterways are so contaminated that they and the ground waters they recharge are public health hazards which have major potential as spreaders of disease.

Our estuaries are being filled, dredged and polluted into ecological deserts. The oceans are no longer fit places for fish.

These are the obvious symptoms that something has been drastically wrong with the environmental policies of this nation and the political system which created those policies.

If, as Thomas Jefferson observed, the combined wisdom of the many is better than the select judgment of the few, we must consider together how we came to the appalling environmental mess in which we find ourselves. It is clear that our political system has failed insofar as environmental protection is concerned. While I have my own views on the matter, it would not serve our purpose at this point to devote much valuable time trying to assess whether this failure has come about through ignorance or through venality. Efforts to correct either are needed, but our task is to go deeper. What must be done to correct ignorance and to block venality must come from more fundamental reform.

But how do we come by reform in government? Massive changes from a series of individually incomplete stpes is the way of it. Each step usually achieves less than was possible. Each step must be tailored to help correct the compromises of the last step which usually can't be predicted in advance. It would take longer and the chance of success would be less to attempt changes in one step.

Some Questions About Our Political System

The desk of a Congressman has a continuing flow of papers across it that raise interesting questions as to our environmental management and the political system to which it is charged.

A case in point is an announcement which arrived while I was dictating these remarks and which reported that my home State of Michigan had received its first federal grant in the amount of $490,000 for comprehensive land-use planning on a statewide basis, including local coordination and participation. And without intelligent land-use management we cannot hope to enhance the environmental values of our waters.

While this federal grant might seem to be a matter for rejoicing, it is really an example of the failure of our political system to act in a timely fashion. This grant graphically illustrates the totally inadequate attention which has been given to the essential task of planning for proper land use. Both the amount of the grant and its tardiness are causes for concern.

It is passing strange that only after almost 200 years of existence as a nation are we getting around to developing comprehensive land-use plans. And it speaks volumes about our system when it is seen that a major state such as Michigan must await a federal grant before it undertakes to develop a sound land-use plan for its citizens. Of course, it remains to be seen if the plans really become an effective basis for public decisions.

A similar demonstration of governmental inadequacy is to be found in the report on our coastal estuaries and wetlands which was filed recently with the Committee on Merchant Marine and Fisheries by the Department of the Interior. The links between rivers and the seas are some of the most critical and most subtle, calling for particularly effective management arrangements.

Enactment of the legislation authorizing the study and report on our estuarine areas represented four years of battling within the Congress. The Department of the Interior expended hundreds of thousands of dollars over a period of two years and then, under pressure from the then Bureau of the Budget, failed to carry through with its mandate to make legislative recommendations to the Congress to identify estuarine areas for preservation and to set out the manner in which these areas would be preserved. The Congress has asked the experts at the Department of the Interior to set out how the system of preservation should work—whether it should be done through the federal government, through state governments or through local units of government and whether it should be through purchase, zoning, easements or lease. Leadership in the development of workable tools for the management of these problems should come from the Executive Branch.

Another example of failure of the political system has been the game played with water pollution abatement programs. Everyone is aware of the environmental damage inflicted upon the magnificent Lake Superior by the Reserve Mining Company. Everyone wonders why Reserve Mining has been permitted to dump taconite tailings into the lake for so long while all of its competitors are compelled to use on-shore disposal for their taconite tailings. A related curious happening took place with regard to Lake Michigan. An Interior Department-sponsored conference to go into the matter of thermal pollution of Lake Michigan was closed to the public until screams from conservationists and concerned citizens forced, through my office, an opening of the conference to the public. One must wonder why the preservation of Lake Michigan must be considered in secret. Why shouldn't a matter as important as a projected flow of hot water into Lake Michigan equal to that of the Mississippi be discussed in open public meeting with full public participation?

One might logically ask, while discussing failures of the political system, why the highway builders, the land developers, the irrigators and others are able to almost totally disregard the adverse effects of their activities on our river systems. Siltation, salt and nutrient pollution and other bad results of land development and highway construction are a major cause of destruction of water quality and environmental values in our river systems.

Why has it taken so long to even begin to develop a national land-use system?

Why have state and local units of government permitted destruction of open spaces upon which our river systems depend?

Why do federal agencies persist in polluting our waters in plain and open defiance of law?

Why has the Executive Branch failed to ask for appropriations for pollution abatement programs which would match what Congress has authorized?

I am afraid that the answer to these questions is to be found in a particular class of inadequacies of our political system. The narrow economic or regional issue is too often given greater weight than is the need for environmental quality. It has been easier to organize effective support for economic development without environmental protection than with it.

444

In this regard, it is interesting to compare our efforts at preserving and enhancing the environment with what is being done in Great Britain. The British recently merged a wide range of activities and placed them under a Minister of the Environment. I had the great pleasure of meeting with Minister Walker a few days ago and he explained that his authority extended to transportation, air and water pollution control and abatement, land use, etc. He told me that it is his function to consider all activities in Britain which affect the environment and to take such actions as are necessary to protect the quality of the human environment. He has the responsibility and he also has the authority necessary to accomplish his mission. He obviously is not afflicted with the curious federal, state and local conflicts which plague our governmental structure; and he is obviously not forced to react to the many political pressures which plague our federal, state and local administrators in taking steps to protect the environment.

There is abundant evidence that the British system is working better than ours in many regards. With substantially more limited resources, the British are moving forward effectively in land-use planning and in cleaning up their air and waterways. They have instituted an environmental management system which is working. We must find a system that will work for us.

I am afraid that our history tells us that politics and politicians have been continually more a brake than an engine in the drive for human progress insofar as the environment is concerned. The evidence of history makes it clear that what has been called "progress.. in the areas of industrial and commercial development within this country has been progress toward environmental disaster. We may make "better things for better living through chemistry," but in the process we have gone a long way towards destroying the Delaware River Basin. In the past, a major steel company could establish its own company town and proceed to make its own rules concerning air and water pollution. Even today electric power producers are permitted to do violence to the Four Corners area in order to supply energy to Los Angeles.

Under an effective system of political decision-making these things would not be allowed to happen.

Basis for More Effective Politics of the Environment

What is our system of political decision-making? Some ideas about this may help in deciding what might be done to make it more effective so that these things would not be allowed to happen. The basic problem is to create self-enforcing, self-regulating mechanisms such that day-to-day decisions of people throughout the system don't lead to such results. But what do we have to work with?

Some Elements of the System. I can't be complete and thorough here. But notice a few things. For one, public decisions in our system are shared—shared between the Congress and the President, between committees of the Congress, between them and the agencies, between the agencies and the groups that take an interest in their work, between factions within those groups, between levels of government, and on and on.

And most of our environmental problems arise because many public and private decisions are made ignoring or almost ignoring the indirect effect of what is immediately at hand. Thus, public policies and programs must be developed to take these indirect effects into account if they are worth considering. The polluter doesn't think of the fish downstream. Somebody has to remind him and force the value of the fish into his calculations. But how do we get this done when so many share in such decisions?

The political and administrative system works in favor of those who have a particular, immediate interest and against interests that may be larger overall but are just that, spread over all. Today is favored over tomorrow. The definite and concrete is favored over the problematical and hypothetical. Those things that mean a lot to a few people are favored over the things that mean a little to the many. This has been called the "producer" bias in the system, but obviously applies to many situations. Environmental values almost always fall in the disfavored class. Why does this bias exist? Because these seemingly more immediate interests have in the past provided a more effective basis for political action—for support of some decisions over others.

It is part of my thesis that the political system can be made to work much better if we provide, throughout, the means to attain a systematized understanding of environmental values and relationships. Understanding by the few, no matter how powerful, can never be enough. Decisions that affect the environment are spread out through the whole society. And I feel that the combined wisdom of the citizenry will operate to force government to take effective action. This will happen more quickly if we improve the means for that wisdom to develop and make itself felt.

It appears from many opinion polls and studies—one taken by the National Wildlife Federation, and others taken by the major professional pollsters—that environmental concerns rank within the first three national issues at all times for the past several years. The problem, then, is to see to it that the political system behaves in its proper role of servant of the public will. It must be the tool for the perfecting of man's relationship with his environment into one of beauty and harmony.

Some Broad Strategy. You may see some basis for my optimism if you reflect a bit on why those favored by the "producer" bias don't completely steal the rest of us blind. For one thing we are all both favored and hurt by this bias in the system. Even those who earn their living working in a pollution-producing factory like to fish and can enjoy just knowing that Grand Canyon is still there. And as the companies and the agencies that we work for grow bigger and more impersonal, we are more and more likely to become active in the organizations that champion these less favored diffused values.

Also, when the "producer" bias in the system goes "too far" more of us are likely to come to the support of those who would shift the balance the other way. "The rules of the game" do shift. The practical definitions of when "too far" happens are subject to change. Environmentally concerned organizations have done much to change our ethical views about the environment. There probably hasn't been a time when there has been a more rapid and profound change than right now. And we need more.

Better understanding of cause and effect is another way to change the rules of the game. When the ordinary citizen understands that there is an alternative to dirty water and spoiled fish habitat he will expect more, demand more from his public officials and the managers of our business establishments. The citizen expects *all* his interests to be protected by those in positions of authority. He often doesn't want to be bothered by the details of particular issues. But he does "throw the rascals out" if he finally realizes they are not doing all that they could. Debate among those who do care about the issues is how he is alerted to this. Thus the potential for conflict is a tool to use in producing changes in the system.

Finally, the rules of the game are changed most effectively when we develop agencies and programs that can effectively represent the neglected interests in the day-to-day business of government. The Congress and the President, the Governors and the Legislators, can only make a limited number of decisions. There must be a better balance between the developmental interests in our agencies and the regulatory-environmental protection interests.

But these agencies face shifting technology, changing perception of environmental effects and system understanding and, certainly, shifting value on the part of the society they serve. For many reasons, internal and external, they have a lot of inertia. Thus, our regulatory arrangements are constantly out of balance, and constant attention is needed to bring them into balance. Government cannot be expected to reform itself. Public education and debate is sorely needed.

447

The many environmentally concerned groups do and must carry much of the burden in achieving needed change. In particular they must provide the kind of lasting interest and support that is needed to maintain the search for balance. The "natural history" of regulatory agencies is instructive. They are often formed at the crest of a wave of public interest. But when the wave falls back, the job is left to the public officials and who is there dealing with them on a day-in-and-day-out basis? Too often it is only those who are supposed to be regulated. Is it any wonder they seem to have more influence than they should?

The interest and support of groups concerned about the environment expresses itself in another way that has the potential for leading to change. With decisions shared by so many there are often many points of leverage and veto. These have been traditionally used against environmental values, but they can be used to its advantage. This is in part because, as I have mentioned, the system has a sensitivity to conflict. The system can be easily overloaded by conflict. Thus, it may take less than a fight on every river project to cause the agencies to find ways to accommodate environmental values and eventually to routinely include procedures that preclude such conflict. Such institutionalization and regularization of environmental responsiveness throughout government must be our goal.

With many interests and widely shared decision-making, the need for coalitions—for the environmental coalition—is great. Overlapping interests must be sought out to join in environmental protection. I am reminded of the success we had in raising the amount available for sewerage grants from the $214 million requested by this administration to $800 million. If you will excuse my sarcasm, a program long oriented to providing funds for our poverty-stricken cities was recognized as also having potential for enhancing environmental quality. More funds were an obvious first step. City officials, civic groups, environmental groups and many others—some rather strange bedfellows—saw a common interest. Now what remains is to see that water quality priorities can be made a basis for giving those grants instead of the first come, first served approach of the past.

The Need for an Expanded Federal Role

Success will be dependent upon many circumstances, and arrival at this point of harmony between man and his environment is going to require immense changes in our political response. Not the least of these changes will be increasing federal responsibility in all environmental areas. This is something which will not easily be done but the process can be seen now in the handling of water and air pollution through federal quality standards, federal enforcement and federal grants.

Many actions that we take to use our rivers are made at the local level. Even federal programs for the development of flood control structures, irrigation storage reservoirs and land drainage draw their support largely from local groups. This kind of development is inherently local. And the environmental aspects of such development as well as waste water management and havitat improvement and other environmental action inspire local interest. It is encouraging to see that this is increasing every day. But what makes environmental concerns often very different from other water concerns is that they also have a large and effective national audience. This means that hanging over every local issue is the possibility of a national response. It also means that the federal responsibility must be well developed and effectively articulated.

It is also true, of course, that correcting an environmental problem may cause that problem to relocate elsewhere if there is not a strong unified national program. But isn't effective support even more to the point? And support for the environment seems easier to get on a national basis.

The federal role in environmental matters must be substantially and quickly enlarged if we are to avoid a crisis which will make our present environmental problems seem small potatoes, indeed.

The Need for a Systems Approach

I have been arguing for a long time for a systems approach to the environment, both insofar as scientific understanding and governmental action.

One must question how 50 states, 3,000 counties and many times that number of cities, villages and townships can be relied upon to come to a uniformly effective systems approach in environmental protection without strong inducements. Equally curious is how we can ever arrive at the hard judgments which must be made on the basis of our present methods for the handling of information, data and the byproducts of research.

Certainly any systems approach to the environment is going to be entirely dependent upon the proper collection, collation, evaluation and storage and retrieval of information and data. Equally important is the blending of the natural sciences, economics, law, sociology and the sciences of human behavior into a total approach to public environmental management problems.

Environmental decisions in the future will have to be dependent not upon the narrow disciplines of yesterday or even the broader disciplinary concepts of today. Rather, decisions will be made upon omnidisciplinary or even nondisciplinary scientific approaches, since the environment successfully defies handling on the basis of a single value or discipline. I am sure the true environmentalist recognizes this fact. Hopefully the dawning will soon come to politicians and our governmental establishment.

The politician's environmental failures are not his alone. The fact is that scientists and environmentalists have failed to draw together and combine products of their various disciplines—natural, social and economic. To take effective action, the politician must have good information and good scientific methods with which to develop a unified systems approach to the environment. But considering the structure of our government this approach must result in a high degree of self-management and self-enforcement.

Some Recent Moves Toward a More Self-Enforcing System

The great wave of public concern sweeping the United States and most of the other developed nations over environmental abuses was until recently unnoticed by many of the most astute politicians. It was only two or three years ago that this situation began to change and since 1969 the tide of public concern has carried forward a number of pieces of environmentally oriented legislation.

A case in point is the National Environmental Policy Act of 1969 (NEPA). The sponsors of this legislation, of which I was one, were greatly surprised to see the manner in which the public rallied behind the proposal and overcame the opposition of the Administration and the dilatory efforts of some of the encrusted poohbahs in the Congressional system.

In signing NEPA into law, the President on New Year's Day 1970 acknowledged this measure to be the first step into the Environmental Decade—the 1970's. This legislation also was the first federal effort to approach environmental matters on a systems basis. It was most appropriate and encouraging that the President came forward with high praise for the legislation. His comments made it clear that a great political lesson had been learned. He in effect reversed his Administration's former opposition to the proposal and accepted the fact that the public was demanding strong action on environmental matters.

To relate that broad and sweeping legislation to the question before us now— river ecology, the impact of man and politics—is not hard. It is obvious that the clear environmental policy statement, the creation of a Council on Environmental Quality to advise the President and to complete an annual environmental report, have a vital place in informing the public and thereby creating the awareness necessary to proper operation of our political system. The absence of this kind of

450

citizen pressure inside the federal system has been one of the major shortcomings of our political system. Its presence means that on the executive side of the government there is a focus for concerns and a means for better resolution of conflict and a more systemized response to the environment.

An important device for good ecological management contained in NEPA is Section 102(2)(C), the provision calling for the filing of environmental impact statements where federal actions or legislation can be expected to have an impact on the human environment. One of the beauties of Section 102(2)(C) is that it tends to be self-enforcing. Such features are essential since no review agency can look over everyone's shoulder. The initial requirement is that the responsible agency file a draft environmental impact statement with the Council on Environmental Quality and make this statement available to other interested federal, state and local agencies and to the public at large. Thus, the initiating agency is forced to give consideration to environmental implications of a proposed action and to study alternatives to that proposed action. Other governmental agencies are put on notice and required to comment on the statement. The public similarly has an opportunity to react to the proposed project or legislation. Following this process of administrative and public exposure, the initiating agency is required to comment on the statement. The public similarly has an opportunity to react to the proposed project or legislation. Following this process of administrative and public exposure, the initiating agency is required to file a final environmental impact statement including proposed project revisions aimed at overcoming environmental problems brought out during the study process. In fact, the environmental impact statement process may point up the folly of a given project and result in its abandonment.

Section 102(2)(C) facilitating access to the courts was a real sleeper which alert conservationists picked up and used to litigate the desirability of the Cross Florida Barge Canal and the proposed Trans Alaska Pipeline. The first project is happily now dead. The second is undergoing further exploration in government, although I must report to you that it probably is not receiving the required honest rethinking imposed by Section 102(2)(C) and other sections of NEPA because of the strong political pressures being applied by powerful economic interests. The history of the supersonic transport program, however, makes it clear that economic political pressure does not always win out over environmental concern.

Thus, a regular tool for systematized decision-making on federal actions has now been placed in the hands of the environmentalists in and out of government. The Council on Environmental Quality is making a manful and honest attempt to lay firm guidelines to assure compliance on the part of federal agencies. Public pressure, particularly in the form of litigation, has helped to encourage compliance. And it is my hope that we will soon have in being an adequately staffed Committee on Environmental Quality which will enable the Congress to exercise a stronger role in assuring compliance with all of the provisions of NEPA and other statutes relating to the environment. This could do what the Countil on Environmental Quality has achieved in the Executive Branch of the government and would add to the self-correcting, self-enforcing character of the decision-making system.

The public also has forced the political system to respond in other ways. Albeit inadequate and cumbersome in their enforcement, there are in being in each of the states water quality standards which have been promulgated in accordance with federal guidelines. Now largely limited to chemical parameters, these standards must be strengthened and extended to all of the important habitat characteristics of the Nation's waterways. And enforcement procedures and financial inducements for compliance must be improved. The federal government has enlarged its budget for construction of waste water treatment facilities, but this program must be extended and enlarged to cover both sanitary and storm sewer systems and inducements to correct nonpoint sources of pollution. Also the cost-sharing should at least be as generous and as sustained as for highways. The air pollution statute has been greatly improved. Environmental values have been built into our airport program. Control and abatement of noise is gaining support.

In the few years since 1969, public opinion has forced real legislative and administrative progress on environmental questions. Previous to that time, a handful of Congressional environmentalists and conservationists had to struggle for public support. In contrast, today, federal, state and local administrators—including the President—have rushed to climb aboard the environmental bandwagon. Hopefully, the public will not be satisfied with verbiage but will continue to insist upon action. There is evidence that it will. I have particular reference to the many new militant and litigacious organizations which have sprung up to carry environmental fights into the courts. New public interest law firms are arriving on the scene daily to serve as attorneys for the environmental movement. I am hopeful that federal environmental class action legislation, such as H.R. 49, of which I am the sponsor, will be enacted and greatly ease the task of these public interest law firms. This, too, would add greatly to the self-correcting nature of the decision-making system.

Much Remains to be Done

Despite these successes in the march towards sound environmental relationships, it would be foolhardy to contend that we have come anywhere close to completing our political job.

As a people, we must substantially revise our economic and social values. We must stop being insensate consumers of more and more things and give greater thought and credence to the fact that we exist in an ecological unit with limited resources and with very real bounds on its ability to withstand the impact of misuse.

I believe that our values are changing, albeit slowly and with considerable strain on the body politic. As a citizen and as a member of the Congress of the United States, I believe it is my duty and obligation to assist in every way possible to reshape our governmental and political institutions so that they will respond to man's growing desire to place himself in harmony with his environment.

To this end, we must all insist on the creation of a national land-use policy which will assure that our resources are utilized in accordance with the general good rather than in compliance with economic or social caprice.

We must develop and implement a national transportation policy which will assure that the movement of people and goods is accomplished in the most economically and environmentally sound manner possible. We can no longer tolerate a situation where noise, air pollution and residential disruption are visited upon the many to convenience the few.

We must create within our great urban areas, particularly in our center cities, the physical, social and economic amenities requisite to a full and complete human existence for all of our citizens.

We must cooperate with other nations to the end that the environmental values and amenities which we seek to attain in our own land are pursued elsewhere. The human capacity to pollute and despoil must in due course be brought under control throughout the world or we must accept the fact that human life on this planet will come to be intolerable in the United States and all other nations.

I have tried to outline some of the modest steps which we are taking within our own political framework to achieve environmental and ecological balance. I have touched briefly upon the vast problems which remain. I hope we and the other peoples of the world have the good sense to achieve the proper solution in time.

SUMMARY

Justin W. Leonard

There seems to be a widespread notion that all the ills of the environment stem from the impact of man, puny newcomer that he is in the evolutionary scheme. But in my Midwest boyhood home it was recognized that even before pioneer settlers had broken the prairie sod, the Missouri ran muddy. One story has it that an eastern tenderfoot visiting Council Bluffs asked a befringed Mountain Man come there to trade his furs if his kind really drank the river water.

"Not exactly," he replied, "but we often cut off a piece with a knife and suck it, like molasses candy."

We have been privileged to receive new information on many of the world's great river systems, and on their problems and possible solutions. We have been reminded that floods which destroy some of man's interests may be essential to others, as with the annual silt deposits of the Nile and the Mekong. As economic development and increasing numbers of people add more wastes to overburdened streams, outcries well up for improvement in water quality and against continued deterioration. And in the concept of improved quality is a comparatively new parameter—esthetics and recreation.

Not new, really. Poets have long been fascinated by living waters. Out of curiosity I checked the *Oxford Dictionary of Quotations*, and found 166 entries for streams as against only sixteen for lakes. What *is* new is the growing insistence that such factors be weighed along with consumptive and pre-emptive uses, power production, navigation, and transport of wastes.

This conference has shown, too, that just as one never looks at the same river twice, so, apparently, no two of us see the same river even when we stand side by side on the bank.

We have been particularly fortunate, therefore, to have three outstanding specialists—as well as a highly articulate chairman—respond to the deceptively innocent query: What *is* a river?

The truly Olympian detachment with which a geologist is able to view the dynamics of river systems against the time scale of geologic epochs seems at first to constitute a completely different conceptual scheme from those developed by biologists to deal with volatile populations of short-lived organisms which may be greatly modified by ephemeral factors, or by flood-control engineers whose long-term plans must still hold a place for the precarious moment when the flash flood laps at the crest of the levee. Robert Curry pointed out that concepts resting on the geomorphological theories of minimum variance and minimization of work (simplified for biologists by the statement that one cannot maximize both efficiency and stability in biological systems) actually define river systems in a most fundamental way. With a grasp of these concepts the biologist may hope to understand the dynamics of a river's energy environments, and the engineer may with greatly improved confidence predict the effects of man's impact on a river system.

In attempting a zoological "definition" (description) Cummins related a wealth of new information stemming from the research of himself and his group. He synthesized this with existing knowledge to present what impresses me as a most useful organization of the seemingly chaotic range of microhabitats in his chart depicting the compartmentalization of running water ecosystems and trophic interactions between various levels of organization. His review of our present knowledge of the dynamics of benthic populations, and Ruth Patrick's later comments on invasion rates, revived a personal memory of my own surprise at a highly localized demonstration of effects of rapid migration and drift. This occurred at Michigan's Hunt Creek Experiment Station in 1940. We connected a newly excavated by-pass channel to the parent stream in early winter after insect oviposition had ceased. Yet within two months the benthic insect fauna of the new channel was nearly identical with that of the natural stream.

Blum rounded out the fruitful discussion with his concluding statement that a botanist sees a river as "a nutrient medium in undirectional motion, less characterized by recycling phenomena than a lake, with complex and diverse chemical and physical parameters which are likely to be fluctuating at different frequencies and in many different ways but generally in such a way as to permit a high level of productivity."

So, guided by men of vision, we turned to the balance of the program with a much more perceptive and penetrating picture of a river than that of the elephant provided by the three blind savants of the old tale.

It is neither feasible nor desirable for a conference summarizer to attempt an "instantaneous replay" of each individual paper. The seven case histories are of themselves enough to insure wide sale and heavy use of the published proceedings.

456

The wealth of factual information each of the rivers treated will have permanent value. I was particularly impressed by the history of the Nile—the world's longest river, the world's most completely documented river in point of time, the world's most completely utilized river—all so appropriate for Africa, land of superlatives.

The statement that "The occupancy of the Illinois River valley for thousands of years by early man probably had little or no impact on the ecosystem of the Illinois river" may well be true in the context its author intended. If, however, there is validity in the view of some ecologists and anthropologists that Illinois was changed from forest to prairie by the deliberate and continued use of fire by early man, the regime of the river doubtless changed, too. One recalls Ruth Patrick's comments on the shift from allochthonous plant detritus to autochthonous phytoplankton as the chief food source as one moves from forest streams to prairie streams. This history of the Illinois River relates pollution, the introduction of the carp, navigation, canal and dam construction, and soil erosion in coherent fashion. American pioneers seem to have missed few opportunities here to exert maximum ecological impact. And the brevity of the time scale was made vivid to me on a very personal level by the written account of one of my great grandfathers. As one of a trio of adventurous youths in their late teens he left Indiana to seek his fortune in the Wisconsin territory the year following Black Hawk's War. And they received assistance in floating their wagons across the river from a band of Indians returning from gathering huckleberries on land long since engulfed by Chicago.

If the late Viennese Waltz King could have viewed the symposium wall map of the Danube he would have been surprised to find it not blue but red; as surprised, indeed, as many of us were to learn that even today, in a drainage area occupied by 70 million people, "only a small part of the sewage (entering the Danube) is purified."

Perhaps it is true that the only real purpose of a river is to help water run downhill. Both Einstein and Blench pointed out that it is not easy to maintain this function in the face of physical forces and human activities. If rivers are the vascular system of our planet, there are many factors operating to develop atherosclerosis.

But we don't seriously contemplate allowing our rivers the luxury of this single use. We elect to consider rivers as one of our prime resources. And since we humans are both drafting the ground rules and calling the plays, we define resources in our own terms. As phrased years ago by Zimmerman, a natural resource has value only as it fills some human need, such as satisfying a want.

457

The earth's iron and nickel core is without value because we haven't learned how to get at it; just as the vast petroleum reserves of our Southwest were without value to the Plains Indians who led precarious lives pursuing the buffalo over the surface.

One of our handicaps, in part, in dealing with water problems has been our slowness in recognizing that our developing culture is changing water from a free commodity to one that is expensive. In their book *Multiple Purpose River Development* (1958), Krutilla and Eckstein noted that "The rationale that any commodity will be allocated according to principles of economic efficiency depends on a number of critical assumptions, namely (1) complete divisibility between inputs and outputs, (2) independence of production and consumption functions, (3) benefits that can be conveniently packaged and quantified in economic terms, and (4) consumer preferences that can be directly compared in the market place."

Recently Tinney and O'Riordan[1] observed:

Water has a two-dimensional utility expressed in terms of quantity and quality. Each may vary over space and time. Certain uses can be expressed in the conventional economic sense of willingness to pay, as for example, water used in agricultural or industrial production processes, where the marginal value product of water can be used to estimate the benefits of water use.

Other types of water use, however, do not satisfy the required market allocation conditions. Certain water resource management benefits are universal in character and cannot be individually packaged or sold. Instead they must be consumed collectively by a whole community. Examples are flood protection, pollution abatement, fishing protection, wildlife preservation, and enhancement of aesthetic attractiveness. Because individuals do not reveal their preferences by a willingness to pay, collective costs and benefits can only be distributed and paid for on a unified basis.

Some water resource management benefits are essentially subjective, personal, and internal in nature. As such, they are difficult to evaluate in monetary terms. These benefits are referred to as extra-market or intangible benefits and include such factors as enhanced recreational environments, preservation of unique ecological habitats, or historical monuments.

----

[1] Tinney, E. Roy, and J. O'Riordan, 1971. Water as a Consumer Commodity, Journal of Soil and Water Cons. 28(3):102-106.

Because these benefits are not easily expressed in terms of a consumer's willingness to pay, they tend to be overemphasized or underemphasized depending on the personal views of decision-makers, political pressures, or the nature of the proposed project.

I was most happy to hear the eloquence of Dr. Crutchfield's exposition of this point, which has long been the *bête noire* of economically illiterate biologists concerned with recreational uses of water.

Obviously, water and the streams which transport it must be amenable to multiple use if they are to constitute a valuable resource. But to me, one of the most encouraging things to emerge from this symposium is recognition of the need for decision-making to become truly interdisciplinary. I do not, God forbid, ask for dilettantism. But the specialist can no longer responsibly fall back on the Philistine attitude that he will do his thing and if *they* are intelligent enough to make use of his golden discoveries, well and good! but if not, it's *their* hard luck. The specialist—and most of us are—must continue to sharpen his individual array of tools. But he must also learn enough of the *corpus* of other specialties to be able to call for help when and from whatever source it is needed, and perhaps most important of all, to accord other specialties respectful consideration. What we must have is not *either-or* but *and-and*.

One of the many things which impressed me in Hammerton's case history of the Nile was that, far too late and with almost criminally inaccurate predictions, government authorities informed some thousands of native villagers that their homes would soon be flooded by the rapidly filling Roseires Reservoir. They were advised to move to a new site which (here is the inexcusable part from the standpoint of the professional) turned out to be one or two meters below water at the upper end of the reservoir. They were flooded out of their new makeshift homes, their hastily replanted crops were drowned out, and their seemingly already crushing misery was augmented by hordes of pestiferous aquatic insects.

Clearly, a situation of this kind should be dealt with by a team of specialists which would include anthropologists and sociologists sensitive to the existing values of the people about to be displaced, as well as people able to predict ultimate reservoir levels with greater accuracy.

That the future holds severe trauma for even greater numbers of people is clear from the prediction that construction of the proposed Jonglei Canal will force a complete shift in the way of life of over 700,000 Nilotic peoples.

Again I find a parallel, albeit of lesser dimensions, in one of my own experiences, this time when I was a member of a multidisciplinary study team looking into pros and cons of the proposed Rampart Dam on Alaska's Yukon River.

Here, as on the Nile, was a natural dam site—an engineer's dream. Over 750 miles inland from the sea, astride the Arctic Circle, the Yukon has cut a narrow, deep notch through a range of hills 1200 to 1500 feet deep. By dropping a (relatively) low-cost concrete plug about 570 feet high and less than a mile wide into this narrow notch, a huge body of water could be backed up: large enough, indeed, to completely inundate the State of Connecticut (an appealing prospect to some of you in this room).

But one of the factors we needed to consider in addition to a host of ecological, meteorological, and economic questions, was that presented by the endemic aboriginals. Nowhere near as numerous as the Nilotic peoples—perhaps nine villages with 1500 or more people would be flooded out and the graves of their ancestors inundated. The migrations of salmon and caribou to which their lives had been long attuned would be cut off or moved out of realistic reach. In addition, another 12000 or more Indians far upstream in Canada's Northwest Territories, who derived about half their livelihood from the salmon run, would be thrown onto government charity.

How were we to evaluate this "people problem"? Where should it rank along with consideration of the pros of economic need (for the new and struggling state) and the cons of environmental alteration? How do we deal with tastes, with attitudes and their possible constructive change? I know what consultants from the fields of sociology and anthropology told us. I know, too, that we were lucky that other more easily quantified parameters have so far determined the course of decision making vis-à-vis the Rampart.

But whether the people whose lives are to be disrupted are Dinka, Athapascans, or white Elizabethan hill people in the Tennessee Valley, the decision-makers should include "people biologists" as well as fish biologists, anthropologists as well as civil engineers, and political spokesmen for those destined to lose as well as for those who are sure they will gain. Only if we can finally overcome our historic penchant for stepping on our own shoelaces can we avoid increasingly disastrous side effects from our major developmental ventures.

But, the final conclusion of almost all speakers—especially those giving river case histories—was essentially hopeful. However bad times have been, they bid fair to get better rather than worse.

SUBJECT INDEX

A

Abstractions, 253-254
Animal production, 45-47
Applied morphometrics, 287-297
  basic principles, 288-289
  examples of applications, 292-297
  history, 287-288
  simple relationships, 290-292

B

Bank score, 311-312
Bed material, 309-310, 312
Benthos, 38-43, 149-151, 197, 222-225,
  230
  depositional assemblages, 41-42
  erosional assemblages, 38-41
  intermediate zone assemblages, 42-43

C

Carp, 148, 161-162
Catchment, 252
Classification of Streams
  cations, 67
  trophic, 67
Columbia river, 77-96, 269
  dams 79-82, 84-93
  dissolved nitrogen, 84-86
  Columbia river development
    program, 81-82
  fish passage, 89-90
  flow, 86-88
  future development, 93-96
  human influences, 78-79, 81
  physical description, 77-78
  previous studies, 80-82
  salmon, 77, 81-93
  temperature, 83-84

Cooperative highways and river research
  programme (Canada), 302
Crustacea, 186, 190, 192, 223

D

Dams, 4, 5, 79-82, 84-93, 148, 154,
  179-181, 189, 198-200, 205, 216,
  296-297
Danube river, 457
  abstraction, 239-240
  chemistry, 241-243
  dams, 238-239
  fisheries and fish production, 244-245
  flood control and land drainage, 237
  history, 247-248
  human influences, 233-245
  hydrology, 233, 241
  navigation, 233-237
  physical description, 233-245
  political frontiers, 248-249
  power generation, 238-239
  radionuclides, 243
  recreation and culture, 246
Delaware River
  fisheries and fish production, 99, 102,
    106-110, 120-124
  history, 99-100, 107
  human influences, 99, 100, 102-103, 105
  hydrology, 103-105
  management, 115-116, 123
  modeling of water quality, 116-123
  physical description, 101-102
  shad, 107-110, 120-123, 125
  socioeconomic and technical problems,
    114-115, 124
  uses, 105-107, 126-127
  water quality, 110-114, 118, 121-122
Deltas, 313
Density currents, 313
Detritus, 37, 44-45

461

Discharge
  definition, 263-264
  effects on fish populations, 266-274
    harvesting, 274
    migration, 267-269
    rearing capacity, 272-274
    spawning, 270-271, 278
  general effects, 264-265
Diversion, 293-295
Diversity, 328-330, 401
Drift, 35, 40

E

Economics of multiple use, 431-440
  analytical problems, 435-437
  constraints on defining systems
    431-432, 434-440
  efficiency, 432-434, 438-439
  non-monetary values, 434-435
  objectives, 438-440
Ecosystem compartmentalization, 34-35
Effects of man, 68, 71
Environmental protection agency, 356

F

Fauna, 456
Fish, 35, 37-40, 45-46, 48. 77, 79-93,
    107-110, 120-123, 125, 147-148, 151,
    154, 156-157, 160, 161-163, 182-186,
    197-199, 201-203, 219-220, 223-228,
    244-245, 266-274, 299, 301, 312, 314,
    334, 336, 338-343, 357-359, 378-379,
    380
  bypassing downstream migrants, 90-93
  dissolved oxygen, 84-86
  fisheries and fish production, 77, 81-82
  flow, 86-88
  passage, 89-90
  salmon, 77, 81-93
  temperature, 83-84
Fisheries and fish production, 253, 256
Flocculation of sediments, 314
Flood control, 253-254
Flora, 37, 40, 42-45, 53-62, 70-73, 111,
    146-147, 151-152, 187-188, 190-196,
    200, 216, 220-222, 229, 316, 327-328,
    378-379, 403-408, 456

algae, 37, 40, 43-45, 53, 57-62, 70-73,
    147, 151-152, 187-188, 190-196,
    220-222, 229, 316, 327, 378, 379,
    403-408
  classification and structure, 59-62
  CO_2, 54
  flow, 56
  light, 54, 57-58
  model streams, 54-57
  periodic phenomena, 58-59
  temperature, 55-56
  theoretical models, 57-58
  vascular plants, 37, 42, 53, 57, 70, 111,
    146, 152, 188, 200, 222, 327, 328
Fluvial geomorphology 9-21
  effects of man, 19-21
  geomorphic limits, 10
  rivers as systems, 10
    hydraulic geometry, 10, 11
    meandering, 14-15
    steady-state balance, 11-16
    thermodynamics, 11-16
    the time factor, 16, 19

H

Helminths, 186, 199, 223-224
Hubbard brook, 25-27
Human influences, 257-260
Hydrodynamics, 9
Hydrologic cycle, 9

G

Geomorphology, 456

I

Illinois river, 457
  agriculture, 155-159
  dissolved oxygen, 136, 148-149, 154
  fisheries and fish production, 138,
    147-148, 158, 160-162
  heavy metals, 137, 152
  history, 131-134, 136-163
  human influences, 131-132, 135-136,
    143-162
  hydrology, 135, 145-147

nutrients, 135, 137, 149, 158, 161
pesticides, 135, 158-159
physical description, 131
recreational use, 160-161
silt, 135, 156
uses, 143-148, 152-153, 156, 159-162
water quality, 135-159
Insects, 35, 37-46, 48-50, 73, 150-151,
    186, 197, 223, 331-333, 335-337, 361

L

Life cycles, 72-73

M

Meandering, 295-296
Mekong river, 2, 3, 5, 455
Mississippi river, 3
Molluscs, 41, 42, 150, 151, 186, 197, 223,
    331, 379-380
Multiple use, 421-430, 458-459
    human population growth, 423-424
    maintenance of industry, 423-424
    management, 424-428
    monitoring, 424-427
    multiple-use conflicts, 428-430
    species preservation, 423

N

National research council (Canada), 304
National water-pollution control act, 3
Nature of sediment, 309-310, 313-315
Navigation, 252-253
Nekton, 35-38
    depositional assemblages, 37-38
    erosional assemblages, 35-36
    intermediate zone assemblages, 38
Nile river, 171-207, 455, 457, 459
    chemistry, 181-182
    dams, 179-181, 189, 198, 200, 205
    fisheries and fish production, 185-186,
        197-199, 201-203
    future outlook, 202-207
    history, 171, 172, 176, 179-180
    human influences, 171, 181, 188-189

hydrology, 174, 176-181, 189
nutrients, 181, 191
physical description, 172-176, 183
previous studies, 175-176, 179, 184-188,
    190-192
uses, 171, 185-186
water quality, 188-189, 190-196, 198-
    199, 206
Nutrients, 389-408
    algal nutrition, 403-408
    bacteria-algae symbiosis, 405-408
    cycling and management, 399-409
    definition, 390
    historical, 389-390
    phosphorus uptake and storage, 403-405
    river ecosystems, 391-392
    sources and sinks, 392-399

O

Orsanco, 2, 360

P

Peace-Athabasca delta, 298-303, 307
Pesticides and industrial wastes, 353-364
    application factor, 358-359
    defining the problems, 353-356, 363-364
    life cycle tests, 359-360
    MATC, 358-359
    toxicity tests, 357-359
Political frontiers, 253, 257
Politics of multiple use, 441-453
    elements of political system, 446
    expanded federal role, 448-449
    failure of political system 442-445
    needs for the future, 453
    problem complexity, 441-442
    recent legislation, 450-452
    strategy for change, 447-448
    systems approach, 449-450
Power generation, 253
Power stations, 319-325, 343-346
    chemical effects, 323-325
    cooling towers, 320, 345
    cooling water, 320-321
    future discharge criteria, 344-345
    future research, 346
    generation, 319, 343-344

463

R

Radionuclides, 367-384
  Clinch river studies, 382-384
  concentration factors, 377-380
  food chain pathways, 373, 377-380
  mineral metabolism, 381-382
  the problem, 367-368
  sorption and dispersion, 374-376
  sources, 368-373
  specific activity, 380
  transport in rivers, 373-376
Recreation and culture, 253, 256
Regulation, 277-280
River Thames
  abstraction, 218-219
  chemistry, 215
  dams, 216
  fisheries and fish production, 219-220,
    223-227, 229
  history, 216-217
  human influences, 216, 217
  hydrology, 215
  physical description, 215
  uses, 217-219, 225-226, 228, 230
  water quality, 217, 228-229
River use and value, 458
River water chemistry
  continentality, 24, 25
  controlling mechanisms, 21-25
  effects of man, 25-27
  precipitation, 23
  salinity increases, 21-22

S

Salmon, 266-273, 276
Shad, 107, 110, 120-123, 125
Solids transport, 267-277
Structure of rivers, 68-71

T

Temperature, 321-323, 344-345
  effects of power stations, 321-323
  natural, 32

Thermal effects on biota, 325, 327-343, 346
  algae, 327
  diversity, 328-330
  fish feeding and growth, 339-340, 342
  fish distribution, 338-339
  fish mortality, 334, 336, 338
  fish spawning, 341-343
  future research, 346
  introduction of exotics, 331
  invertebrates, 328-331
  life histories, 332-334
  tolerant species, 330
  vascular plants, 327-328
Tropical relationships, 43-45, 67-72,
  223-225
Tropical rivers, 2-5
  Brahmaputra, 4
  compared with temperate, 3, 4, 5
  clams, 4
  Mekong, 2, 3, 5
  Parana, 4

V

Vegetation, 275-276

W

Wash-load, 310-313, 315
  deposition, 310-311, 313-314
  transport, 310
Waste disposal, 253, 255
Water quality, 110-114, 118, 121-122, 217,
    228-229, 241-243, 274-276, 312, 314,
    315-316, 321-324, 344-345, 359,
    361-363, 389-408
  dissolved oxygen, 110-112, 118-122
  heavy metals, 397, 399-401
  nutrients, 112-113, 241-242, 323,
    389-408
  oxygen, 229, 241-243, 314, 324, 361
  pesticides, 361
  pH, 361
  silt, 276
  temperature, 275, 321-323, 344-345
  toxic materials, 315

toxicants, 359, 362
transparency, 316
Waterfowl and wildlife, 138-139, 147, 157-158, 160-161, 203-204, 299-301, 307

Y

Yukon river, 460